GENOMICS AND PROTEOMICS

Functional and Computational Aspects

GENOMICS AND PROTEOMICS

Functional and Computational Aspects

Edited by

Sándor Suhai

Deutsches Krebsforschungszentrum
Heidelberg, Germany

Springer Science+Business Media, LLC

Library of Congress Cataloging-in-Publication Data

Genomics and proteomics : functional and computational aspects / edited by Sándor Suhai.
p. cm.
Includes bibliographical references and index.

1. Gene mapping--Data processing. 2. Nucleotide sequence--Data processing. 3. Proteins--Analysis--Data processing. I. Suhai, Sándor.

QH445.2 .G473 2000
572.8'633'0285--dc21

00-022525

Proceedings of the International Symposium on Genomics and Proteomics: Functional and Computational Aspects, held October 4–7, 1998, in Heidelberg, Germany

ISBN 978-1-4757-8174-8 ISBN 978-0-306-46823-0 (eBook)
DOI 10.1007/978-0-306-46823-0

Originally published by Kluwer Academic / Plenum Publishers, New York in 2000
Softcover reprint of the hardcover 1st edition 2000

http://www.wkap.nl/

10 9 8 7 6 5 4 3 2 1

A C.I.P. record for this book is available from the Library of Congress

PREFACE

Genome research will certainly be one of the most important and exciting scientific disciplines of the 21st century. Deciphering the structure of the human genome, as well as that of several model organisms, is the key to our understanding how genes function in health and disease. With the combined development of innovative tools, resources, scientific know-how, and an overall functional genomic strategy, the origins of human and other organisms' genetic diseases can be traced. Scientific research groups and developmental departments of several major pharmaceutical and biotechnological companies are using new, innovative strategies to unravel how genes function, elucidating the gene protein product, understanding how genes interact with others—both in health and in the disease state.

Presently, the impact of the applications of genome research on our society in medicine, agriculture and nutrition will be comparable only to that of communication technologies. In fact, computational methods, including networking, have been playing a substantial role even in genomics and proteomics from the beginning. We can observe, however, a fundamental change of the paradigm in life sciences these days: research focused until now mostly on the study of single processes related to a few genes or gene products, but due to technical developments of the last years we can now potentially identify and analyze all genes and gene products of an organism and clarify their role in the network of life processes. This breakthrough in life sciences is gaining speed worldwide and its impact on biology is comparable only to that of microchips on information technology.

The main purpose of the International Symposium on Genomics and Proteomics: Functional and Computational Aspects, held October 4–7, 1998 at the Deutsches Krebsforschungszentrum (DKFZ) in Heidelberg, was to give an overview of the present state of the unique relationship between bioinformatics and experimental genome research. The five main sessions, under the headings: expression analysis; functional gene identification; functional aspects of higher order DNA-structure; from protein sequence to structure and function; and genetic and medical aspects of genomics, comprised both computational work and experimental studies to synergetically unify both approaches.

The content of this volume was presented mostly as plenary lectures. The conference was held at the same time as the Annual Meeting of the Gesellschaft für Genetik (GfG). It is a great pleasure to thank Professor Harald zur Hausen and the coworkers of DKFZ for their help and hospitality extended to the lecturers and participants during the meeting. We would also like to thank the European Commission and the companies BASF AG, BASF-LYNX Bioscience AG, Bayer AG, BIOMEVA GmbH, Boehringer

Mannheim GmbH, Hoffmann-La Roche Ltd., Knoll AG, Merck KGaA, and Schering AG for the funding of the symposium. The organizers, Annemarie Poustka, Hermann Bujard, and Sándor Suhai, profited greatly from the help of the scientific committee, Claus Bartram, Jörg Hoheisel, Fotis Kafatos, Jörg Langowski, Peter Lichter, Jens Reich, Manfred Schwab, Peter Seeburg, and Martin Vingron. Furthermore, the editor is deeply indebted to Anke Retzmann and Michaela Knapp-Mohammady for their help in organizing the meeting and preparing this volume.

Sándor Suhai

CONTENTS

1

. . . AND COUNTING

DNA-Microarrays

Jörg D. Hoheisel*

Functional Genome Analysis
Deutsches Krebsforschungszentrum
Im Neuenheimer Feld 506
D-69120 Heidelberg
Germany

In recent years, emphasis in genome research has moved away from the more descriptive presentation of the rather static sequence fundaments of an organism toward the evaluation of the dynamic processes taking place within a living cell on the level of nucleic acids (and beyond). This adds another dimension of complexity, since the entire organism has to be re-analysed very many times over with probes generated under different environmental conditions or taken from different (tissue) parts. The observed scale of fluctuation is somewhat surprising although this is not news as such. The genomic approaches only bring home this message more clearly and convincingly, because it is reflected in the puzzling composition of the information obtained. Toward a comprehensive understanding, rather elaborate and fast methods are therefore essential and accurate numbers need to be determined. The last issue is critical, since already subtle variations can precipitate enormous consequences, especially in regulative processes. Many presentations at the recent *Symposium on Genomics and Proteomics* dealt with methodologies capable to perform this sort of analyses, at least in principle, and highlighted the perspectives and challenges ahead.

The term "DNA-microarray" stands for the currently most prominent and promising type of technology in this respect. By simultaneously analysing the hybridisation behaviour of probe molecules at very many different sequences, it combines simplicity

* Tel.: +49-6221-424680, Fax: -424682, e-mail: j.hoheisel@dkfz-heidelberg.de

Genomics and Proteomics, edited by Sándor Suhai.
Kluwer Academic / Plenum Publishers, New York, 2000.

of the assay with the high throughput required for genomic approaches. A simple look at the numbers of relevant publications (Figure 1) published during the last few years illustrates both the increased awareness of the array-based approaches and the actual start of data production by such means (for review see Nature Biotechnol. 17, 1999), although a considerable number of relevant publications is missing because of search-intrinsic restrictions to only certain types of manuscripts and journals. Also, there are currently indeed still more reviews and forecasts on the subject than reports on actual data, yet this is bound to change very soon.

The potential range of microarray applications is as wide as is the field of life sciences and commerce. Thus, there is not a single one technique for all applications—nor will there ever be one—but a rather wide spectrum of array types, adapted to the particular needs. Rather than decrease, this variety will increase with the number of applications (and companies getting involved) at least for some time, since certain techniques are well suited for one kind of analysis while less fitting another. Also, there are many new areas of application out there either not yet being worked at or, most likely, not even thought of today, in a development similar to PCR, when from a single basic principle very many derivatives evolved. One field, for example, yet virtually unexplored by microarray techniques is the analysis of the information encoded in the DNA structure rather than sequence. It has been demonstrated that not only functional information is genetically encoded that way but, in addition, that even short term memory effects are possible (e.g., Pohl, 1987). Another example is the determination of the methylation status of DNA, important for both structure and function (Olek et al., 1996).

As with many scientific developments during their initial phases, the microarray techniques are still full of pitfalls and problems. It has been shown that mutational analyses of the p53-gene can be carried out at higher accuracy than by sequencing, the current gold standard (Ahrendt et al., 1999), but this does not hold true for many other

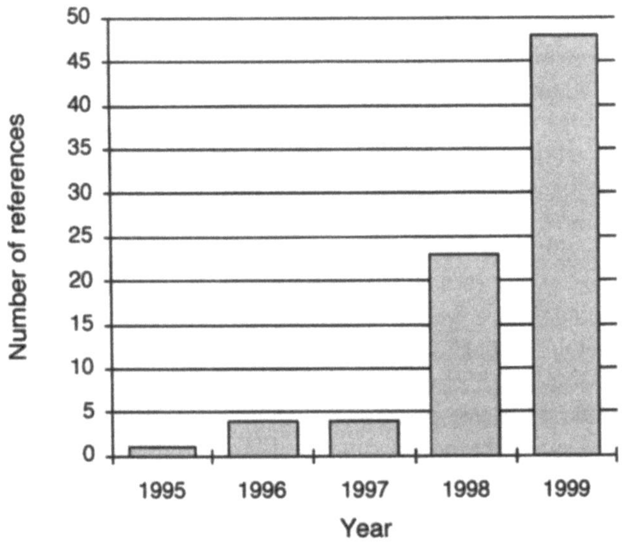

Figure 1. Number of hits when searching *Medline* for manuscripts dealing with applications of *DNA-arrays*, *microarrays* and *DNA-chips*. The value for 1999 is an extrapolation based on the number published in the period January to March and probably an underestimate of the eventual total.

applications. In addition, the field is lacking standardisation. There are as many ways of how to perform measurements as there are laboratories, differing not only because of the different systems and pieces of equipment employed but also from a variety of factors such as the lack of broadly defined controls or widely agreed protocols for probe preparation. Quality assessment is already a critical issue within a single laboratory. A direct comparison of data from different sources is currently difficult to achieve, and in some instances impossible.

Apart from technical developments toward improved data quality, another and in any case also parallel approach is the accumulation of results, even if of different quality or obtained from different sources. Because of the amount of data, a statistical evaluation will become possible at some stage. For the complexity of the analysed specimen, statistical approaches are prerequisite anyway for genomic studies. The production of a set of transcript profiles recorded at some 300 conditions on the complete gene repertoire of yeast, for example (Brown, 1999), illustrates the power of this approach. The resulting matrix of 300 conditions on 6000 genes, however, is already challenging in terms of evaluation even with the help of bioinformatics tools. To develop and optimise even relative simple tasks such as a user-friendly presentation of the data, let alone directed data mining, will be engaging software developers for many years to come, since ever more sophisticated algorithms will be needed to deal with the sheer mass of data and the extraction of the relevant information.

Another development already taking shape is the extension of the basic methodology created for nucleic acids to other molecule classes (e.g., Büssow et al., 1998). Studies on the interaction between biomolecules will have to be carried out in a highly parallel manner, because of the extreme complexity of their relationships within and between cells. Only by such means, sufficient data will be gathered in order to even get a glimpse at cellular functioning and its regulation. Thus, apart from being an important tool in their own right, DNA-microarrays are a forerunner currently establishing basic features and analysis strategies which will be taken advantage of during years to come.

REFERENCES

Ahrendt, S.A., Halachmi, S., Chow, J.T., Wu, L., Halachmi, N., Yang, S.C., Wehage, S., Jen, J., and Sidransky, D. (1999) Rapid sequence analysis in primary lung cancer using an oligonucleotide probe array, Proc. Natl. Acad. Sci. U.S.A. **96**, 7382–7387.

Brown, P.O. (1999) Watching the yeast genome in action, Curr. Genetics **35**, 173. Presentation at the XIX International Conference on Yeast Genetics and Molecular Biology.

Büssow, K., Cahill, D., Nietfeld, W., Bancroft, D., Scherzinger, E., Lehrach, H., and Walter, G. (1998) A method for global protein expression and antibody screening on high-density filters of an arrayed cDNA library, Nucleic Acids Res. **26**, 5007–5008.

Olek, A., Oswald, J., and Walter, J. (1996) A modified and improved method for bisulphite based cytosine methylation analysis, Nucleic Acids Res. **24**, 5064–5066.

Pohl, F.M. (1987) Hysteretic behaviour of a Z-DNA-antibody complex, Biophys. Chem. **26**, 385–390.

2

OBTAINING AND EVALUATING GENE EXPRESSION PROFILES WITH cDNA MICROARRAYS

Michael Bittner,[1] Yidong Chen,[1] Sally A. Amundson,[2] Javed Khan,[1] Albert J. Fornace Jr.,[2] Edward R. Dougherty,[3] Paul S. Meltzer,[1] and Jeffrey M. Trent[1]

[1]Cancer Genetics Branch
National Human Genome Research Institute
[2]National Cancer Institute
National Institutes of Health
Bethesda, Marland 20892
[3]Computer Assisted Medical Diagnostic Imaging Laboratory
Department of Electrical Engineering
Texas A&M University

1. INTRODUCTION

The ability to detect the RNA products of transcription by hybridization with nucleic acid probes of known sequence is a long-standing and central capability of molecular biology. Until recently, the primary focus of this kind of experimentation has been careful examination of the mRNA levels of one or a few genes per experimental series. Experiments frequently examined the steady state levels of a message in cells from different tissues or different pathologic states, the temporal transcription pattern of a gene during processes such as development, or the response to some defined stimulus. Recently, the products of genomics research have provided a strong impetus to develop methods that allow evaluation of the message levels of many genes simultaneously. Several technologies that enable one to develop profiles of gene transcription have been developed. As a result of initial reports of the results of such profiling, there is considerable interest in understanding what is required to carry out such experiments, what range of information can be gathered in these experiments, and what analytical methods can be applied to the results obtained. A very broad review of this field has been presented in a

Genomics and Proteomics, edited by Sándor Suhai.
Kluwer Academic / Plenum Publishers, New York, 2000.

supplementary issue of the journal Nature Genetics.[1] The following review will focus on the underlying concepts, methodologies, and current capabilities of gene expression profiling carried out by means of cDNA microarrays.

2. INFORMATION FROM GENE EXPRESSION

2.1. The Problems of Determining Gene Function and Control

Efforts have been launched worldwide to produce gene maps, lists of genes and complete genome sequence data for a number of organisms. At present, public and private efforts have resulted in complete genome sequences for 17 organisms, including the eukaryotes *Saccharomyces cerevisiae*[2] and *Caenorhabditis elegans*.[3] Parallel efforts that seek to develop clones and sequences (ESTs) based on sampling the sets of expressed mRNAs are also proceeding at a significant rate. Roughly 2.1 million such sample sequences have been deposited in public databases. Due to the collaborative efforts of the IMAGE Consortium,[4] the National Center for Biotechnology Information[5] and a number of companies supplying molecular biological reagents, both sequences and cloned DNA for somewhat more than 1.2 million human ESTs can be obtained. The development of high-throughput capabilities to clone and sequence nucleic acids has far eclipsed the capability to conduct more definitive biochemical studies of the functions and controlling inter-relationships of this emerging cohort of genes. Clearly, a variety of approaches to the analysis of gene function which can exploit the outputs of large-scale, highly-parallel analysis are desirable as aids to sensible orchestration of such further research.

2.2. Gene Functions, Controls, and Genomic Data

Rather than assign functions to already known genes, gene discovery has traditionally worked from an explicitly or implicitly defined function towards the gene that encodes the protein responsible for carrying out that function. A large repertoire of gene isolation tools have been developed which exploit ways of conditioning the expression of genes, methods for making the survival of a cell or the production of an easily detected marker dependent on some form of gene-dependent complementation, and combinations thereof. Recent genomics approaches invert these schemes, finding genes based solely on their presence in a particular tissue or cell type. This form of gene discovery frequently provides neither a suggestion of what the newly identified gene does nor hints as to how it is regulated.

2.2.1. Biological System Properties Provide Analytical Opportunities. The ability to study changes in gene expression for many genes simultaneously has been widely viewed as a possible way to extract information about what uncharacterized genes do, and how they are regulated. There are good reasons why this may be a workable approach. The rationale is best stated within the context of the current understanding of complex, adaptive systems. In the past sixty years, through use of the increasing computation and simulation capabilities supplied by computers, it has become possible to begin to model complex systems such as economies, societies, global weather systems, ecological systems and biological systems. These systems are all composed of very large numbers of interactive components having individual capabilities and propensities. These systems can exhibit complex behavior as a result of the enormous number of possible inter-component interactions. The resulting large set of possibilities make it hard to predict exactly what the system will do as a result of interactions between the modules and the local (non-system) environment, even when a great deal is known about the properties and behavior of individual system components.

The characterization of some of the features of construction and operation of these systems provides insights, which should facilitate the use of expression data to study the function and control of the component parts of biological systems. One of the key aspects of the construction of complex systems is their modularity. A very concise description of this aspect is presented by H. A. Simon in a lecture on the hierarchic nature of complex systems.[6] In general, he points out that complex systems are composed of largely independent subsystems, each of which operates to achieve its individual goals. Within a subsystem, the interactions between members are widespread and frequent. The interactions between subsystems are less frequent, involve far fewer members of the participating modules, and are most frequently geared to adjust the net inputs or outputs of the subsystems through feedback loops. This mode of interaction allows the subsystems to operate largely independently of each other. At the organism level, homeostatic circuits based on cross talk between organs, such as the interplay between heart, blood vessels and kidneys in the regulation of blood pressure and blood volume, provide familiar examples of this form of organization.

For those functions where the modules act with the most independence, it is possible to gain a strong sense of what portion of the work done by the whole system is attributable to that particular module. Such analysis by decomposition is a familiar tool for biologists. Many of the fields of study within biology are organized along the lines of the observed hierarchy of assembled functional units of macromolecules, organelles, cells, organs, bodies and ecologies.

Expression profiling is well suited to the study of modular action at the cellular level. Those cellular subsystems that have been characterized to greater or lesser extents, such as those responsible for intermediary metabolism, energy production, control of the cell cycle, and DNA replication employ a wide variety of control strategies. An important component of these controls is variation in the level of the mRNAs specifying the protein components of subsystems. Transcription is clearly not the only way to modulate the presence or activity of a gene, and exactly how comprehensive a picture of regulated change is obtained by observation of transcriptional regulation is certainly debatable. Still, given the ubiquity of control at the level of mRNA abundance, it is reasonable to assume that at least some of the relevant modulation will be seen as changes in the quantities of mRNA of that module's components.

2.2.2. Simple Interpretation Strategies. If alteration in message abundance proves to be a sufficiently rich source of information, then the most basic approach to interpreting the changes will be to look for two distinctive forms of change. The changes which occur as a consequence of adjustment of a subsystem, such as the adjustment of intermediary metabolism in response to a change from fermentation to respiration,[7] will reflect the very tight interactivity between the parts of that functional module. Concerted changes of many genes that cooperate to achieve a particular function will be observed. When such coordinated behavior is observed during a variety of adjustments of that subsystem, the implication will be that the co-varying genes are components of that functional entity. While this would be an inexact specification, it would certainly be a useful preliminary categorization of an uncharacterized gene.

The second discernible form of change that should emerge from expression profiles is the type resulting from signaling between subsystems. In this case, change in the level of a gene product will precede the alteration of the level of components of a number of subsytems. Well known examples of proteins whose action causes widespread change in the cell are p53, an early component of the cell's response to DNA damage,[8] and myoD, an early regulator of the muscle differentiation program.[9] Temporal profiling as cells respond to a stimulus or execute a differentiation program may well identify genes that are integral to the initiation and propagation of these actions.

While such reasoning argues that it will be possible to obtain information about gene function and control, there remain questions as to how readily this can be achieved in practice. The extraction of data from profiles will have to deal with the confounding effects of the size and complexity of biological systems. The expected compartmentalization of the changes observed to be covariant will be blurred by the way the cell is constantly running many dynamic, tightly interlocked processes in parallel. It remains to be seen how much data, and of what precision will be required to allow potent inference of function and control.

3. LARGE-SCALE METHODS OF STUDYING GENE EXPRESSION

A very appealing aspect of expression profiling as an approach is that detection schemes for gene expression studies can be either hybridization or sequencing based, both of which can be carried out in highly parallel, large scale formats, exploiting the sequences and clones resulting from genomics projects. Sequencing-based approaches to this form of study include sequencing of cDNA libraries[10,11] and serial analysis of gene expression (SAGE).[12] Hybridization methods have evolved from early membrane-based, radioactive detection embodiments[13] to multi-gene versions of this methodology,[14,15] and thence to highly parallel quantitative methods using fluorescence detection. These recent techniques use either preformed cDNAs printed to a glass surface[16] or oligonucleotides synthesized in situ by photolithographic methods[17] as the known sequence detectors.

In prior hybridization-based approaches to detecting expression levels, mixtures of cellular RNA were either immobilized as an unfractionated pool or else electrophoretically fractionated and immobilized as continuous, size-separated fractions. The specific mRNA gene products were detected by the use of radioactively labeled, known sequence nucleic acid probes. Thus, even if RNA from a number of sources were immobilized on a single matrix, one could only extract information about the abundance of a single gene in the course of a single experiment. By inverting the immobilized and free components of such an experiment, the abundance of many mRNAs can be evaluated in a single experiment. Large numbers of known sequence probes are immobilized as an array of detection units, and the pool of RNAs to be examined is labeled and then hybridized to the detectors. When the detectors used in this format are cDNAs, the experiment is termed a cDNA microarray analysis of gene expression.

4. ANALYSIS OF GENE EXPRESSION WITH cDNA MICROARRAYS

4.1. cDNA Array Detectors

cDNA arrays are typically prepared by printing small (2–5 nanoliter) volumes of solutions of DNA (100–500 μg/ml) onto glass microscope slides. The slides are chosen for their uniform thickness, flatness, and low intrinsic fluorescence. Coatings are applied to the slide to enhance its hydrophobicity, limit the spread of the printed droplet of DNA solution, and increase its capability to retain DNA following chemical or photo cross-linking. Some of the coatings in common use are poly-L-lysine, amino silanes, and amino reactive silanes.[16,18] A simple approach is to use poly-L-lysine coated slides, and to UV cross-link the DNA to the coated surface. The use of coatings which leave charged amines on the surface of the slide requires that a chemical passivation step be included after cross-linking, so that the labeled DNA introduced at the hybridization step does not have a strong electrostatic tendency to bind to the slide. This is can be achieved by reacting the amines with succinic anhydride in a buffer composed mostly of organic solvent.[18] The transfer of DNA solution to the slide surface is commonly accomplished by the use of a pen-like device which is dipped into the source DNA solution, filled by capillary action, and then contacted with the slide surface to transfer a few nanoliters of solution. Printing speed and precision are achieved by using using highly accurate industrial robots to move the pens.

cDNA arrays provide great flexibility in the choice and production of the defined sequence probes to be printed on the slide. In essence, any DNA complementary to an mRNA can be used as a hybridization detector. Practical considerations tend to shape the choice of which DNA detectors to use, and how to prepare them. One limit to the performance of flurochrome based detection systems is the tendency of flurochromes to bind to a wide variety of hydrophobic substances. For this reason, it is very useful to prepare the DNA for arraying in a method that facilitates easy purification away from cellular debris. A simple method currently in use is to prepare purified template DNA from cells and then to use PCR amplification followed by ethanol precipitation, gel filtration or both to prepare relatively pure DNA for printing.

The choice of template source and PCR strategy vary with the organism being studied. In organisms with smaller genomes and infrequent introns, such as yeast and prokaryotic microbes, purified total genomic DNA serves as template and sequence specific oligonucleotides are used as primers. In dealing with large genomes and genes with frequent introns such as human and mouse, cloned ESTs and primers directed to the plasmid sequences adjacent to the cloning insertion site are used. A further consideration is the necessity of matching the target of hybridization to the portion of the message that will serve as template in the message labeling reaction. If reverse transcription from the polyA termini and incorporation of fluor-tagged nucleotides are used to produce labeled cDNA targets, the labeled products will be complements of the 3′ end of the messages, usually extending 600 to 1000 bases from the priming site. Where available, ESTs provide well-matched complements for such labeled species, as ESTs in the public banks are typically 600 to 2000 base pair copies of the 3′ ends of genes.

For all organisms, the ability to efficiently select genes to be placed on an array is limited by the genomics and informatics infrastructures that have been developed for that organism. While it is desirable to represent every gene from an organism on an array detector, this is currently only possible for organisms with small, simple, completely sequenced genomes. The only multicellular eukaryote for which a complete gene array could be built in the near future is *C. elegans*, which has somewhat more than 19,000 genes inferred from the genomic sequence. Yet even for this model organism relatively sparse EST holdings may impede rapid progress. In the case of the even larger mouse and human genomes a complete complement of genes has not been defined, and thus arrays necessarily represent only a sampling of the full set of genes. While it is possible to array uncharacterized, redundant gene sets arising directly from cDNA cloning, this approach is seldom used. The expense in time and materials incurred in printing cDNAs onto arrays makes it worthwhile to expend the effort to develop highly characterized, non-redundant gene sets for printing.

4.2. Labeled cDNA Representations of the mRNA Pools

The known gene probes immobilized on microarrays are hybridized to fluor-tagged cDNA copies of the message pools of the cells to be analyzed. Fluor-tagged representations can readily be produced with a single round of reverse transcriptase (RT) extension from an oligo dT primer hybridized to the polyA termini of mRNAs in the message pool. Alternatively, mRNA may be purified by selection on an oligo dT matrix, and then used as template for oligo dT primed RT extension, though this reduces the amount of labeled cDNA which can be obtained from a given amount of starting cells due to handling losses during selection. Care must be taken to obtain quite pure RNA for labeling and hybridization, as the performance of fluorescence assays is easily degraded by impurities such as lipid or protein. Many of the protocols described in the microarray literature specify RNA purifications that allow the RNA to be purified to the final, useable form before concentrating by precipitation steps. A likely reason for this common feature is that early precipitation steps could form aggregates of the RNA with cellular components such as carbohydrate which would not be easily disaggregated in subsequent steps, and would contribute strongly to non-specific binding to the slide surface.

The cDNA copies of the message pools to be compared are made fluorescent by inclusion of fluor-tagged nucleotides in the RT reaction. The best fluor-tagged nucleotides characterized for this purpose to date are dUTP conjugates of the cyanine dyes Cy3 and Cy5. While only incorporated at rates of 1 to 2% (of total nucleotide incorporated), these flurochromes have high extinction coefficients and quantum yields, and reasonable photostability. In addition, their absorption and extinction maxima are roughly 100 nm apart, facilitating optical filtration, and their absorption peaks are in spectral regions accessible with a variety of lasers.

For organisms with sizeable genomes such as mouse and human, there is a requirement for large amounts of labeled cDNA to produce reliable fluorescent signals from low abundance transcripts. Figure 1 displays the number of transcripts of a specified abundance which would be found in the column of liquid above an immobilized cDNA probe in a typical cDNA microarray hybridization as one varies the amount of total RNA used to generate the labeled cDNA. This number can serve as a crude estimate of the amount of a particular transcript that could be captured during hybridization. The volume from which labeled molecules can be captured is limited by the low rate of diffusion of sizeable nucleic acids ($D_0 \sim 10^{-7}$ to $10^{-8}\,cm^2$/second),[19] and by the low likelihood of significant mixing by thermal convection during an isothermal hybridization. The corresponding density of flurochromes (per $100\,\mu^2$) resulting from a 100% capture rate of the local labeled species is also plotted in this Figure, illustrating the practical consequence of this limitation to the hybridization. Using these suppositions, a species of mRNA present at one copy per cell (approximately 1 transcript in 10^5) would be expected to yield approximately 10 flurochromes per $100\,\mu^2$ pixel if $100\,\mu g$ of total RNA were converted to labeled cDNA and hybridized to the array. With any assay noise, this would be at the lower end of the capabilities of fluorescent detection. As can be seen in Figure 2, the normal amount of the low abundance gene CDKN1A, which may be present at 1 copy per cell, is just detectable.

The requirement for significant amounts of RNA to detect the bulk of the transcript species, which are estimated to be present at 1–20 copies per cell,[20] is currently a

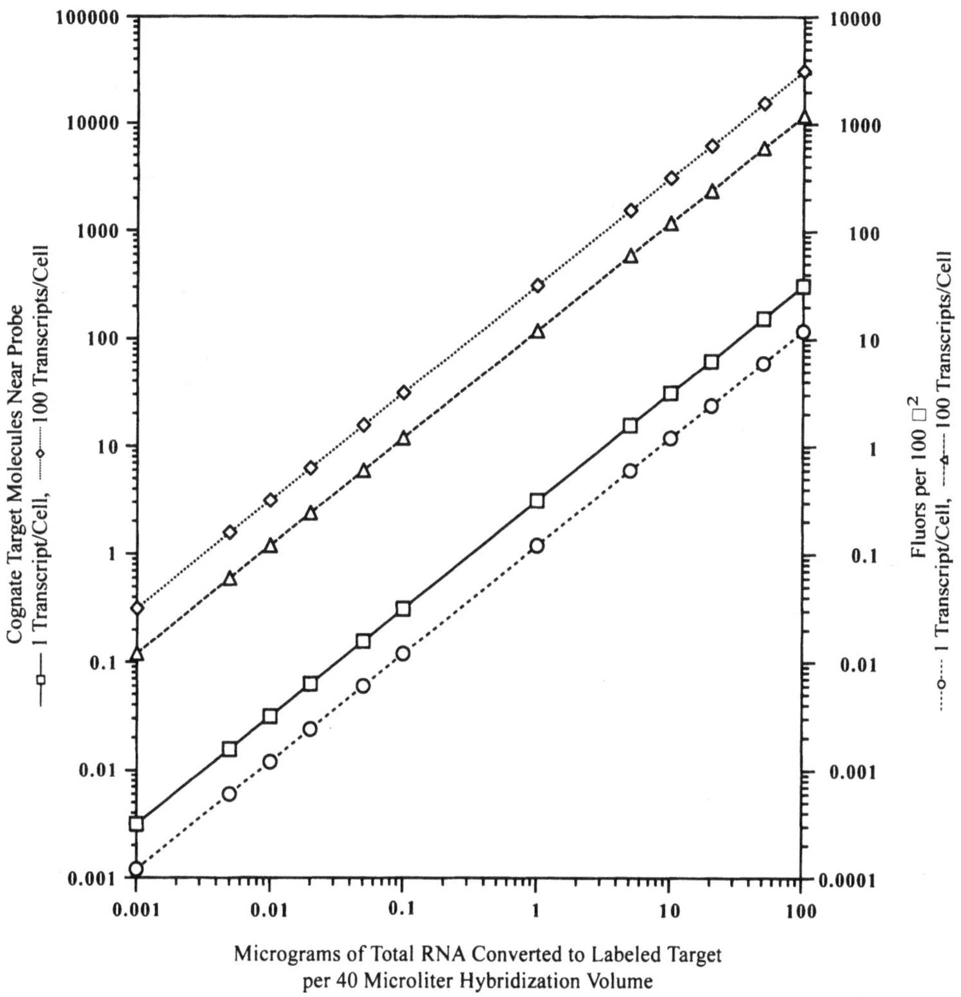

Figure 1. Calculated yields of flurochrome deposition on a hybridization detector. The amount of flurochrome that a probe could capture was calculated using the following set of assumptions. The amount of total RNA extracted from 10^8 cells is 1.25 milligrams. All mRNA is recovered in extraction and converted by reverse transcriptase to cDNA with an average length of 600 bases. Fluor-tagged nucleotides are introduced into the cDNA transcripts at a rate of 2 per 100 bases. All those targets in the column of liquid immediately above the probe at the start of the hybridization reaction will reach the probe and hybridize to it.

significant limitation of the technique. A variety of methods to reduce the RNA requirements for signal production are being analyzed. Methods that would circulate the labeled cDNA efficiently over the hybridization area would bring more probe molecules into contact with their cognate targets. Amplification methods based on using phage RNA polymerase copying of cDNA products have been developed[21] and exploited[17] for use in arrays. Amplification methods in which detectable molecules are precipitated onto the site of immobilized probe by typical histochemical methods have also been adapted for use with arrays.[22]

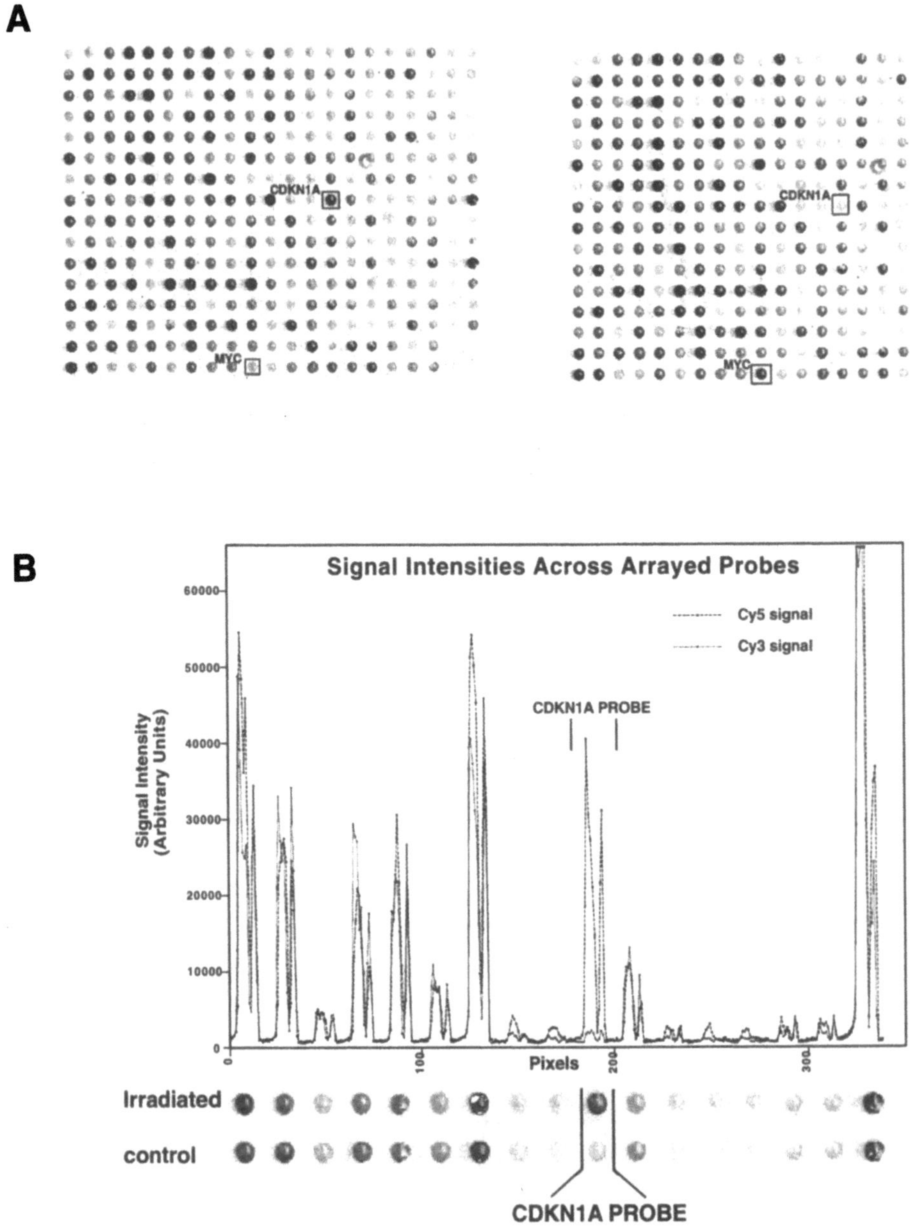

Figure 2. Detecting equivalent and disparate message levels with a cDNA microarry. Panel A is the pseudo-colored image of a portion of a microarray to which fluorescent cDNA representations of the mRNA pools of radiation treated and untreated ML1 cells were hybridized. Treated cells were harvested 4 hours after receiving 20 Gy of gamma irradiation.[24] Fluorescent intensities of the treated cells and the untreated control cells were placed in the red and green image channels respectively. The two most differentially expressed genes detected in this experiment were MYC, which is high in the control cells and CDKN1A ($p21^{CIP1,Waf1}$), which is high in the irradiated cells. Panel B is a detailed intensity plot of the control and irradiated fluorescent intensities at the immobilized probes surrounding CDKN1A. (From reference 1 by permission of the publisher.)

5. DATA EXTRACTION AND ANALYSIS

5.1. Image Analysis

5.1.1. Intensity Evaluation. The fluorescent intensity associated with each probe spot is determined from images taken with a confocal scanning microscope adapted to scan large areas at moderate resolution (100–400 μ^2 pixels). This provides approximately 50 to 200 samplings of the intensity at each immobilized probe site. The regularity of placement of the detectors which robotic spotting provides coupled with the very sharp images resulting from confocal imaging makes it possible to use many available, well-developed image analysis tools and methods. Approaches such as adaptive detection of local background and morphological modeling allow accurate detection even of weak signals.[23]

cDNA microarray analysis is carried out as a comparative hybridization between two samples. This is both convenient for the goal of detecting changes in expression patterns between samples, and necessary for obtaining the most accurate evaluation of the relative message levels in the samples. By simultaneously hybridizing a reference cDNA pool, derived from the reference cell line, and the test cDNA pool, internal normalization of the data from each immobilized probe is achieved. Analysis of comparative hybridization is greatly simplified by the large excess of probe hybridization sites over labeled target. In this situation, target molecules are not competing for sites at the immobilized probe, and hybridization is proportional to the pool size of each target in each sample. An example of results from such a comparative hybridization of flurochrome labeled cDNA probes from gamma irradiated and unirradiated cells[24] is presented in Figure 2. Panel A shows a portion of the microarray image where the target cDNA fluorescence from gamma irradiated cells is presented in the red color channel and the target cDNA fluorescence from unirradiated cells is presented in the green color channel. The greatest differences in message level detected in this array are for the genes CDKN 1A ($p21^{Cip1/Waf1}$), which is much more abundant in the irradiated cells and therefore appears red, and MYC, which is more abundant in the unirradiated cells and therefore appears green.

5.1.2. Data Normalization. A large number of scalar efficiencies affect the fluorescent intensities present in the two image channels. Variations in the amount of message used to produce the labeled targets, efficiencies of incorporation of the fluor-tagged nucleotides, absorbtion and quantum efficiencies of the flurochromes, the strengths of the illuminating lasers, the transmission efficiencies of the interference filters, and the wavelength dependent sensitivities of the photomultipliers all effect the observed signal intensities. During image acquisition, bulk normalization of these scalar efficiencies is carried out by adjusting the photomultipliers' sensitivities so that the intensity at most probes is nearly equal. The degree of matching which can be obtained is demonstrated in Panel B of Figure 2, which shows the fluorescent intensity profile obtained by sampling a line of image pixels, which run through the center of immobilized probes in the vicinity of the probe for CDKN1A. As would be expected, when a wide sampling of genes is made, most of the genes will show similar levels of transcripts, even in similar cells responding to different stimuli, or dissimilar cell types.

In addition to this kind of bulk normalization, more refined normalization, based on the mean intensities of all or a subset of the probes can be carried out on the extracted image data. In comparisons between closely related cells, the bulk of the genes surveyed will have very close levels of expression, and normalization based on all genes will produce a good estimate of the best normalization, and of the expected variance in expression levels between genes. As the cells become more and more dissimilar, more genes show dissimilar levels of expression. In this case, it is useful to normalize with a subset of genes whose functional level is more likely to be comparable between cell types, so called housekeeping genes. The use of such subsets allows finer discrimination of what expression levels are similar and different between dissimilar cells through more accurate determination of the minimum expected variance between expression levels.[23]

An example of the tendency of expression profiles to broaden as cells become more different is presented in Figure 3. A comparison of a cell line against itself produces a very tight distribution of intensity values around the 1:1 diagonal, while a comparison of two different cell lines shows a much broader distribution of values around the diagonal. The mean intensity values for a subset of 88 housekeeping genes, while noticeably more distributed in the case of different cell lines are still less highly varied than the entire gene set.

Examination of hybridizations with a very high degree of concordance also illustrates a technical difficulty in the evaulation of median intensities at the lower limits of detection, where the intensity of the signal is very close to the background intensity. Small differences in the levels of non-specific assay backgrounds localized on either the immobilized probe, or the local background, produce an artificial difference in the observed mean intensities, and distorts the distribution of ratios derived from low intensity data. Caution in the interpretation of this segment of the data is clearly required.

5.1.3. Statistical Estimation of Expression Differences. In any analysis of the difference between expression ratios of differing cells, a measure of how statistically significant the observed differences are is a critical aid to interpretation. As previously mentioned, it is possible to construct a significance test on the basis of the observed level of variance between sets of genes expected to be invariant between the samples.[23] In practice, the observed variances of mean intensities from 1:1 of a subset of genes are used to calculate a probablility density function for the ratios. From this function, it is possible to estimate the extent of variance required to state at a specified level of confidence that a gene is not within the same distribution as genes that are invariant.

When this kind of variance analysis is performed on the data sets shown in Figure 3, the distributions of ratios observed are shown in the histograms of Figure 4. A curve representing the ratio distribution predicted by the variances of the housekeeping gene set is sketched over the histograms. In the case of the same cell comparison, the coefficient of variance (CV) of the housekeeping genes was small, 11.2%, and a 99% confidence interval for inclusion in the presumptive 1:1 distribution ranged from 0.65 to 1.53. For the more disparate cells, the CV of the housekeeping set was 17.6%, leading to a broader 99% confidence interval of 0.49 to 2.02. The very tight confidence interval predicted for the same cell comparison underestimates the effect of the observed ratio distortion of the lowest intensity genes seen in Figure 2, and thus a number of genes having ratios between 1.53 and 2 are presumably incorrectly identified as outside of the 1:1 distribution. A method of analysis which recognizes the increased difficulties of correct prediction of the weakest signals and broadens the confidence interval for these regions, needs to be developed.

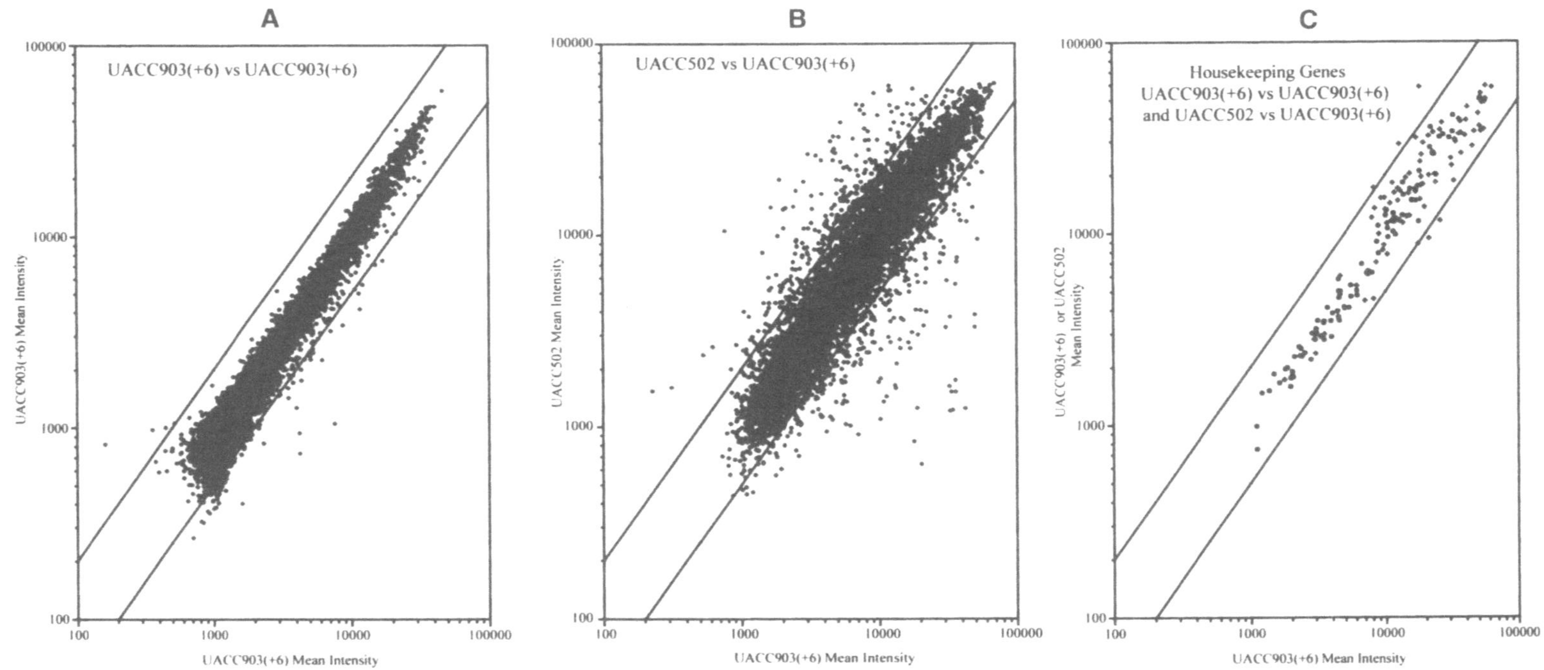

Figure 3. Scatterplots of mean probe intensities obtained when identical and different mRNA pools are used to produce target species. Panel A depicts the mean intensities of comparatively hybridized, fluorescently labeled cDNAs both derived from the tumorigenicity suppressed[32] melanoma cell line UACC903(+6) to an 8067 detector cDNA microarray. Panel B shows the mean intensity distribution of a hybridization of UACC903(+6) and a tumorigenic melanoma cell line UACC502. Panel C shows the mean intensity distributions of the housekeeping genes from the 903(+6) against itself (dots) and 903(+6) against UACC502 (crosses) shown in A and B. The solid lines are drawn at intervals of twofold change from equivalent fluorescent intensity.

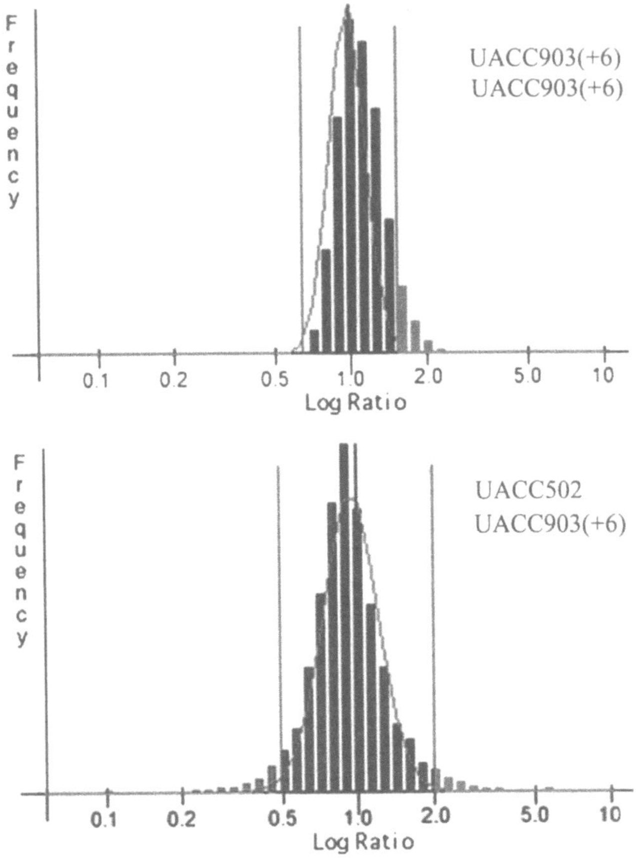

Figure 4. Histograms of the ratio distributions of genes when identical and different mRNA pools are used to produce target species. The data from Figure 3 are plotted to show the frequency distribution of clones having a particular ratio. A curve showing the predicted distribution of ratio frequencies based on the behavior of an 88 gene subset of the 8067 genes on the array is plotted in gray. Vertical lines represent the boundaries of a 99% confidence limit calculated on the basis of the distribution of the housekeeping genes.

5.2. Assay Reliability

For any new technology it is necessary to determine the reproducibility of the determinations, and to test the accuracy of the measurements against other means of carrying out the same measurement.

5.2.1. Reproducibility. Since each microarray experiment provides data from a large number of detection events, simple replication of an experiment provides sufficient data for a detailed analysis of reproducibility. Figure 5 shows the concordance of observed ratios between two separate measurements of the change in expression pattern of a cell line responding to ionizing radiation damage, as a function of the average mean intensity for the detection of that gene. In this set of experiments, the fluorescent dyes used to tag the irradiated and unirradiated cDNAs were switched between experiments to exaggerate any dye-specific variances.

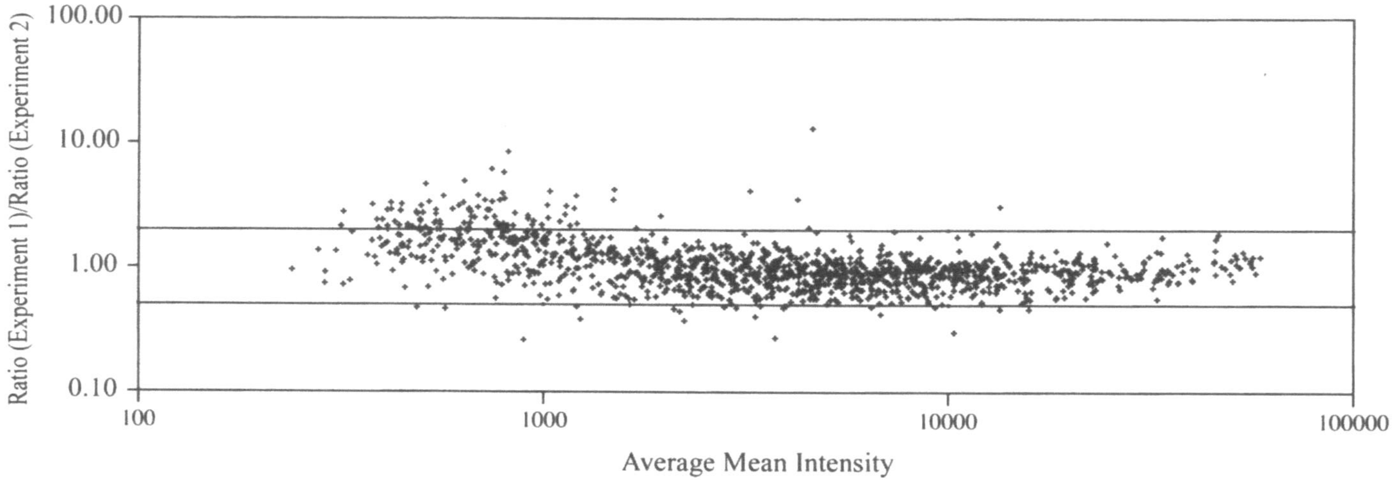

Figure 5. Inter-experiment ratio comparison. The ratios of the gene expression ratios for 1238 genes determined in duplicates of the experiment described in Figure 2 are plotted versus the mean fluorescent intensity of the unirradiated control for each gene. The solid lines are drawn at intervals of twofold deviation from ratio equivalence.

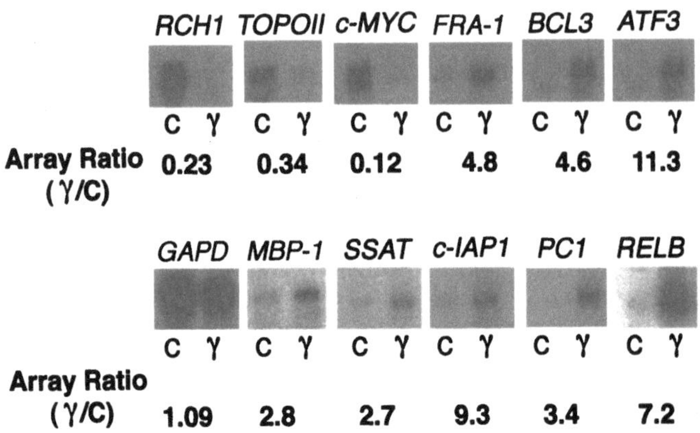

Figure 6. Comparison of array ratio determinations to Northern blots. Northern blots of mRNA prepared as described in Figure 2 were used to assay the agreement between Northern and array estimates of mRNA abundance. Blot lanes containing 1 μg of untreated control mRNA from cell line ML-1 (C) or 1 μg of mRNA from gamma irradiated ML-1 cells (γ), were probed with labelled EST PCR product identical to the DNA immobilized on the cDNA array as a detector for that gene. (From reference 24 by permission of the publisher.)

5.2.2. Accuracy. Microarray determinations of differences in expression have been approached with due skepticism by both the few labs having immediate access to the technology and the many who have obtained microarray data through collaborations. The general finding has been that changes on the order of two fold or more, observed in genes whose level of expression is several times the minimal detectable level can readily be detected as changes in the direction specified by Northern blotting, quantitative dot blot, or quantitative PCR. Sufficient reported data is not available to provide a strong assessment of the numerical agreement between the ratios determined by array and by other means, however the data available would indicate agreement to within approximately a factor of 2. Figure 6 and Table 1 provide example data sets of comparisons between expression ratios determined by arrays, Northern blots and quantitative dot blots.

6. ANALYSIS OF MULTIPLE DATA SETS

Few of the objectives of studying a large sampling of cellular gene expression can be met by a single comparison between a pair of samples. If a study of the cell's response to a change in environment or genetic composition is undertaken, then multiple samples over a time course are required to examine the complete cellular reaction. When attempting to discern the molecular commonalties and differences of a complex disease such as cancer, many examples of cancers diagnosed as the same need to be compared to determine the extent of commonality and the range of variations likely to be encountered. If the goal of the study is to examine the interconnections which constitute a particular cellular system, then measurements of the expression response of known members of that system in cells where other known and suspected members do not exhibit normal functionality would be advantageous. Each of these types of investigation will generate data sets too large to be systematically evaluated by simple inspection. Computational aids are therefore required to apply analytic methods and data filtration and then organize the presentation of the data in ways that highlight the various patterns being examined.

Table 1. Comparison of quantitative ratio estimates from cDNA microarrays and membrane blots[a]

Gene	Array	Hybridization	Control (mean intensity)
MYC	0.13	0.11	17,913
GADD153	1.5	1.8	6,939
MCL1	2	2.5	722
BCL-XL	2.4	1.7	313
BAK	2.7	1.9	341
MDM2	3.4	12	328
GADD45	5.8	32	516
CIP1/WAF1	44.5	91	743
RCH1	0.25	0.54	24,932
TOPOII	0.36	0.4	37,173
SATB1	0.48	0.61	30,364
BCL7A	1.8	1.2	4,284
ERCC2	1.8	0.6	621
IL-TMP	2.7	1.4	383
S/S NAC	2.8	5.5	4,160
MRC-OX	3.6	1.5	290
PC1	3.6	8.9	343
BCL3	4.9	9.7	584
FRA-1	5.1	4.5	672
RELB	7.6	28	398
IAP	9.8	3.4	132
ATF3	12	6.3	571
beta-actin	0.88		51,500

[a]RNAs from control and irradiated cell line ML-1 were prepared as described in Figure 2. Determinations of the expression ratios of the genes in this table were carried out by either cDNA microarray assay or by a quantitative dot blot. For the dot blot, serial dilutions of RNA were fixed to nylon filters and hybridized to either a probe for the target gene of interest or a polyU probe. The signal intensities were then determined by using the non-saturated protion of the curve, and the relative signal of the gene of interest was normalized to the amount of poly A in each sample, as determined by polyU hybridization.[33] The mean signal intensities (arbitrary units) detected for the control sample for each of these genes, as well as that of the beta actin gene are included for reference. (From reference 24 by permission of the publisher.)

Many of the analyses reported produce either subsets of the data which are the products of a particular filtering operation, or produce simplified representations of the relationships of the parent samples after a large set of expression values has been distilled into a much smaller set of numeric descriptors.

6.1. Characterization by Similar Expression Patterns in Subsets of Genes

Many variations on the theme of finding similar patterns of gene expression between cells are possible. The simplest form of this approach is to reduce the number of genes under consideration by filtering for a given magnitude of change and for a given percent of times when change exceeding this magnitude is observed. The extent of change can be an arbitrary or statistically defined magnitude. Such searches are computationally simple, and can readily be carried out with any of a number of inexpensive commercial programs. By simply looking for genes which change at the same time and in the same direction, it is possible to find new candidates for inclusion in well studied biological processes such as the shift of yeast metabolism from glucose to ethanol metabolism.[7]

More sophisticated forms of this mode of analysis couple intuitive representations of patterns of change with the well-developed mathematical methods of cluster analysis.[25,26,27] Such visualization leaves no doubt that the genes depicted are behaving in a very orderly fashion and are responding as part of a larger, integrated system. Both temporal expression patterns and patterns associated with cell types can be detected in this fashion. Studies at this level are very well suited to extending knowledge of cellular mechanisms by identifying genes whose expression profiles suggest that they play a role in a well established pathway. In the studies cited, new candidate participants in known cellular processes were observed, and interesting ways of linking expression data to other forms of data, such as promoter elements were shown to be applicable.

The use of such correlation/visualization tools to organize the presentation of data obviously gives the highest specificity of prediction in cases where the stimulus applied produces only a small number of expression changes. In this setting, new genes involved in that response may be rapidly identified and investigated by looking for other indications of co-regulation, such as common promoter elements.

6.2. Characterization on the Basis of Similar Expression States

Another approach to asking questions about the relationship of expression patterns and the behavior of the cell is to look at the overall differences in expression between cells. Such a study has been carried out on alveolar rhabdomyosarcoma (ARMS), a cancer having a very characteristic cytogenetic translocation, which fuses two genes to form a chimeric transcription factor.[28] The new gene contains a PAX DNA recognition domain and a FKHR transcription domain.[29] Expression profiles comparing 7 rhabdomyosarcoma lines and 6 other cancer lines against a control cell line were obtained. A simple visual measure of the relatedness of the ARMS lines relative to each other and to other types of cancers can be seen in Figure 7. This figure shows 12 scatterplots comparing the ratios for each gene between one of the ARMS lines, RMS13, and all of the other samples.

Taking a Pearson correlation coefficient for each of the possible pairwise combinations of cell lines can produce a quantitative measure of this similarity. The output of this analysis is a matrix of measurements from 0 to 1, where low values denote highly dissimilar expresssion profiles and high values closely matched profiles. Two informative ways of displaying this form of correlation are a multidimensional scaling plot and a hierarchical clustering dendogram, Figure 8. In the multidimensional scaling plot, the similarity of the cell types is represented as a map distance in a two-dimensional plot. The distance between cell types is adjusted to be as close to one minus the Pearson coefficient as possible. In such a map, cell types that are close to each other have similar expression profiles. The hierarchical clustering dendrogram representation uses a similar comparison metric, and clusters cell types in order of decreasing similarity. In both of these cases, the similarity of the ARMS cell lines, and their aggregate dissimilarity to other cancer cell lines is clear. It is worth noting that the most similar non-rhabdomyosarcoma line is TC71, a Ewing's sarcoma line, another cancer originating from muscle tissue. Such a finding suggests that efforts to compare profiles across

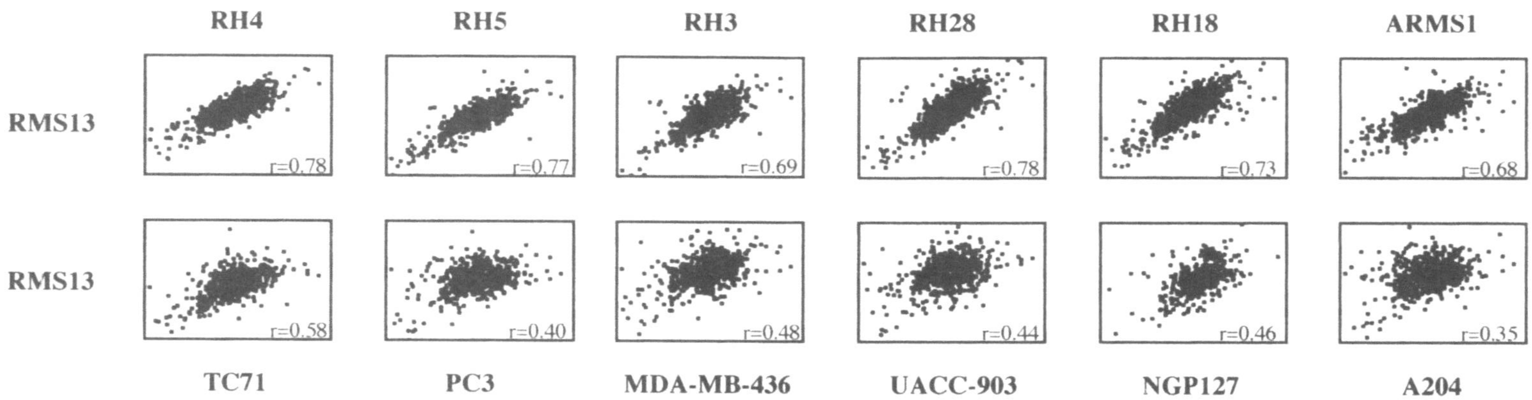

Figure 7. Comparison of expression ratios between an alveolar rhabdomyosarcoma cell line (RMS13) and twelve other cell lines. The top row of scatterplots shows the comparison of gene expression ratios between RMS13 and six other alveolar rhabdomyosarcoma cell lines. The bottom row of scatterplots compares expression in RMS 13 versus expression in six other cancer cell lines. A microarray containing 1238 ESTs was used in the hybridizations. (From reference 28 by permission of the publisher.)

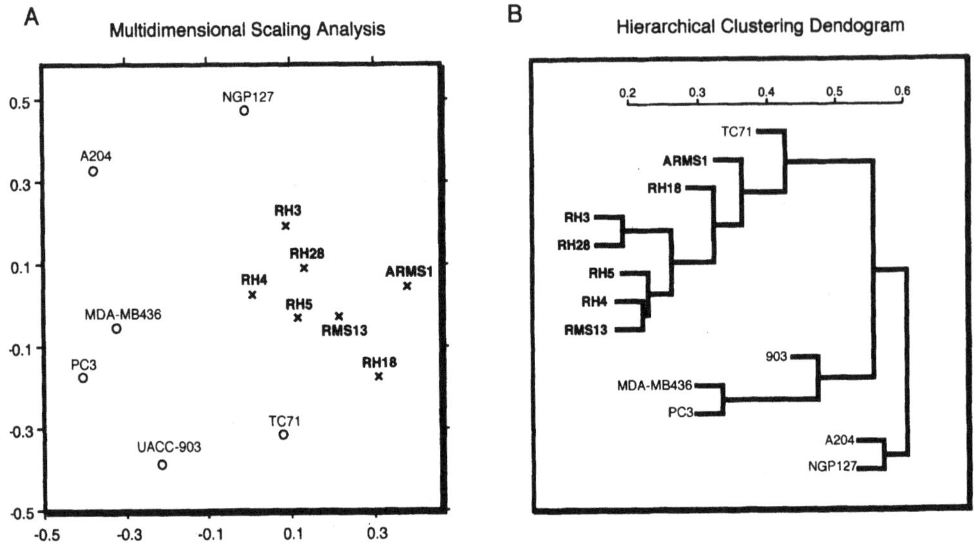

Figure 8. Graphical representations of the cumulated differences in mRNA levels between cells lines. A Pearson's correlation coefficient was calculated for each possible pairwise comparison of the 13 cell lines described in Figure 7. For the calculation, the data was filtered to include only ratio values from genes for which the mean intensity exceeded 2000 units for one of the cell lines, to avoid the inaccuracies associated with very low level detection (Figure 3). The values from the correlation matrix are then used to produce a multidimensional scaling (MDS) analysis, Panel A, or a hierarchical clustering dendogram (HCD), Panel B. In the MDS plot, the distances between the cell lines represent the best two-dimensional fit to 1-Pearson's Coefficient. Cell lines with identical expression patterns map to the same point, while a distance of 1 separates cell lines with entirely different expression patterns. HCD shows the clusters that arise from assembling the most closely related pairs, and then producing a dendogram that displays these clusters in order of decreasing similarity. In both panels, alveolar rhabdomyosarcoma cell lines are in dark, bold type and other cancer cell lines are in light, regular type. (From reference 28 by permission of the publisher.)

cancer types may disclose profile similarities arising from strictures imposed by the tissue of origin.

An exciting prospect for the application this form of profiling analysis is the study of cancers that do not exhibit such genetic uniformity. An early goal will be to attempt to discern subclasses, each of which is characterized by similar expression profiles. The possibility of finding subclasses which correlate tightly to responsiveness to therapy has been one of the most widely recognized clinical opportunities for expression profiling. A number of groups will undoubtedly begin to try and incorporate expression profiling in clinical trials in the near future.

6.3. Statistical Prediction of Expression Behavior in Varied Cell Contexts

While correlative analysis of expression data will undoubtedly provide new and valuable insights, the inherent ambiguity of correlation when applied to a very complex system suggests the need for complementary analytical tools. It is increasingly evident that the control of transcription is accomplished by mechanisms that readily interpret a large variety of inputs.[30,31] A sense of the extreme variety of responses which different cells mount to a fairly simple stimulus, genomic damage, can be gained by examining Table 2. This Table catalogs expression changes for a series of 12 genes across 12 cell lines

Table 2. Visualizing contextual effects on gene expression[a]

p53 wild type		*RCH1*	*BCL3*	*FRA1*	*REL-B*	*ATF3*	*IAP-1*	*PC-1*	*MBP-1*	*SSAT*	*MDM2*	*CIP1/WAF1*	*BAX*
ML-1	IR	0.4	**5.6**	**6.9**	**57.0**	**9.2**	**2.1**	**19.5**	**3.7**	**3.4**	**12.0**	**31.0**	**9.5**
	MMS	0.7	0.8	0.8	1.8	**7.3**	0.7	1.4	0.9	1.5	**9.2**	**11.3**	**2.1**
Molt4	IR	0.5	1.1	1.4	**9.5**	**13.7**	1.2	**10.0**	1.1	1.0	**4.5**	**31.0**	**5.8**
	MMS	1.4	1.0	**3.7**	0.5	**32.0**	1.3	1.3	0.6	1.2	1.1	**12.6**	0.9
SR	IR	0.5	0.9	1.7	**5.6**	**6.1**	**2.9**	**3.1**	**2.0**	1.2	**8.9**	**8.7**	1.2
	MMS	1.2	0.4	1.1	1.6	**10.3**	1.2	1.0	0.6	1.2	**3.2**	**6.3**	0.5
A549	IR	0.6	1.1	1.1	1.4	1.7	1.2	1.8	0.8	1.1	**4.8**	**7.5**	1.9
	MMS	0.8	0.5	1.9	0.5	**20.0**	1.2	1.4	0.8	1.1	1.6	**2.8**	1.1
	UV	0.8	0.6	1.5	0.6	**4.4**	0.6	1.5	0.5	1.4	0.5	**5.7**	1.3
MCF7	IR	0.4	1.8	**9.5**	**4.4**	1.6	1.0	0.6	1.1	1.1	**2.5**	**22.4**	**2.9**
	MMS	1.0	0.6	**4.8**	1.4	**22.0**	0.8	1.3	0.9	0.9	**2.8**	**5.0**	1.4
	UV	0.9	1.5	**24.7**	1.0	**3.9**	0.5	1.2	1.0	1.0	**2.3**	**9.8**	1.6
RKO	IR	0.7	**2.2**	0.7	**3.4**	**5.3**	**3.0**	**3.5**	1.7	1.4	**2.5**	**4.0**	1.7
	MMS	0.8	1.2	0.7	1.5	**39.0**	1.4	0.7	1.4	1.2	0.9	**6.4**	0.7
	UV	0.8	1.7	0.9	0.8	**4.6**	1.1	1.1	1.1	1.4	1.4	**2.7**	1.0
p53 mutant													
CCRF-CEM	IR	0.5	**3.8**	1.0	**13.4**	**10.7**	1.3	**11.9**	1.7	0.7	1.0	1.8	0.9
	MMS	0.8	0.8	1.2	0.5	**26.0**	0.9	0.8	1.1	0.5	1.2	0.7	1.2
HL60	IR	0.5	**6.2**	1.9	**8.3**	**3.2**	1.0	**5.2**	1.6	**2.2**	1.0	**2.0**	0.5
	MMS	0.9	1.1	**2.7**	1.3	**9.4**	1.2	1.7	0.9	0.8	**3.6**	**5.9**	0.7
K562	IR	0.7	1.4	1.2	1.6	0.9	1.0	1.1	1.3	1.0	1.2	1.4	1.0
	MMS	1.0	1.7	1.2	0.8	**37.0**	0.9	0.8	0.8	1.3	1.2	ND	0.9
H1299	IR	0.7	0.7	0.7	**2.1**	1.0	1.3	**4.7**	1.1	0.9	1.0	1.3	1.1
	MMS	0.9	0.7	0.6	1.0	**21.0**	0.5	1.9	0.9	0.7	0.6	**6.3**	0.9
	UV	0.6	0.6	0.5	1.2	**3.8**	0.5	**3.8**	0.7	0.8	0.9	**2.2**	1.5
RKO-E6	IR	0.2	**3.1**	0.5	**7.2**	1.4		**4.6**	1.4	1.0	1.6	1.5	1.3
	MMS	0.3	1.3	1.1	0.8	**42.0**	1.1	0.8	1.3	0.8	0.9	**2.3**	0.7
	UV	0.2	1.0	0.7	0.7	**2.8**	0.5	0.6	0.7	0.9	0.5	**2.8**	1.0
T47D	IR	0.7	1.8	0	**4.8**	1.5	1.6	1.9	1.1	0.9	0.8	**2.2**	0.7
	MMS	0.8	0.8	0	1.6	**38.0**	1.5	0.7	0.8	0.9	0.7	**3.9**	0.8
	UV	1.1	1.3	0	1.9	**3.5**	1.5	**3.3**	0.5	1.2	0.9	**2.7**	0.8

[a]A set of genes found to have altered expression levels following exposure to ionizing radiation were characterized for their responsiveness to three forms of genotoxic stress in a panel of cancer cell lines. The relative amounts of mRNA from a cell line four hours after exposure to ionizing radiation (IR), methyl methane sulfonate (MMS) or ultraviolet radiation (UV) versus an untreated control are shown. Ratios were determined by the blot method described in Table 1.

exposed to different genotoxic stresses. All of the genes respond with a strong change in expression level in the cell line ML-1 when it is exposed to ionizing radiation, as seen in the first row of the Table. All of the genes also change expression in at least one other cell line and treatment, however the variation in responsiveness across the different cells and treatments is quite high.

This type of data suggests a possible way to approach the thorny problem of finding specific gene interactions via expression data. By examining expression across a wide sampling of cell types and varied stimuli, it should be possible to find relationships in which knowledge of the states of a set of genes will accurately predict the state of another gene. Finding such relationships in the very large sets of data that would be required to expose minimal predictive sets will be computationally challenging. Even sifting a set of four genes capable of accurately predicting a fifth gene from a set of five hundred genes assayed under several hundred conditions is a daunting task using the best tools now available in probabilistic multivariate analysis.

7. CONCLUSION

Control of gene transcription is one of the key ways in which cells control their activities. As it becomes possible to obtain a more panoramic view of the consequences of transcriptional control, new and interesting questions can be framed at many different levels. At the simplest levels we can ask whether such studies will readily identify molecular targets for therapeutics. At a higher level, we can ask whether the overall picture of gene expression can become the basis for more refined stratification of complex diseases into uniform subtypes. At the most general level, it will become possible to experimentally examine the workings of a control system that is more robust and more highly integrated than any human designed system.

The pursuit of the organizational principles of a complex adaptive system as complicated as a cell seems likely to accelerate the growth of the study of complexity. The availability of suitable experimental data will sharpen the collaborative efforts between theoretical and experimental biologists and those mathematicians, engineers and computational scientists already involved the study of such systems. Basic biological concepts such as the ability to evolve systems by variation and selection are now beginning to have serious impacts in engineering and computation. It seems likely that the powerful analytic tools that have been developed in mathematics, engineering and computation will likewise provide biologists with radically new ways to study living systems.

REFERENCES

1. B. Phimister (ed.), "The chipping forecast," Nat Genet 21, no. 1 Suppl (1999).
2. A. Goffeau et al., "Life with 6000 genes," Science 274, no. 5287(1996):546, 563–7.
3. The C. elegans Sequencing Consortium, "Genome sequence of the nematode C. elegans: a platform for investigating biology.," Science 282, no. 5396(1998):2012–8.
4. G. Lennon et al., "The I.M.A.G.E. Consortium: an integrated molecular analysis of genomes and their expression," Genomics 33, no. 1(1996):151–2.
5. G.D. Schuler et al., "A gene map of the human genome," Science 274, no. 5287(1996):540–6.
6. Herbert Alexander Simon, The sciences of the artificial, 3rd ed. (Cambridge, Mass.: MIT Press, 1996).
7. J.L. DeRisi, V.R. Iyer, and P.O. Brown, "Exploring the metabolic and genetic control of gene expression on a genomic scale," Science 278, no. 5338(1997):680–6.
8. S.A. Amundson, T.G. Myers, and A.J. Fornace, Jr., "Roles for p53 in growth arrest and apoptosis: putting on the brakes after genotoxic stress," Oncogene 17, no. 25(1998):3287–99.
9. H.M. Blau, "Regulating the myogenic regulators," Symp Soc Exp Biol 46(1992):9–18.
10. M.D. Adams et al., "Complementary DNA sequencing: expressed sequence tags and human genome project," Science 252, no. 5013(1991):1651–6.
11. K. Okubo et al., "Large scale cDNA sequencing for analysis of quantitative and qualitative aspects of gene expression [see comments]," Nat Genet 2, no. 3(1992):173–9.
12. V.E. Velculescu et al., "Serial analysis of gene expression" Science 270, no. 5235(1995):484–7.
13. J.C. Alwine, D.J. Kemp, and G.R. Stark, "Method for detection of specific RNAs in agarose gels by transfer to diazobenzyloxymethyl-paper and hybridization with DNA probes," Proc Natl Acad Sci USA 74, no. 12(1977):5350–4.
14. L.H. Augenlicht et al., "Patterns of gene expression that characterize the colonic mucosa in patients at genetic risk for colonic cancer," Proc Natl Acad Sci U S A 88, no. 8(1991):3286–9.
15. G. Pietu et al., "Novel gene transcripts preferentially expressed in human muscles revealed by quantitative hybridization of a high density cDNA array," Genome Res 6, no. 6(1996):492–503.
16. M. Schena et al., "Quantitative monitoring of gene expression patterns with a complementary DNA microarray," Science 270, no. 5235(1995):467–70.
17. D.J. Lockhart et al., "Expression monitoring by hybridization to high-density oligonucleotide arrays," Nat Biotechnol 14, no. 13(1996):1675–80.

18. J. DeRisi et al., "Use of a cDNA microarray to analyse gene expression patterns in human cancer," Nat Genet 14, no. 4(1996):457–60.
19. Charles Tanford, Physical chemistry of macromolecules (New York: Wiley, 1961).
20. J.O. Bishop et al., "Three abundance classes in HeLa cell messenger RNA," Nature 250, no. 463(1974):199–204.
21. J. Phillips and J.H. Eberwine, "Antisense RNA Amplification: A Linear Amplification Method for Analyzing the mRNA Population from Single Living Cells," Methods 10, no. 3(1996):283–8.
22. J.J. Chen et al., "Profiling expression patterns and isolating differentially expressed genes by cDNA microarray system with colorimetry detection," Genomics 51, no. 3(1998):313–24.
23. Y. Chen, E.R. Dougherty, and M.L. Bittner, "Ratio-based decisions and the quantitative analysis of cDNA microarray images," J. Biomed Optics 2, no. 4(1997):364–74.
24. S.A. Amundson et al., " cDNA microarray hybridization reveals complexity and heterogeneity of cellular genotoxic stress responses.," Oncogene in press (1999).
25. M.B. Eisen et al., "Cluster analysis and display of genome-wide expression patterns," Proc Natl Acad Sci U S A 95, no. 25(1998):14863–8.
26. V.R. Iyer et al., "The transcriptional program in the response of human fibroblasts to serum," Science 283, no. 5398(1999):83–7.
27. P.T. Spellman et al., "Comprehensive identification of cell cycle-regulated genes of the yeast Saccharomyces cerevisiae by microarray hybridization," Mol Biol Cell 9, no. 12(1998):3273–97.
28. J. Khan et al., "Gene expression profiling of alveolar rhabdomyosarcoma with cDNA microarrays," Cancer Res 58, no. 22(1998):5009–13.
29. J.A. Biegel et al., "Detection of the t(2;13)(q35;q14) and PAX3-FKHR fusion in alveolar rhabdomyosarcoma by fluorescence in situ hybridization," Genes Chromosomes Cancer 12, no. 3(1995):186–92.
30. H.H. McAdams and L. Shapiro, "Circuit simulation of genetic networks," Science 269, no. 5224(1995):650–6.
31. C.H. Yuh, H. Bolouri, and E.H. Davidson, "Genomic cis-regulatory logic: experimental and computational analysis of a sea urchin gene [see comments]," Science 279, no. 5358(1998):1896–902.
32. J.M. Trent et al., "Tumorigenicity in human melanoma cell lines controlled by introduction of human chromosome 6," Science 247, no. 4942(1990):568–71.
33. M.C. Hollander and A.J. Fornace, Jr., "Estimation of relative mRNA content by filter hybridization to a polythymidylate probe," Biotechniques 9, no. 2(1990):174–9.

3

LARGE SCALE EXPRESSION SCREENING IDENTIFIES MOLECULAR PATHWAYS AND PREDICTS GENE FUNCTION

Nicolas Pollet,[1] Volker Gawantka,[1] Hajo Delius,[2] and Christof Niehrs[1]

[1]Division of Molecular Embryology
[2]Division of Applied Tumorvirology
Deutsches Krebsforschungszentrum
Im Neuenheimer Feld 280
D-69120 Heidelberg
Germany

1. INTRODUCTION

The genome of a given organism is considered in biology as the fundamental invariant (Monod, 1970). It is virtually the same throughout lifetime and, to a lesser extent, over generations. In contrast, genetic information is expressed in complex and ever-changing temporal and spatial patterns throughout development and differentiation. The description and analysis of these patterns is crucial to elucidate the functions of genes and to understand the network of genetic interactions that underlies the process of normal development.

While the study of the expression pattern of a gene is a prerequisite to understand its physiological function, the characterisation of the expression of most known genes is incomplete. As a consequence it is almost impossible to compare gene expression patterns, and there is no specialised public databases available storing the data. At the same time, genome science has to bridge the gap between DNA sequence and function. To date, the study of cDNA copies of mRNAs have proven to be the most efficient way for large scale gene identification and analysis. The additional information as to where and when a mRNA is present will be essential to help elucidating gene function. Databases of gene expression are needed as a resource for the emerging field of functional genomics. Yet, most of existing methodologies used to characterise gene expression are not amenable to systematic studies using large number of samples.

Genomics and Proteomics, edited by Sándor Suhai.
Kluwer Academic / Plenum Publishers, New York, 2000.

The generation of the expression data for large numbers of genes should be a means of placing newly characterised sequences into context with respect to their sites of expression, to study the correlation between gene expression and function, and to correlate the expression profiles with regulatory sequences.

Genetic analysis of development in invertebrates such as *Drosophila* or *Caenorhabditis elegans* has proven to be a powerful approach to study developmental mechanisms (Miklos and Rubin, 1996). For example most of the genes known to be involved in hedgehog, dpp/BMP and wnt signalling pathways were identified through classical genetic screens in *Drosophila*. The characterisation of these genes and their vertebrate homologues has greatly advanced our understanding of cell signalling pathways that regulate development.

Genetic screens, however, have significant limitations. Genes with subtle loss-of-function phenotypes or genes whose function can be compensated for by other genes or pathways are unlikely to be found. These two classes of genes may represent the majority of genes in the fly, since it is estimated that two-thirds of *Drosophila* genes are not required for viability (Miklos and Rubin, 1996). In addition, screens designed to identify specific phenotypic defects often do not recover genes with pleiotropic roles during development, since the requirement for gene function in one developmental process can mask its requirement in another.

To identify all classes of developmentally important genes, expression-based and other molecular screens are needed to supplement classical genetic screens. In *Drosophila*, the most productive of these screens to date have used P element-based enhancer traps, but P element insertion is not random and enhancer trap screens are biased toward identifying genes that are favoured for insertion by P elements (Spradling et al., 1995; Kidwell, 1986). In a screen based on in situ hybridisation, 80% of the genes found were not previously described, underscoring the potential of this approach (Kopczynski et al., 1998).

We present a large-scale screen for genes that are expressed in specific tissue or cell-types during embryonic development in *Xenopus* (Gawantka et al., 1998). The approach used is a high throughput procedure of whole-mount in situ hybridisation to mRNA, followed by sequence analysis. The results have been compiled in a publicly available database, Axeldb.

2. STRATEGY

Spatial and temporal embryonic expression profiles of the genes represented in a neurula stage cDNA library were determined by RNA in situ hybridisation to whole-mount *Xenopus* embryos (Harland, 1991). This developmental stage was selected because most of the genes expressed during gastrulation are still transcribed, and genes involved in neurogenesis are already active.

RNA probes were prepared from individual, randomly picked cDNA clones and screened on albinos embryos at stages gastrula, neurula and tailbud. This enabled to characterize gene expression at the critical phases of mesoderm regionalization, neurogenesis and organogenesis.

When a restricted expression pattern was observed, it was described in a semi-quantitative way and pictures of stained embryos taken. The corresponding cDNAs were partially sequenced.

3. RESULTS

3.1. Expression Pattern Analysis

Of 1765 clones screened, 449 (26%) represent genes expressed in specific patterns during embryogenesis (Figure 1), whereas 51% of the cDNAs showed ubiquitous pattern of expression and 23% did not produce detectable staining in the embryo.

A wide variety of temporal and spatial expression patterns were observed, (examples in Figure 2). The frequency of gene expression at different stages and in various tissues is summarised in Table 1. The most prominent figure is the increase in the complexity of gene expression patterns as development proceeds, and notably in the central nervous system (82% of genes at stage 30) and in the tailbud region. In *Xenopus* embryos, the expression in endoderm can not be reliably assessed due to the limitations of the whole-mount procedure, where penetration of tissues rich in yolk is a problem (Harland, 1991).

The comparison of expression pattern led to the identification of four groups of genes with shared, complex expression pattern that we refer as synexpression groups (Table 2).

The Bmp4 group consists of six members (two isolated in this study) which all encode components of the BMP signalling pathway as studied in early dorso-ventral patterning of mesoderm, including ligands, receptor and downstream components of the pathway (Hogan, 1996). The expression pattern of these genes is similar to the growth factor itself, and *Bmp4* indeed coordinately induces them.

The genes in the endoplasmic reticulum (ER) group are highly expressed in tissues active in secretion. Genes of this group act in the early steps of secretion (Rothblatt et al., 1994), either in translocation (e.g. translocon subunits) or in protein folding in the ER (protein disulphide isomerase). The common regulatory mechanism of this group is unknown but it suggests a transcriptional feedback between the secretory load of a cell and the expression of key components involved in protein translocation across the ER.

The Delta1 group includes mostly bHLH genes that are expressed in a characteristic pattern of this ligand of the Notch receptor, including the central nervous system and the forming somites (Chitnis et al., 1994). The possibility that members of this group function in the Notch pathway has been confirmed by functional analysis of two novel members of this group (C. Kintner, E. Bellefroid, T. Pieler, pers. commun.). The shared expression is likely due to Delta1 responsive elements in the gene's promoters.

The largest synexpression group identified is the chromatin group. Characteristic for these genes is their repression in tissues becoming postmitotic. Most of these genes are known to encode chromatin proteins (e.g. histones, HMG proteins), or genes indirectly interacting with chromatin such as ornithin decarboxylase, a key enzyme in spermidin synthesis. The common regulatory mechanism of this group is also unknown but it is likely cell cycle related.

3.2. Sequence Analysis

For each differential expression pattern observed, we sequenced the 5′ and 3′ ends of the corresponding cDNA. By sequence analysis, we could identify redundant sequences and clones, and concluded that 273 genes were identified. The most abundant cDNA clones found were identified as being derived from genes coding high mobility protein, histone H3 and 16S mitochondrial RNA. The results of sequence similarity

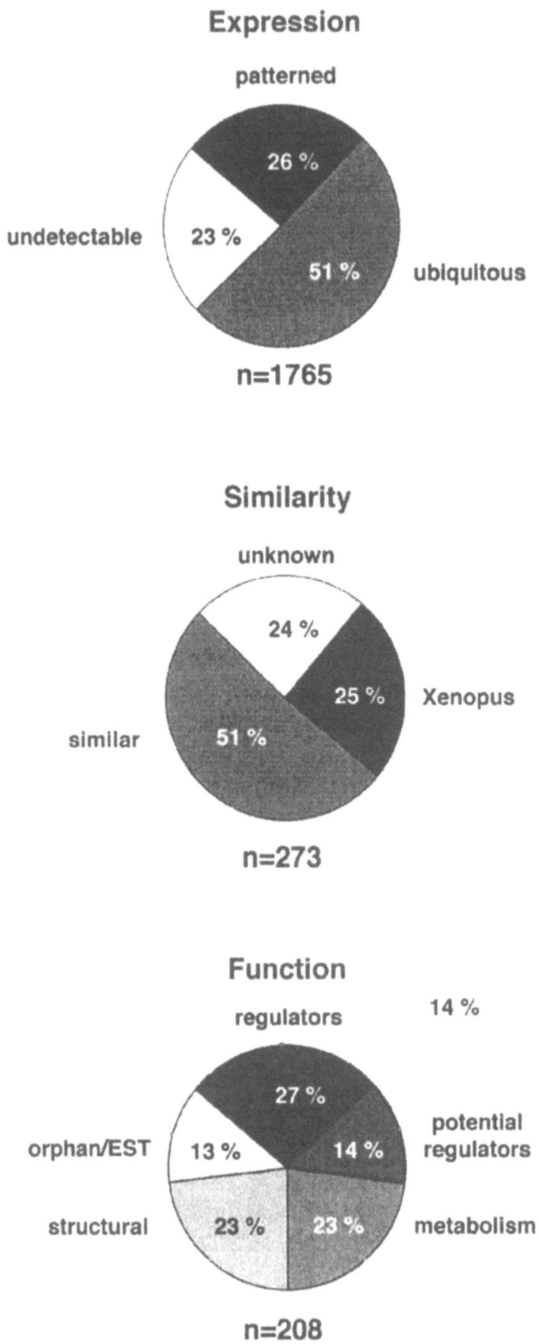

Figure 1. Overview of expression and sequence informations. Classification of the clones according to gene expression pattern, sequence similarity and predicted function (top, middle and bottom respectively). Values are given as percentages of total number of cDNAs examined (n = 1765), the number of unique, differentially expressed genes (n = 273) and the number of unique, differentially expressed genes with sequence similarity (n = 208).

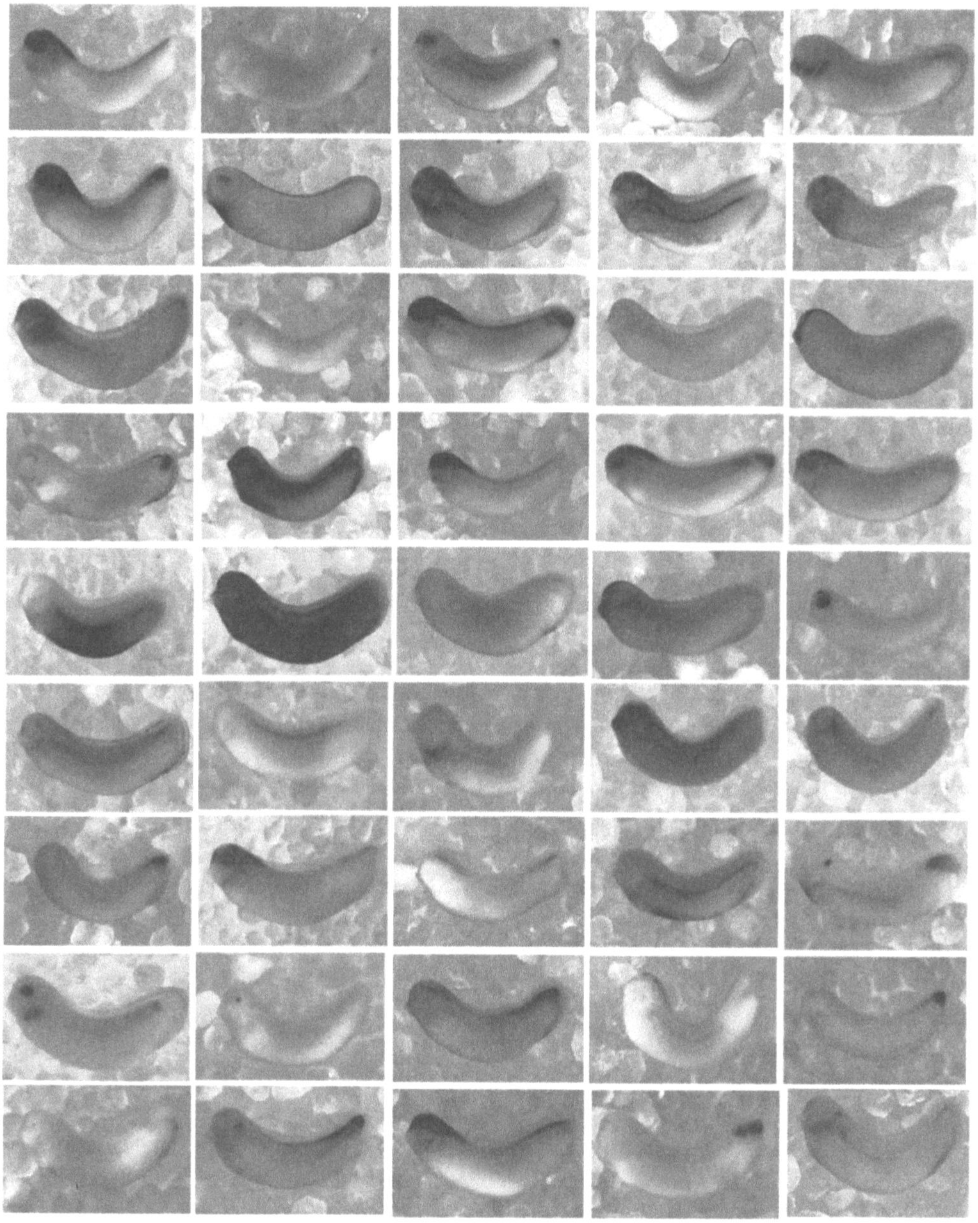

Figure 2. Expression of a subset of genes. Whole-mount in situ hybridizations of tailbud embryos are shown in lateral view, anterior is to the left and dorsal to the top. The gene names from top to bottom and left to right are: 2.9, 2.15, 3.14, 5A18, 5C21, 5D9, 5E23, 5I4, 6A5, 6D6, 6D16, 8C1, 8C9, 8F9, 9B4, 9C8, 9D1, 10A5, 10C6, 11A10, 11E2, 12A4, 12F1, 12F11, 13F8, 13H2, 14E5, 16E2, 17A1, 17C3, 17G2, 19F1.1, 19G2.1, 21E1.1, 22F11.1, 23E9.1, 23F2.1, 23G1.2, 25A6.1, 26C1.1, 26C10.1, 26E7.1, 30F5.2, 32B3.1, 32B12.2 (Pollet et al. 1999).

Table 1. Frequency of gene expression

	n	%
Developmental stage		
Gastrula	220	81%
Neurula	253	93%
Tailbud	269	99%
Tissues at tailbud stage		
Brain	197	72%
Spinal cord	179	66%
Eye	178	65%
Ear vesicle	192	70%
Nasal vesicle	164	60%
Epidermis	175	64%
Cement gland	84	31%
Hatching gland	115	42%
Notochord	92	34%
Somites	111	41%
Pronephros	145	53%
Lateral plate	109	40%
Blood	105	38%
Visceral arches	174	64%
Proctodeum	147	54%
Tailbud	179	66%

searches, both at the nucleotidic and proteic level are outlined in Figure 1. We made a classification of those genes with attributable function (Figure 1) and observed that 27% represent regulators (growth or transcription factors, receptors, signal transducers).

3.3. Data Availability

A Xenopus laevis database (Axeldb) was developed with the aim to compile the expression patterns, the DNA sequences and associated informations coming from this study. We used ACEDB (A *Caenorhabditis elegans* database) as our database management system (Durbin and Thierry-Mieg, 1991). ACEDB is publically available and widely used in many genomic centers, its basic data model is easy to tailor, and it comes with powerful data visualization capabilities. We modified the basic ACEDB data model by adding objects with information specific for expression patterns, synexpression groups and expression domains. ACEDB provides a convenient framework for browsing and manipulating the integrated results, as well as a scriptable access and a web interface (Stein and Thierry-Mieg, 1999). Access to Axeldb can be made in two ways. First, a web interface is available at the URL: http://www.dkfz-heidelberg.de/abt0135/axeldb.htm (Figure 3). Second, data (including pictures) and models for the UNIX version of ACEDB are available at the ftp server ftp.dkfz-heidelberg.de in outgoing/abt0135/axeldb. Users can query the database through class objects: clone, expression pattern, expression domain, tissue and through sequence similarity searches.

4. CONCLUSION

We used a whole-mount in situ hybridisation based screen in *Xenopus* embryos to identify differentially expressed genes during early development. The expression profiles

Table 2. Synexpression groups[a]

BMP4 GROUP: ***dorsal eye, ventral branchial arches, posterior dorsal fin edgelproctodeum***		
not isolated	Bmp4	TGFb growth factor
not isolated	XBMPRII	BMP type II receptor
not isolated	Smad6	signal transduction, inhibitory smad
not isolated	Smad7	signal transduction, inhibitory smad
9C8	Xvent2	homeobox transcription factor
5E23	putative transmembrane protein	transmembrane protein
DELTA1 GROUP: ***central nervous system, eyes, tailbud, forming somites***		
not isolated	XDeltal	Notch receptor ligand
5D9	protein with ankyrin repeats	protein/protein interaction
8C9	HES5 related	bHLH transcription factor
11A10	HES5 related	bHLH transcription factor
10C6	HES1 ralated	bHLH transcription factor
ER-IMPORT GROUP: ***strong in cement gland, pronephros, notochord; weak ubiquitous***		
27H8.1	SEC61 a	subunit of ER protein conducting channel
25C5.1	SEC61 b	subunit of ER protein conducting channel
1.16	SEC61 g	subunit of ER protein conducting channel
22A8.1	translocon associated protein b	subunit of translocon
3.40	translocon associated protein g	subunit of translocon
9C5	protein disulfide isomerase	ER-located enzyme
18F9	no homology	–
CHROMATIN GROUP: ***not in cement gland, notochord, anterior somites; strong in all other regions***		
30F11.1	histone H2A	chromatin-associated protein
21H2.1	histone H3	chromatin-associated protein
26E1.1	HMG1	chromatin-associated protein
27C9.2	HMG2	chromatin-associated protein
11G6	HMG14	chromatin-associated protein
12G2	HMG17	chromatin-associated protein
22C2.2	thyroid rec. intractor (HMG)	chromatin-associated protein
5B20	NAP1	chromatin-associated protein
16H8	NO38	chromatin-associated protein
5F8	modifier 2 protein	chromatin-associated protein
19C7.1	prothymosin a1	chromatin-associated protein
32C10.1	hnRNP U	chromatin-associated protein, splicing
5C2	CArG-binding factor A-related	transcription factor
14E10	CArG-binding factor A-related	transcription factor
19E1.1	NF45	transcription factor
5J20	hnRNP K	transcription factor, RNA/ssDNA-binding
23G4.1	protein arginine N-methyltransferase	hnRNP/histone methylase
32E11.2	ornithine decarboxylase	polyamine biosynthesis, chromatin structure
29A11.2	hnRNPA1	nuclear shuttling protein, splicing
32F8.1	SRP 20	splicing factor
22F1.1	Smt 3	suppressor for centromere mutant MIF2 (yeast)
29C5.2	EST	–
10A8	EST	–
26F2.1	no homology	–
27G8.2	no homology	–

[a]Sequence similarities of cDNAs belonging to the synexpression groups are listed. A brief description of the expression pattern is given in the headline of each group. Clone ID, sequence similarity and putative function are listed. Within a group clones are sorted according to related function.

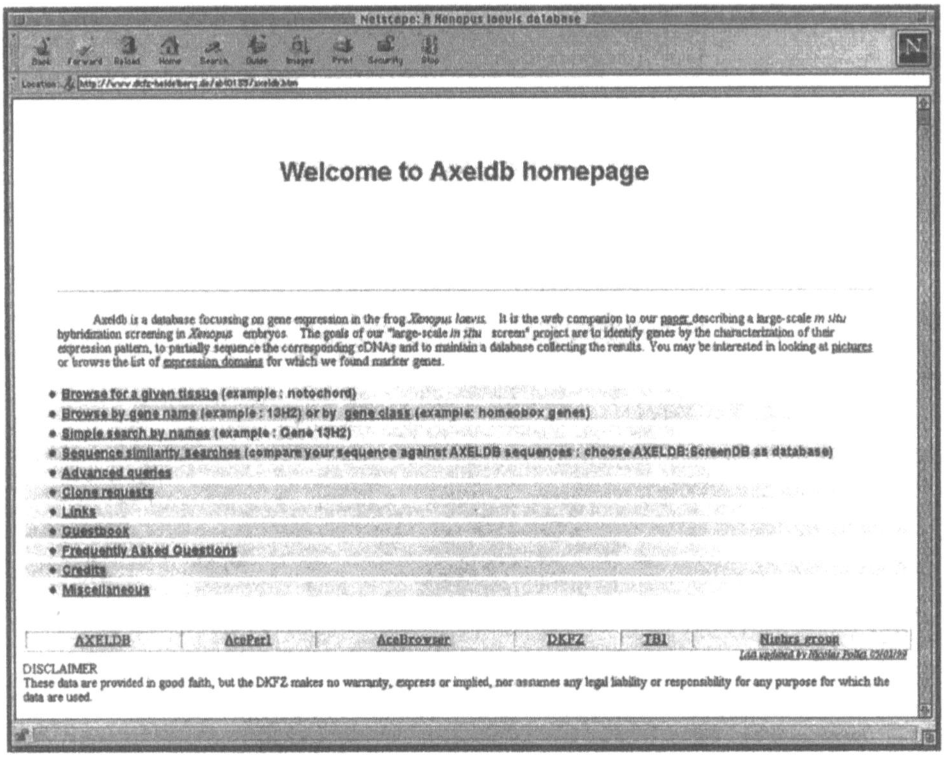

Figure 3. The Axeldb homepage. The URL is <http://www.dkfz-heidelberg.de/abt0135/axeldb.htm>.

of 273 genes and their associated sequence information is available on a public database, Axeldb.

By comparing expression profiles, we identified groups of genes with shared, complex expression pattern which also share function. These synexpression groups predict molecular pathways involved in patterning and differentiation. Within groups, strong predictions can be made about the function of genes without sequence similarity. These results indicate that large scale expression screening is an alternative to identify molecular pathways and elucidate gene function of unknown genes.

A great advantage of the in situ screen is the immediate availability of the cloned cDNA, which readily allows a gain-of function test by microinjection of synthetic mRNA in Xenopus. By this approach two novel homeobox genes discovered in this screen could be implicated in dorso ventral mesoderm patterning (Gawantka et al., 1995; Onichtchouk et al., 1996).

Using filter-arrayed cDNA libraries, robotic processing of DNA and RNA probes and automated whole-mount in situ hybridization, gene expression screening can be largely automated (our unpublished results). Hence, there is the perspective of carrying out a saturating analysis of embryonic gene expression.

ACKNOWLEDGMENTS

We gratefully acknowledge C. Blumenstock, K. Hirschfeld, R. Nitsch and R. Pfister for technical assistance. Our many thanks goes to H. Schmidt and M. Vingron for their invaluable bioinformatic support. This study was supported by a grant from the DFG to C.N. and a Marie Curie Fellowship to N.P.

REFERENCES

Chitnis, A., Henrique, D., Lewis, J., Ish, H.D., and Kintner, C. (1995) Primary neurogenesis in Xenopus embryos regulated by a homologue of the Drosophila neurogenic gene Delta. Nature 375, 761–6.

Durbin, R. and Thierry-Mieg, T. (1991) A *C. elegans* database. Documentation, code and data available from anonymous FTP servers at lirmm.lirmm.fr, cele.mrc-lmb.cam.ac.uk and ncbi.nlm.nih.gov.

Gawantka, V., Delius, H., Hirschfeld, K., Blumenstock, C., and Niehrs, C. (1995). Antagonizing the Spemann organizer: role of the homeobox gene Xvent-1. Embo J. 14, 6268–6279.

Gawantka, V., Pollet, N., Delius, H., Vingron, M., Pfister, R., Nitsch, R., Blumenstock, C., and Niehrs, C. (1998) Gene expression screening in Xenopus identifies molecular pathways, predicts gene function and provides a global view of embryonic patterning. Mech. Dev. 77, 95–141.

Harland, R.M. (1991) In situ hybridization: an improved whole-mount method for Xenopus embryos. Methods Cell Biol 36, 685–95.

Hogan, B.M. (1996) Bone morphogenetic proteins:multifunctional regulators of vertebrate development. Genes Dev. 10, 1580–1594.

Kidwell, M.G. (1986) in *Drosophila*: A practical approach, ed Roberts, E.D. (IRL, Washington DC), pp 59–83.

Kopczynski, C.C., Noordermeer, J.M., Serano, T.L., Chen, W.-C., Pendleton, J.D., Lewis, S., Goodman, C.S., and Rubin, G.M. (1998) A high throughput screen to identify secreted and transmembrane proteins involved in *Drosophila* embryogenesis. Proc. Natl. Acad. Sci. USA 95, 9973–9978.

Miklos, G.L. and Rubin, G.M. (1996) The role of the genome project in determining gene function: insights from model organisms. Cell 86, 521–529.

Monod, J. (1970) Le hasard et la nécessité. Editions du Seuil, Paris.

Onichtchouk, D., Gawantka, V., Dosch, R., Delius, H., Hirschfeld, K., Blumenstock, C., and Niehrs, C. (1996) The Xvent-2 homeobox gene is part of the BMP-4 signaling pathway controling dorsoventral patterning of Xenopus mesoderm. Development 122, 3045–3053.

Rothblatt, J., Novick, P., and Stevens, T. (1994) Guidebook to the Secretory Pathway, pp. New York: Oxford University Press.

Spradling, A.C., Stern, D.M., Kiss, I., Roote, J., Laverty, T., and Rubin, G.M. (1995) Gene disruptions using P transposable elements: an integral component of the *Drosophila* genome project. Proc. Natl. Acad. Sci. USA 92, 10824–10830.

Stein, L.D. and Thierry-Mieg, J. (1998) Scriptable access to the Canorhabditis elegans genome sequence and other ACEDB databases. Genome Res. 8, 1308–1315.

4

THE GLEAN MACHINE

What Can We Learn from DNA Sequence Polymorphisms?

Daniel L. Hartl,[1] E. Fidelma Boyd,[2] Carlos D. Bustamante,[1]
and Stanley A. Sawyer[3]

[1]Department of Organismic and Evolutionary Biology
Harvard University
Cambridge, Massachussetts 02138
[2]Department of Geographic Medicine and Infectious Diseases
Tupper Institute, Tufts University Medical School
Boston, Massachussetts 02111
[3]Department of Mathematics
Washington University
St. Louis, Missouri 63130

ABSTRACT

Nucleotide sequences contain hidden information about the forces for conservation and variation that shaped their evolutionary history. To glean sequences for hidden information motivates the study of similarities in sequence among orthologous and paralogous coding sequences, and also gives impetus for improved methods of phylogenetic estimation and hypothesis testing. Variation within populations is also evidential for evolutionary history. Within coding sequences, different patterns of variation are often observed between nonsynonymous nucleotide substitutions, which cause amino acid replacements, and synonymous nucleotide substitutions, which do not. For some coding sequences these differences are consistent with an evolutionary scenario featuring greater functional constraints on amino acid sequences than on nucleotide sequences. We have developed a sampling theory of selection and random genetic drift for interpreting the numbers of wildtype and variant nucleotides found among the polymorphic sites present in sequences of multiple alleles of a gene. This sampling theory has been used to interpret the patterns of intrapopulation polymorphism of 28 genes in *Escherichia coli* and *Salmonella enterica*, each gene exhibiting greater than 50 polymorphic sites among the alleles examined. Many of these genes have an excess of singleton amino acid

Genomics and Proteomics, edited by Sándor Suhai.
Kluwer Academic / Plenum Publishers, New York, 2000.

polymorphisms, relative to the number of singleton synonymous polymorphisms. (A singleton polymorphism is one in which the sample is monomorphic except for a single variant.) In 22/28 genes, there is a greater proportion of singleton nonsynonymous polymorphisms than the proportion of singleton synonymous polymorphisms, and in 8 genes this excess is statistically significant. This pattern is consistent with a model in which most amino acid polymorphisms are slightly deleterious and hence present in samples at lower than expected frequencies. Furthermore, the sampling distribution of polymorphic synonymous nucleotide sites implies selection for optimal codon usage and enables estimation of the magnitude of the selection coefficients.

1. INTRODUCTION

The fundamental tenet of Darwinian evolution, expressed in modern genetic terminology, asserts that hereditary variation *within populations* gradually becomes converted into genetic differences *between species*. The primary processes that effect the transformation of within-species variation to among-species variation are generally agreed to be mutation, migration, natural selection, and random genetic drift. Population geneticists have expended considerable effort to understand the dynamics of these evolutionary forces.[1] For many decades the theoretical advances were handicapped by the lack of an adequate observational database. The advent of high-throughput DNA sequencing has helped to remedy this situation. The ability to sequence entire genomes, though still relatively small genomes, has given tremendous impetus to comparative sequence analysis, and the insights into the evolutionary processes that may be gleaned from sequence comparisons. The dazzling data from whole genomes has inevitably overshadowed genetic variation within species. But within-species genetic variation is inextricably related to between-species genetic differences. They lie on different sides of Darwin's evolutionary equation.

Much can be learned about microevolutionary processes from analyzing genetic variation present within populations. Although the individual microevolutionary forces of mutation, migration, natural selection, and random genetic drift are ordinarily too small (or if not too small, then too statistically confounded) to be estimated from observations of natural populations over a few generations. In evolutionary time there is a sort of "integration" over the microevolutionary forces, so that their effects become magnified. In favorable cases, the configuration of genetic polymorphisms in contemporary populations can be used to infer the direction and magnitude of the evolutionary forces that shaped the polymorphisms. These inferences require a suitable sample of genetic variation from natural populations, an appropriate theory of genetic change in populations, and a sampling theory that connects the actual sample of data to the underlying population theory. In this paper we give one example of such an approach, extending previous work in refs. 2 and 3. Alternative approaches to evolutionary inference from DNA sequences are exemplified in refs. 4, 5, 6, and 7.

2. GENETIC POLYMORPHISMS MODELED AS A POISSON RANDOM FIELD

Consider an indefinitely large set of homologous nucleotide or amino acid sequences, each of length L, corresponding to alleles of a single gene in a haploid organism. For each nucleotide or amino acid site in each of these sequences, we define a mutation rate μ, scaled by the length of the sequence L and also by the effective population number N. Expressed in terms of the usual mutation rate, μ_o, defined as the probability of a new mutation per mutable site (in this case per nucleotide or per amino acid) per generation, the definition of μ is

$$\mu = \mu_0 \times L \times N$$

For each new mutant at a nucleotide or amino acid site, we define a selection rate, γ, scaled again according to the effective population number. In terms of the conventional selection coefficient s, which defines the difference in fitness between the wildtype and mutant genetic types, γ is given by

$$\gamma = s \times N$$

For the sake of convenience we assume that the selection rate is the same for all new mutants, or at least for all new mutants that have any chance of becoming polymorphic in the population. Our estimates of γ are therefore some sort of average taken across different mutable sites and across evolutionary time. More realistic models could incorporate a probability distribution for the selection rate, but this is unnecessary for present purposes.

Let $X\{i, k\}$ be the population frequency of descendants of the ith mutation that are present in the population exactly k (an integer) generations after its occurrence. Each trajectory $X\{i, k\}$ is assumed to be stochastically independent of all the others, and the initial condition is $X\{i, 0\} = 1/N$ for each i because each mutation, at the time of its occurrence, is unique and present exactly once in the population. Each new mutation starts an independent selection-drift process, and no additional mutants occur at the same site until the ultimate fate of the original mutant (loss or fixation) has been realized. In practical terms this means that, at any one time, each nucleotide or amino acid site may have at most two alternative forms in the population. Note that the $X\{i, k\}$ are completely independent processes, and there are no constraints on the sum of the $X\{i, k\}$. These properties define the Poisson random field model of molecular evolution, which is distinct from the infinite-alleles model of multiple alleles at a single locus[8] whose frequencies must necessarily sum to 1, and also distinct from the infinite-sites model[9] in which the analogs of the $X\{i, k\}$ are not independent.

Under the standard Fisher-Wright selection-drift model (without mutation), each of the discrete processes $X\{i, k\}$ can be approximated by a diffusion equation with drift and diffusion coefficients $\gamma p(1-p)$ and $p(1-p)$, respectively. The technical requirements for converging to this diffusion are detailed in ref. 2. Under the diffusion approximation, it can shown that the limiting probability density of the frequency p of polymorphic mutant alleles is given by

$$2\mu \frac{1-e^{-2\gamma(1-p)}}{1-e^{-2\gamma}} \frac{dp}{p(1-p)} \quad \text{for} \quad \gamma \neq 0 \tag{1}$$

or by

$$2\mu \frac{dp}{p} \quad \text{for} \quad \gamma = 0 \tag{2}$$

The diffusion approximation also leads to convenient expressions for the flux of fixations of the mutant alleles,[2] and while these have important implications for the theory of genetic polymorphism and divergence, they are not relevant for present purposes. Note

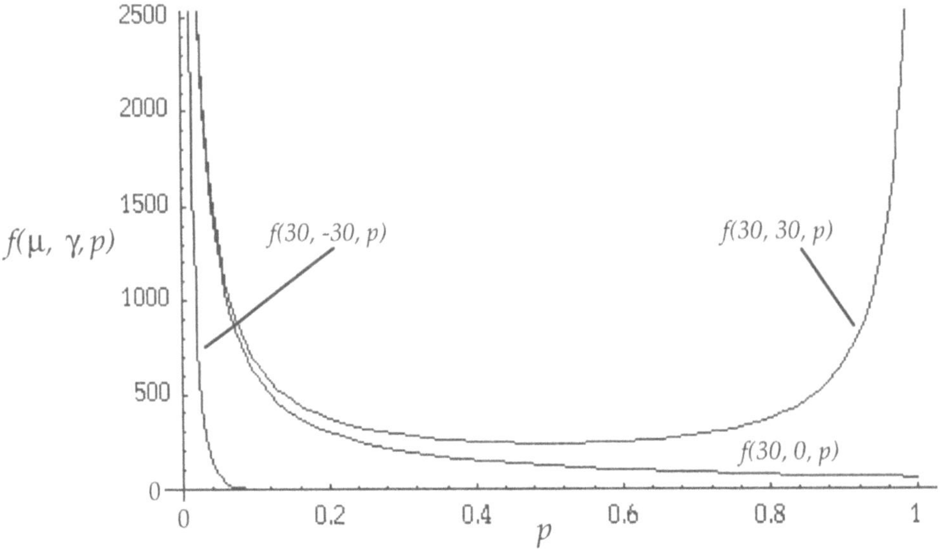

Figure 1. Plots of $f(\mu, \gamma, p)$ from Equations 1 ($\gamma \neq 0$) and 2 ($\gamma = 0$).

that Equations (1) and (2) are not integrable at $p = 0$, which implies that the population always contains a very large number of rare mutant alleles.

In Equation (1), $\gamma > 0$ corresponds to a favorable mutant allele (positive selection) and $\gamma < 0$ to a deleterious allele (negative selection). In case $\gamma = 0$ the mutant allele is selectively neutral. Curves of the probability densities for specific values of μ and γ are shown in Figure 1. The parameter values have been chosen to emphasize the point that, compared to the neutral case, $\gamma < 0$ leads to a relatively much greater proportion of polymorphisms in which the mutant allele is rare. This is expected on intuitive grounds, because a deleterious mutant allele is less likely to reach an intermediate frequency than a neutral allele. Likewise, when $\gamma > 0$ (positive selection), there is a relatively greater pile-up of mutant allele frequencies near $p = 1$.

3. INTRASPECIFIC POLYMORPHISMS OF GENES IN ENTERIC BACTERIA

In this paper we use the Poisson random field model to analyze sequences from the enteric bacteria *Escherichia coli* and *Salmonella enterica*, all extracted from publicly accessible databases. The data are summarized in Table 1, which includes all reported studies of multiple allelic sequences that include 50 or more polymorphisms. The total amount of sequence from all genes combined is a little over 600 kb. The data are tabulated by species (*Sen* for *S. enterica*, *Eco* for *E. coli*), gene symbol, protein product, number of alleles studied, number of nucleotides sequenced for each allele, and total number of polymorphisms. The polymorphisms are classified as "silent" polymorphisms (synonymous nucleotide polymorphisms in the coding sequence that code for the same amino acid) or else as amino acid polymorphisms (nonsynonymous nucleotide polymorphisms that code for different amino acids).

Table 1. Intraspecific DNA polymorphisms in *E. coli* and *S. enterica*

Spp	Gene	Protein	No. alleles	No. sites	Total kb	Total polys	Synon polys	AA polys
Sen	*spiR*	Transcriptional regulator	11	434	4.8	69	44	25
Sen	*mdh*	Malate dehydrogenase	19	849	16.1	127	113	14
Eco	*cls*	Cardiolipin synthase	10	1461	14.6	50	46	4
Sen	*aceK*	Isocitrate dehydrogenase kinase	16	1743	27.9	306	251	55
Eco	*trpB*	Tryptophan synthase B	25	1194	29.9	98	90	8
Sen	*gnd*	Gluconate dehydrogenase	66	1311	86.5	270	186	84
Eco	*trpA*	Tryptophan synthase A	25	807	20.2	70	59	11
Sen	*invE*	Invasion protein	19	1119	21.3	158	140	18
Eco	*gnd*	Gluconate dehydrogenase	33	1338	44.2	370	313	57
Sen	*fimA*	Major pilin protein (fimbrin)	17	602	10.2	99	62	37
Sen	*icd*	Isocitrate dehydrogenase	16	1164	18.6	208	201	7
Eco	*putP*	Proline permease	12	1467	17.6	106	99	7
Sen	*invA*	Invasion protein	16	1951	31.2	291	260	31
Eco	*phoA*	Alkaline phosphatase	8	1416	11.3	58	50	8
Sen	*fliC*	Phase I flagellin	42	1527	64.1	252	186	66
Eco	*trpC*	Anthranilate isomerase	25	1359	34.0	93	63	30
Eco	*icd*	Isocitrate dehydrogenase	17	1212	20.6	68	63	5
Sen	*spaP*	Secretory pathway protein	16	675	10.8	81	74	7
Sen	*putP*	Proline permease	15	1467	22.0	206	186	20
Eco	*fimA*	Major pilin protein (fimbrin)	7	555	3.9	51	31	20
Eco	*aceK*	Isocitrate dehydrogenase kinase	16	1722	27.6	159	143	16
Sen	*spaO*	Secretory pathway protein	16	912	14.6	160	98	62
Eco	*sfaA*	Major S-pilin protein	5	537	2.7	86	38	48
Sen	*spaN*	Secretory pathway protein	16	1026	16.4	200	83	117
Sen	*spaM*	Secretory pathway protein	16	441	7.1	62	36	26
Sen	*gapA*	Glyceraldehyde-3 P dehydrog'ase	16	924	14.8	109	95	14
Eco	*atpA*	Membrane-bound ATP synthase	6	546	3.3	134	22	112
Sen	*invH*	Invasion protein	17	444	7.5	83	38	45

What can we glean from these data? There are a few obvious points. First, the number of polymorphisms per kb of DNA has a wide range, from a low of 2.7 polymorphisms per kb in *E. coli* anthranilate isomerase to a high of 40.9 per kb for *E. coli* membrane-bound ATPase. The average is 9.7 polymorphisms/kb, and the median is near the 8/kb observed in *E. coli* malate dehydrogenase and *E. coli* gluconate dehydrogenase. There is less variation in the number of silent polymorphisms per kb, ranging from 1.9/kb (*E. coli* anthranilate isomerase) to 14.2/kb (*E. coli* major S-pilin protein), with a mean and median both very close to 6.1 silent polymorphisms/kb. The greatest variation is observed in the proportion of all polymorphisms that are amino acid polymorphisms. The low is 3% for *S. enterica* isocitrate dehydrogenase and the high 84% for *E. coli* membrane-bound ATPase. The average is 25% amino acid polymorphisms, whereas the median is 15–16% found for *E. coli* gluconate dehydrogenase and *E. coli* tryptophan synthase A. The proportion of silent polymorphisms may vary because of differential selection for optimal codon usage bias.[3,10,11,12] It may vary also because of local differences in mutation rate or other factors perhaps related to population structure. The proportion of all polymorphisms that are amino acid polymorphisms may reflect differing levels of selective constraint, rendering amino acid replacements so deleterious that they are not found as polymorphisms in reasonably sized samples. In some cases amino acid polymorphisms may be promoted by diversifying

Table 2. Sample configurations of polymorphisms in 7 alleles of the *gnd* gene in *E. coli*

Configuration of sample	Fourfold degenerate synonymous sites	Twofold degerate synonymous sites	Amino acid replacements
(7, 0)	92	50	249
(6, 1)	15	19	7
(5, 2)	5	7	
(4, 3)	4	5	
(5, 1, 1)		3	
(4, 2, 1)		3	
(3, 3, 1)		2	
(3, 2, 2)		2	
(4, 1, 1, 1)		1	

selection (see ref. 13 for evidence regarding *E. coli fimA*, and also ref. 14 for an unrelated example).

There is another aspect of the polymorphisms that is not apparent in Table 1, which has to do with the sampling configurations of each of the mutants. The meaning of the term "sampling configuration" is best made clear by an example, and for simplicity we shall use the data on polymorphisms of *E. coli* gluconate dehydrogenase among 7 natural isolates (see ref. 4 for the details). For each aligned nucleotide or amino acid in the sequence, the sample configuration is a list, in order of decreasing magnitude, of the most common type in the sample, the second most common type in the sample, the third most common type in the sample, and so forth. Thus the sample configuration (7, 0) in Table 2 refers to all nucleotide or amino acid positions that are monomorphic in the sample, since there are only 7 aligned sequences. Note that the nucleotide sites have been classified as fourfold synonymous (in which any nucleotide codes for the same amino acid), twofold synonymous sites (in which either pyrimidine, or either purine, codes for the same amino acid), or amino acid replacements. The sample configuration (6, 1) refers to sites at which the majority nucleotide or amino acid is present in 6 members of the sample and the minority in 1 member, the sample configuration (5, 2) represents a 5–2 majority–minority split, and so on. Sample configurations with three or four nonzero entries are those in which, at the particular site in the sample, there are three or four genetic variants simultaneously segregating.

More generally, in a sample of aligned allelic sequences, each of length L, a sample configuration of the form $(r - 1, 1)$ is called a "singleton" configuration. It is through such singleton configurations that the data in the sample can be connected with the distribution of population allele frequencies exemplified in Figure 1. This is done by means of the sampling formula explained in the next section. The motivation for examining singletons is apparent from comparisons of the frequencies for silent and amino acid polymorphisms in Table 2. Among fourfold or twofold degenerate silent polymorphisms, the proportion of singletons is 15/24 = 62.5% and 19/42 = 45.2%, respectively. Among amino acid polymorphisms, the proportion of singletons is 0/7 = 0%. The difference is statistically significant. Why are there so few nonsingleton amino acid polymorphisms?

4. A SAMPLING THEORY FOR POLYMORPHISMS IN POISSON RANDOM FIELDS

If r sequences, each of size L, are sampled at random from the Poisson random field described earlier, then each site having population frequency p of the mutant allele will yield a binomial sample because we are assuming that at each site at most two variants can segregate simultaneously. Because the values of p across all sites are given by the limiting probability density of the frequency p in Equation (1), it follows that the expected number of sites yielding exactly k mutant and $r - k$ nonmutant type has a Poisson distribution with mean

$$M_k = 2\mu\int_0^1 \frac{1-e^{-2\gamma(1-p)}}{1-e^{-2\gamma}}\frac{1}{p(1-p)}\binom{r}{k}p^k(1-p)^{r-k}\,dp \quad \text{for} \quad 1 \le k \le r-1 \qquad (3)$$

For sites with $\gamma = 0$, an analogous integral follows from Equation (2). Furthermore, under the Poisson random field model, the expected numbers given by Equation (3) are independent Poisson random variables for $1 \le k \le r - 1$.

It follows from Equation (1) that the expected number of sites in the sample that have singleton configurations is given by $M_1 + M_{r-1}$ and that the expected total number of polymorphic sites in the sample is given by $\sum_{k=1}^{r-1} M_k$. The ratio of these two quantities is the proportion of singletons among all polymorphisms in the sample. This ratio obviously depends on the value of γ, and the form of this dependence is shown in Figure 2. Note that information about the sign of γ is preserved, even though, for each polymorphic site, we have no way of knowing which variant is the mutant and which is the nonmutant. This is because Equation (1) is not symmetric in γ, even when p is replaced with $1 - p$. The main point of Figure 2 is that sites with a smaller value of γ will have a larger proportion of singletons as a fraction of all polymorphisms at the site. The discrepancy is most pronounced when γ is negative, corresponding to detrimental mutations. Thus one possible explanation for the excess of singleton polymorphisms in Table 2 is that, relative to silent polymorphisms, more amino acid polymorphisms are slightly detrimental. This does not imply that silent polymorphisms have $\gamma = 0$, but only that the γ for silent polymorphisms is larger than that for amino acid polymorphisms.[3] More generally, Equation (3) can be used to obtain maximum likelihood estimates of μ and γ taking into account the entire set of sample configurations.[3] Nevertheless the implications for singletons are appealingly intuitive: new mutations that are more detrimental are less likely to achieve appreciable population (and sample) frequencies than new mutations that are less detrimental.

5. TREASURE YOUR SINGLETONS

William Bateson is usually credited with having coined what used to be every young geneticist's mantra, "treasure your exceptions." We may paraphrase this as "treasure your singletons," because Figure 2 suggests that the proportion of singletons may be used to test whether amino acid polymorphisms in a gene are subjected to different selective forces than adjacent synonymous polymorphisms present in the same gene. The test is a test for homogeneity in a conventional 2×2 contingency table with the layout shown in

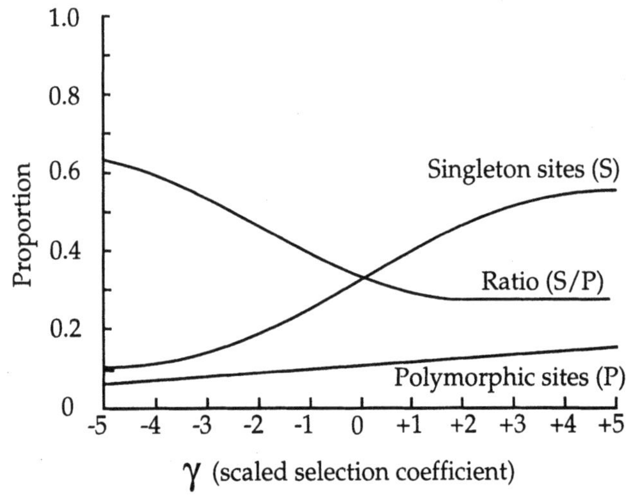

Figure 2. Poisson random field expectations for the number of polymorphisms (P), the number of singleton polymorphisms (S), and their ratio, for various values of the scaled selection coefficient γ.

Figure 3 for the *E. coli* gene for gluconate dehydrogenase (*gnd*). Why the gene for gluconate dehydrogenase? Because this gene is completely typical in terms of the proportion of all sites that are polymorphic and also in terms of the proportion of all polymorphisms that are amino acid polymorphisms (Table 1). If we define p_1 to be the proportion of all amino acid polymorphisms that are singletons, and p_2 to be the proportion of all silent polymorphisms that are singletons, then the null hypothesis of the test is H_0: $p_1 = p_2$. Singificance in the direction of $p_1 > p_2$ might suggest that amino acid replacements that become polymorphic are, on average, subjected to stronger negative selection than are synonymous nucleotide substitutions that become polymorphic.

What is the power of such a test? This depends on the value of p_1/p_2 as well as on the marginal totals. We have carried out power simulations for datasets that have various numbers of total polymorphisms and various values of p_1/p_2, with the stipulation that the relative proportions of amino acid and silent polymorphisms are those observed for the *gnd* gene in *E. coli.* The results are illustrated in Figure 4. With 100 or more total polymorphisms, and a proportion of amino acid polymorphisms of 15%, the power of the test is at least 40%, even for values as small as $p_1/p_2 = 2$. In Table 1, 15 among the 28 genes have more than 100 polymorphisms, and several others have close to 100. For *E. coli gnd*, by the way, the estimated value of $p_1/p_2 = 2.2$.

	Singleton	Nonsingleton
Amino acid polymorphisms	22	35
Silent polymorphisms	55	258

Figure 3. Layout of a 2 × 2 contingency table to test for an excess of singleton amino acid polymorphisms.

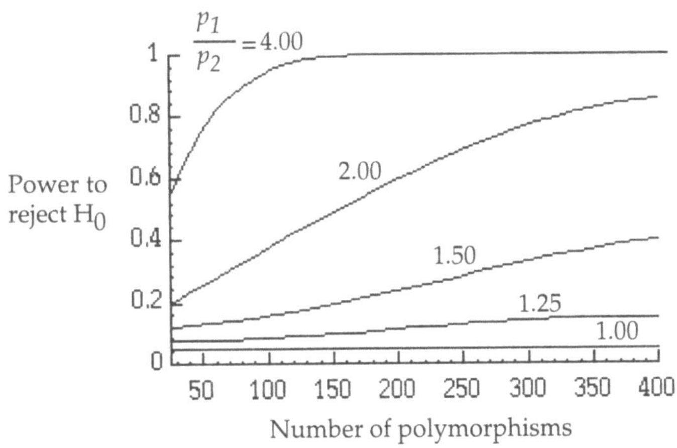

Figure 4. Power of the contingency chi-square test in Figure 3, for fixed values of p_1/p_2, and when the marginal totals of amino acid polymorphisms and silent polymorphisms are maintained in proportion to those observed for the *gnd* gene in *E. coli.*

6. RESULTS AND CONCLUSIONS

We have carried out a homogeneity test for a 2×2 singleton versus nonsingleton table like that shown in Figure 3, for all of the sequences in Table 1. The results are shown in Table 3 and graphically in Figure 5. In Table 3 the genes are ranked in descending ordered of p_1/p_2, which is the same order as given in Table 1 and also the left-to-right order of the genes in Figure 5. The P values in the far right column were computed using Fisher's exact test. The result is that 8 of the top 9 entries have P values (corrected for multiple comparisons) of $P < 0.15$, indicating and excess of singleton amino acid polymorphisms.

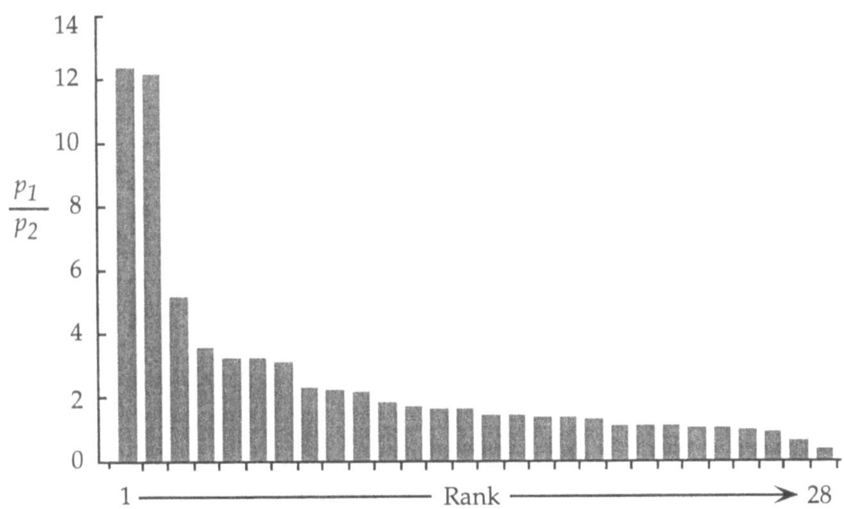

Figure 5. Ratio of the proportion of amino acid polymorphisms that are singletons (p_1) to silent polymorphisms that are singletons (p_2), for the sequences in Table 1, in order of rank.

Table 3. Tests for an excess of singleton amino acid polymorphisms

Spp	Gene	Protein	Total polys	AA polys / Total	p_1/p_2	P
Sen	*spiR*	Transcriptional regulator	69	0.36	12.32	0.003
Sen	*mdh*	Malate dehydrogenase	127	0.11	12.11	<0.0001
Eco	*cls*	Cardiolipin synthase	50	0.08	5.11	0.003
Sen	*aceK*	Isocitrate dehydrogenase kinase	306	0.18	3.51	0.003
Eco	*trpB*	Tryptophan synthase B	98	0.08	3.21	0.005
Sen	*gnd*	Gluconate dehydrogenase	270	0.31	3.21	<0.0001
Eco	*trpA*	Tryptophan synthase A	70	0.16	3.06	0.003
Sen	*invE*	Invasion protein	158	0.11	2.22	0.228
Eco	*gnd*	Gluconate dehydrogenase	370	0.15	2.20	0.001
Sen	*fimA*	Major pilin protein (fimbrin)	99	0.37	2.09	0.106
Sen	*icd*	Isocitrate dehydrogenase	208	0.03	1.79	1.000
Eco	*putP*	Proline permease	106	0.07	1.66	0.417
Sen	*invA*	Invasion protein	291	0.11	1.57	0.704
Eco	*phoA*	Alkaline phosphatase	58	0.14	1.56	0.272
Sen	*fliC*	Phase I flagellin	252	0.26	1.41	0.464
Eco	*trpC*	Anthranilate isomerase	93	0.32	1.40	0.117
Eco	*icd*	Isocitrate dehydrogenase	68	0.07	1.33	1.000
Sen	*spaP*	Secretory pathway protein	81	0.09	1.32	1.000
Sen	*putP*	Proline permease	206	0.10	1.27	0.717
Eco	*fimA*	Major pilin protein (fimbrin)	51	0.39	1.09	1.000
Eco	*aceK*	Isocitrate dehydrogenase kinase	159	0.10	1.08	1.000
Sen	*spaO*	Secretory pathway protein	160	0.39	1.05	1.000
Eco	*sfaA*	Major S-pilin protein	86	0.56	0.98	1.000
Sen	*spaN*	Secretory pathway protein	200	0.59	0.98	1.000
Sen	*spaM*	Secretory pathway protein	62	0.42	0.92	1.000
Sen	*gapA*	Glyceraldehyde-3 P dehydrog'ase	109	0.13	0.85	1.000
Eco	*atpA*	Membrane-bound ATP synthase	134	0.84	0.56	0.214
Sen	*invH*	Invasion protein	83	0.54	0.34	0.238

What are the magnitudes of the selection coefficients that can account for the statistically significant p_1/p_2 values? The values of γ along with their 95 percent condfidence intevals have been estimated from the full maximum likelihood equations for the Poisson random field.[2,3] These are shown in Table 4, along with the corresponding values of the conventional selection coefficient s, calculated under the assumption that $N = 1.8 \times 10^8$ for *E. coli* and *S. enterica.* (This approach does not work for *E. coli* cardiolipin synthase because, in this case, there are no nonsingleton amino acid polymorphisms.) At least for the *S. enterica* genes *spiR* and *mdh*, the estimated values of γ are suspiciously small, relative to the 95 percent confidence intervals; this observation, taken together with the exceptionally high values of p_1/p_2, suggests that the excess of singleton amino acid polymorphisms may result from diversifying selection rather than slightly deleterious mutations. For the *mdh* gene, the discrepancy could also be due to some nonequilibrium population structure, since the amino acid polymorphisms fail to give a satisfactory fit to the Poisson random field model. For the rest of the genes in Table 4, the γ values range from -1.1 to -6.7, which may be compared to a value of $\gamma = -1.3$ for selection against nonoptimal synonymous codons in *E. coli gnd*[3], which is a gene with moderate codon usage bias.

What biological information have we gleaned from this analysis using Poisson random fields? First, given sufficient numbers of polymorphisms, there is often enough

Table 4. Estimates of selection intensity against polymorphic amino acid mutants

Spp	Gene	Protein	p_1/p_2	γ	s
Sen	*spiR*	Transcriptional regulator	12.32	-0.20 ± 3.48	-1.1×10^{-9}
Sen	*mdh*	Malate dehydrogenase	12.11	-1.26 ± 2.35	-7.0×10^{-9}
Eco	*cls*	Cardiolipin synthase	5.11	—	—
Sen	*aceK*	Isocitrate dehydrogenase kinase	3.51	-1.07 ± 1.51	-5.9×10^{-9}
Eco	*trpB*	Tryptophan synthase B	3.21	-2.45 ± 3.36	-1.4×10^{-8}
Sen	*gnd*	Gluconate dehydrogenase	3.21	-6.70 ± 2.60	-3.7×10^{-8}
Eco	*trpA*	Tryptophan synthase A	3.06	-3.51 ± 3.77	-2.0×10^{-8}
Eco	*gnd*	Gluconate dehydrogenase	2.20	-4.23 ± 1.90	-2.4×10^{-8}

statistical power to distinguish possible selective effects acting on amino polymorphisms from those acting on synonymous polymorphisms of nonoptimal codons present in the same gene. Additional power can be gained by including divergence between species.[2,6,15] The interspecific comparisons require that a sufficiently closely related species can be identified. By "sufficiently closely related species" we mean one the has diverged sufficiently recently in time that few nucleotide or amino acid sites have been hit by multiple mutations, since saturation effects can bias the tests in the direction of detecting selection, including spuriously suggesting positive selection for amino acid replacements (see Figure 2, p. 29, in ref. 16). Most genes in *E. coli* and *S. enterica* are sufficiently near saturation for silent-site differences that statistical tests comparing polymorphism with divergence may be biased.[17]

In the present analysis of intraspecific polymorphisms, in a significant number of cases there is an excess of singleton amino acid polymorphisms, relative to singleton synonymous polymorphisms, which may suggest that most polymorphic amino acid mutants are slightly detrimental. (We again emphasize the fact that extremely high values of p_1/p_2 as observed for the *S. enterica* genes *spiR* and *mdh* may be suspect.) Third, when there is evidence for slightly detrimental selection acting on mutant amino acid replacements, for example in the last five entries in Table 4, the the intensity of selection is on the order of a small multiple of the effective population number. These intensities of selection, acting across evolutionary time, are large enough to leave their mark on the sample configurations of polymorphic sites, but they are at least six orders of magnitude smaller than any difference in growth rate that can be detected in competition experiments in the laboratory.[18,19] It should be emphasized that the failure to identify a statistically significant excess of singleton amino acids does not necessarily imply that amino acid polymorphisms are as relatively weakly selected as nonoptimal synonymous codon polymorphisms. Note in Table 3 that 22 of the 28 values of p_1/p_2 are greater than 1, which suggests slightly deleterious amino acid polymorphisms for at least some of these genes that are not detected as statistically significant because of the limitations on the power of the test apparent in Figure 4.

If many polymorphic amino acid replacements are slightly detrimental, where are they located in the three-dimension structure of the molecule? Unfortunately, for most of the proteins for which there is extensive polymorphism data (Table 1), the protein structures have not been worked out; and for most of the protein structures that have been worked out, little polymorphism data exists. Nevertheless, there are a few proteins with both types of data, and we have just begun to examine where in the molecule the polymorphisms may tend to lie. Already there are some interesting hints. For example, in the case of *E. coli* anthranilate isomerase (E.C. 5.3.1.24), an enzyme involved in the pathway of tryptophan biosynthesis, there is a threefold difference ($G = 5.44$, $P < 0.02$) in the ratio of amino acid polymorphisms to silent polymorphisms between those residues that are accessible to solvent (on the outside of the protein) versus those that are not. On the other hand, the silent polymorphisms themselves are randomly distributed along the polypeptide chain, as expected since they do not alter the amino acids.

One possible explanation for the nonrandom distribution of amino acid polymorphisms is that replacement mutations in residues that are on the inside of the protein may tend to disrupt the stability of the molecule, making these mutations sufficiently deleterious that they will be eliminated from the population by purifying selection and not be observed as polymorphisms. In contrast, many replacement mutations that are on the outside of the protein may have so little effect on protein stability that they can be neutral or slightly detrimental, allowing them to be maintained as polymorphisms, at least long enough to be included in samples the size of those in Table 1. If this hypothesis is correct, we should expect to see a relatively greater proportion of amino acid polymorphisms in solvent-accessible parts of other enzymes and soluble proteins. Perhaps by bringing polymorphism data and structural data together, we can gain insight into the forces that shape protein evolution.

ACKNOWLEDGMENTS

This work was supported by grants from the U. S. National Institutes of Health (DLH), The Howard Hughes Medical Institute (CDB), and the National Science Foundation (SAS).

REFERENCES

1. Hartl D.L. and Clark A.G. *Principles of Population Genetics.* Sinauer Associates (Sunderland, MA, 1997).
2. Sawyer S.A. and Hartl D.L. 1992 Population genetics of polymorphism and divergence. *Genetics* 132:1161–1176.
3. Hartl D.L., Moriyama E.N., and Sawyer S.A. 1994 Selection intensity for codon bias. *Genetics* 138:227–234.
4. Sawyer S.A., Dykhuizen D.E., and Hartl D.L. 1987 A confidence interval for the number of selectively neutral amino acid polymorphisms. *Proc. Natl. Acad. Sci. USA* 84:6225–6228.
5. Hudson R.R., Kreitman M., and Aguadé M. 1987 A test of neutral molecular evolution based on nucleotide data. *Genetics* 116:153–159.
6. McDonald J.H. and Kreitman M. 1991 Adaptive protein evolution at the *Adh* locus in *Drosophila. Nature* 351:652–654.
7. Templeton A.R. 1996 Contingency tests of neutrality using intra/interspecific gene trees: The rejection of neutrality for the evolution of the mitochondrial cytochrome oxidase II gene in the Hominoid primates. *Genetics* 144:1263–1270.
8. Ewens W.J. 1972 The sampling theory of selectively neutral alleles. *Theor. Popul. Biol.* 3:87–112.
9. Watterson G.A. Estimating species divergence times using multi-locus data. *Population Genetics and Molecular Evolution* (Ohta T. Aoki K., eds.) Springer-Verlag, (Berlin, 1985), 163–183.
10. Bulmer M. 1991 The selection-mutation-drift theory of synonymous codon usage. *Genetics* 129:897–907.

11. Eyre-Walker A. and Bulmer M. 1993 Reduced synonymous substitution rate at the start of enterobacterial genes. *Nucleic Acids Res.* 21:4599–4603.
12. Eyre-Walker A. and Bulmer M. 1995 Synonymous substitution rates in enterobacteria. *Genetics* 140:1407–1412.
13. Boyd E.F. and Hartl D.L. 1998 Diversifying selection governs sequence polymorphism in the major adhesin proteins FimA, PapA, and SfaA of *Escherichia coli. J. Mol. Evol.* (in press)
14. Riley M.A. 1993 Positive selection for colicin diversity in bacteria. *Mol. Biol. Evol.* 10:1048–1059.
15. Akashi H. 1995 Inferring weak selection from patterns of polymorphism and divergence at "silent" sites in Drosophila DNA. *Genetics* 139:1067–1076.
16. Ayala F.J., Chang B.S.W., and Hartl D.L. 1993 Molecular evolution of the *Rh3* gene in *Drosophila. Genetic* 92:23–32.
17. Guttman D.S., and Dykhuizen D.E. 1994 Detecting selective sweeps in naturally occurring *Escherichia coli. Genetics* 138:993–1003.
18. Hartl D.L. and Dykhuizen D.E. 1981 Potential for selection among nearly neutral allozymes of 6-phosphogluconate dehydrogenase in *Escherichia coli. Proc. Natl. Acad. Sci. USA* 78:6344–6348.
19. Hartl D.L. and Dykhuizen D.E. The neutral theory and the molecular basis of preadaptation. *Population Genetics and Molecular Evolution* (Ohta T. Aoki K., eds.) Japan Scientific Societies Press, (Tokyo, 1985), 107–124.

5

AUTOMATIC ASSEMBLY AND EDITING OF GENOMIC DATA

B. Chevreux, T. Pfisterer, and S. Suhai

Deutsches Krebsforschungszentrum Heidelberg
Im Neuenheimer Feld 280
69120 Heidelberg

1. DEFINING THE PROBLEM

Shotgun sequencing a genome for subsequent reconstruction is comparable to assembling a jigsaw puzzle. Genomes, of course, are much more complex than your average jigsaw puzzle: they tend to be about 500 to 5000 pieces, printed on both sides, with many vital pieces possibly missing. Some of the pieces are dirty or unrecognizable, and several pieces from another puzzle have been mixed in. Additionally, a few pieces themselves appear to have been assembled by a very impatient two-year-old with a pair of scissors and a bottle of glue. Talk about fun.

This comparison might be somewhat far fetched, but it reflects approximatly the reality with which biologists are confronted to. Shotgun sequencing is a process which reconstitutes small and sometimes faulty DNA fragments into their true DNA sequences. The devil is in the details, however. Biologists must constantly work around these less-than-ideal conditions when sequencing and assembling contiguous DNA blocks, or contigs. If the gathered readings (reads) were 100% error free, then a multiplicity of problems would not occur. But in reality, the extraction of the data is a physical process and subject to stochastic error probabilities, which, in combination with the highly repetitive properties of DNA, tends to impede the formation process in an awesome way.

The common method for assembling DNA consists of using fault-tolerant algorithms which produce a basic sequence which then has to be reviewed and manually corrected by human experts. Incorrectly assembled sequences must be dismantled and reassembled at different places. For large scale sequencing projects, this method is very slow and inefficient and represent the most important bottle-neck.

Genomics and Proteomics, edited by Sándor Suhai.
Kluwer Academic / Plenum Publishers, New York, 2000.

As an alternative, we present a package based on a new integrated assembler and editor which can greatly shorten the time involved in this process. The assembler determines all potential error regions and uses a rule-based decision system, comparable to a human operator, to fall back to the original data if it is necessary to resolve discrepancies. Their interaction allows the automatic detection and prevention of wrong decisions. The integrated automatic editor provides the knowledge and expertise of a human expert, helps the assembler during the assembly process, and removes standard errors from the resulting assembly.

2. THE DATA

We give a short oversight over the type of data we are working with and draft a possibility to classify the error types in a single read.

2.1. Data File Formats and Quantities

Every base in a sequence calls for a certain amount of data that has to be extracted in a read. Among others are the original electrophoresis signals from the ABI or ALF sequencing machines, probabilities for the different bases, and various administrative data. These are stored in SCF (Standard Chromatogram Format) format which can be read by the assembler and the editor. When rounding generously, each base of a sequence needs about 100 bytes for support in an SCF file.

In effect, even a middle-sized project with 1000 reads of about 1000 bases each results in 100 megabytes of SCF data files which must be considered for a correct assembly.

This does not even include additional data won by further analyses like sequencing vector removal, tagging of repetitive sequences etc. The Staden package stores these in EXP (EXPeriment) file format which is read by the assembler.

Both the assembler and the editor can read and write assembled contigs or projects using the CAF format (Common Assembly Format)* which is a complete textual description of an assembly. This enables the communication between the assembler and the editor to use either shared memory structures or CAF files. Exchange with other proprietary formats has to be realized by file format conversion.[5]

2.2. Sources of Errors

As mentioned briefly above, the base material from which the output sequences are derived tend to be error prone. Therefore, it is necessary to know exactly what data results from a shotgun sequencing process and what types of errors it might contain. These errors can be classified into three distinct categories.

2.2.1. Primary Errors. We identify errors on the chemical level as primary errors. Base mutation within the sequence and the formation of chimeras during the polymerase chain reaction (PCR) fall into this category. These errors occur before the sequencing and characterize themselves by the fact that the signals—possibly of outstanding quality—are false. They cannot be detected using the data of an individual read.

---

* http://www.sanger.ac.uk/Software/CAF/

2.2.2. Secondary Errors. Secondary errors are caused by read operations. Chromatography is a chemical process subject to stochastic oscillations, which can cause sub-optimal signal quality. This becomes visible in the under- and over-oscillations of the signals, non-separated curves, and signal peaks or dropouts. These in turn raises errors in the interpretation process (base calling) of these signals. Secondary errors can be reduced by using better improved chemistry[9] or they can be "repaired" in some cases by human experts at the read level, but in most cases other reads covering the same place in the sequence have to be consulted to support decision making.

2.2.3. Tertiary Errors. Data gathered from a sequencing process must be worked over before assembly step can begin. The most important task consists of the removal of the sequencing vectors. Those vectors make assembly much more difficult to perform correctly, if not impossible. Due to primary and secondary errors, algorithms cannot always cut off the sequencing vectors accurately. Unmarked residues of these vectors in a sequence are called tertiary errors. These always occur at the start or end of a sequence. They cannot be detected on the signal level as errors, but only in combination with the knowledge of the bases contained within a sequencing vector.

3. DEFINITIONS

We define an alphabet A^g which contains all the characters needed for an assembly,

$$A^g = \{A, C, G, T, N, *\} \tag{1}$$

where the letters A, C, G, and T stand for their respective bases, N for a unspecified base and "*" for a gap enclosed by two bases in a sequence.

As we will see later we also need a character "∇" to semantically show positions which are not covered by the read without altering the content respective the meaning of the sequence. Thus the new alphabet reads now

$$A^G = \{A, C, G, T, N, *, \nabla\} \tag{2}$$

A sequence S^G is an ordered succession of characters originating from the alphabet A^G

$$S^G = (s_1, \cdots, s_n) \quad \text{with} \quad n = \|S^G\| \quad \text{and} \quad s_i^G \in A^G \tag{3}$$

where $|S^G|$ denominates the length of the sequence -excluding ∇—and $\|S^G\|$ the length of the sequence including ∇.

An alignment L can therefore be described as a vector of sequences or directly as a two-dimensional array.

$$L = \begin{pmatrix} S_1^G \\ \vdots \\ S_k^G \end{pmatrix} \text{ or else } L = \begin{pmatrix} s_{11}^G & \cdots & s_{1n}^G \\ \vdots & \ddots & \vdots \\ s_{k1}^G & \cdots & s_{kn}^G \end{pmatrix} \tag{4}$$

As can be seen, all sequences must have the same length. This is why the end-gaps (∇) exist: they allow the positioning of a single sequence within an alignment without changing something to it.

The numeric result of a comparison of two elements s_1^G and s_2^G of a sequence S^G is called score: score(s_1^G, s_2^G). The score of a column in an alignment is the sum of scores of the permutation of elements in this column

$$\text{score}(s_1, \cdots, s_k) = \sum_{j=1}^{k} \sum_{l=j}^{k} \text{score}(s_j, s_l) \tag{5}$$

The score of an alignment of k sequences S is the sum of scores of all the columns in this alignment.

$$\text{score}(S_1, \cdots, S_k) = \sum_{i=1}^{\|S_k^G\|} \sum_{j=1}^{k} \sum_{l=j}^{k} \text{score}(s_j, s_l) \tag{6}$$

The coverage of a column is the number of characters of this column belonging to the alphabet A^g. This means that end-gaps ∇ do not count as coverage.

4. THE ASSEMBLER

As we stated before our model is based on the desequentialisation of assembly and editing steps whilst reconstructing a DNA sequence. By integrating both packages up to a certain point we expect to obtain better results than current approaches. The foundation for this framework is laid by the assembler which is explained in this section.

4.1. Working Principles

The primary objective of an assembler is to provide a structural frame of assembled sequences. Up to now two very different approaches are used to tackle down this problem. A greedy solution will try to assemble as much as possible regardless of possible errors. A quality based solution will assemble only those sequences that fit together with almost no errors at all. Our integrated approach allows us to go a third way: we start to assemble high quality parts of an assembly first and gradually incorporate lesser quality, always checking back to the electrophoresis signals to inhibit misassemblies at error locations.

Ideally an assembler must be able to do many things. It has the overview of all potential matches between any of two sequences S_1 and S_2, it builds contigs by performing multiple alignments on several sequences (preferably including base qualities even at this step), and it discerns wrongly inserted sequences in an align L and removes them, trying to insert them elsewhere as appropriate. Its result is an alignment with as few errors as possible and with a consensus that, with high probability, does not contain any errors.

This ideal case fails in practice at its basic condition: there is no algorithm known capable of aligning any number of sequences within a justifiable amount of time and memory expenditure. Wang and Jiang[15] have shown in their study on the computational complexity of multiple sequence alignment that this problem is NP-complete.

The most common solution to this problem is instead of processing n sequences at the same time, an alignment is built up by pairwise alignments of an already existing consensus to a new sequence. This solution is computable in a finite time, but sub-optimal for several reasons. One of the more important reasons is that errors occuring at an early stage of the assembly influence the rest of the process.

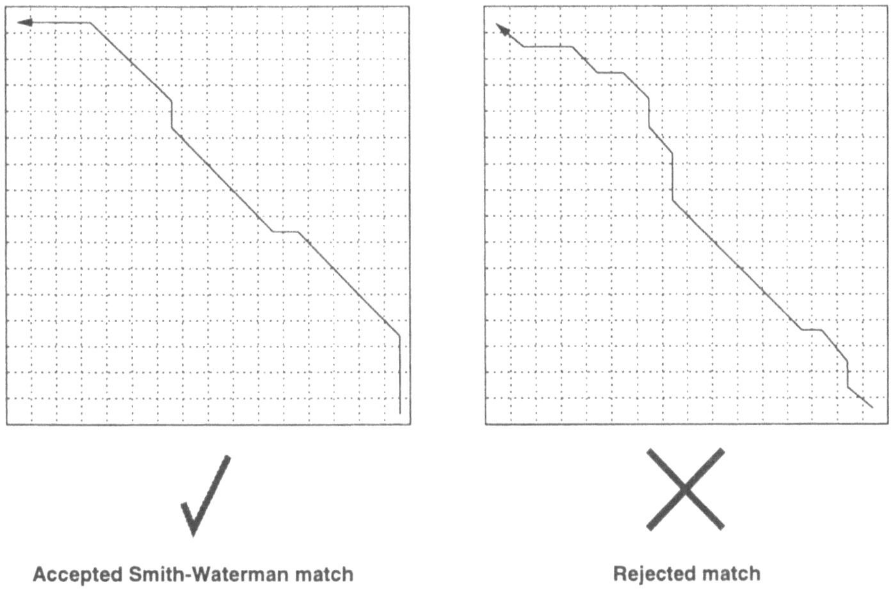

Figure 1. Two potential matches being compared by Smith-Waterman. The left align has almost no errors so that quality and weight of the alignment is computed. The right alignment resulted from a spurious match from the scanner and is rejected.

4.2. Realisation

In order to find all potential matches, the first step consists of comparing each sequence S with every other one in forward and reverse direction. The scanner finds any potential match within a user definable maximum error threshold (usually about 10–20% error rate) in any of two sequences if they correspond in a minimum number of bases. It is irrelevant how the errors came into the sequences and of what type—insertion, deletion, base change—they are.

All potential matches get a much more closer inspection in the second step. This examination is performed by a standard Smith-Waterman alignment algorithm. We pursue two different purposes by doing this: spurious matches resulting from the first step are sorted out and—at the same time—an alignment of these sequences is computed together with its quality respective weight. The quality respective weight of an alignment is calculated from a combination of the score(S_1, S_2) of the alignment and the number of spotted errors—on the base level—in the overlap. Long overlaps with very few errors get, due to their higher score, a stronger weight than short overlaps with no errors at all. The two sequences forming this dual alignment are taken as possible candidates for a multiple alignment in a contig to be constructed later in the process if their weight lies above a given threshhold.

All pairs of candidates build up one ore several non-directional weighted graphs. We draw paths through these graphs which help us building contigs by pairwise alignment as each distinct graph represents one contig. For starting the buildup in each graph we choose the sequence with the best weights to all its neighbours as anchor point. This ensures a long and—most likely—relatively error free sequence as a good starting point.

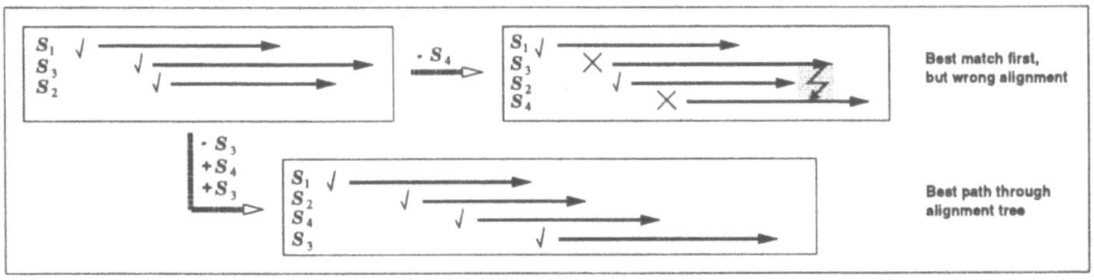

Figure 2. A contradiction arises when the sequences S_1 to S_4 get incorporated sequentially into a contig. This contradiction—recognized on the character level and not resolved by base alternatives on the signal level—leads to the building of an alternative pathway S_1 S_3 S_2 and S_4 resulting into an error free contig.

However, algorithms for computing a minimum spanning tree—as presented for example in[14]—cannot be applied when searching for a solution for the best path through the graphs, since each read taken into a contig influences the ongoing assembly process. Paths are therefore build up interactively with a look ahead technique and each pair of sequences newly taken into an alignment is immediately checked for consistency errors. Figure 2 shows a typical case where this strategy leads to success.

Strictly speaking, a contig is an alignment of several sequences with additional information like its consensus, direction of the sequences etc. The contig behaves semi-intelligently by checking whether or not too many non-explainable errors result in its consensus when a new sequence is added to it. If this is the case, the sequence is rejected by the contig and the pathfinder algorithm has to search for alternatives through the graph. The contig relies on knowledge-based routines for signal analysis and processing in order to resolve errors in its consensus as thoroughly as possible. These routines are provided by the automatic editor and scrutinize the signal for possible alternatives at the location of a spotted error. If the error can be resolved by an alternative, it is explainable for the contig and the read is not rejected.

This is an important advantage of our newly developed system: unsolvable errors (misaligns) are detected during the assembly process based on the original signal data. Alternative positioning of the sequences can then be tried out in cases like this. This is a crucial leap forward in the quality improvement of alignments. The functionality of the assembler and its interaction with the automatic editor are shown in Figure 3.

4.3. Planned Improvements

We are currently developing an improved assembly verification system to handle highly repetitive sequences in a stricter way than it is currently done. The next step on schedule will be to enlarge the clipped sequences in an assembly. Within each sequence a significant part of the data has been clipped away because of apparently poor quality of the electrophoresis signals. Once a raw assembly with high quality has been built, these low quality parts of each sequence can be gradually incorporated into the assembly to enlarge the coverage as a higher coverage immediately leads to a lesser error probability per base in the consensus.

As an extension to this, we also plan to use lower quality only sections of contigs to allow the joining of contigs when the expert system is able to solve most discrepancies occuring there.

5. THE EDITING PROBLEM

As we have seen the sequences obtained from electrophoresis can—and probably will—contain errors. We had to take this into account when we assembled the data fault tolerantly but we did not correct the faults at this stage of the process. This is usually done afterwards during the editing and the so called finishing of the sequences. The majority of errors in the sequences show up as a discrepancy in the alignment; i.e. not all bases in a column of the assembly are identical. A highly skilled and labour intensive task is to adjudicate between those conflicting readings. Hence, DNA sequencing productivity could be improved if the necessary time for checking and editing these readings could be reduced. In the following we briefly discuss some previous work in the field and present our solution to tackle the decision problem by modeling the knowledge used by the expert. We use a distinct fault hypotheses generation task to be able to find solutions for problems with multiple faults.

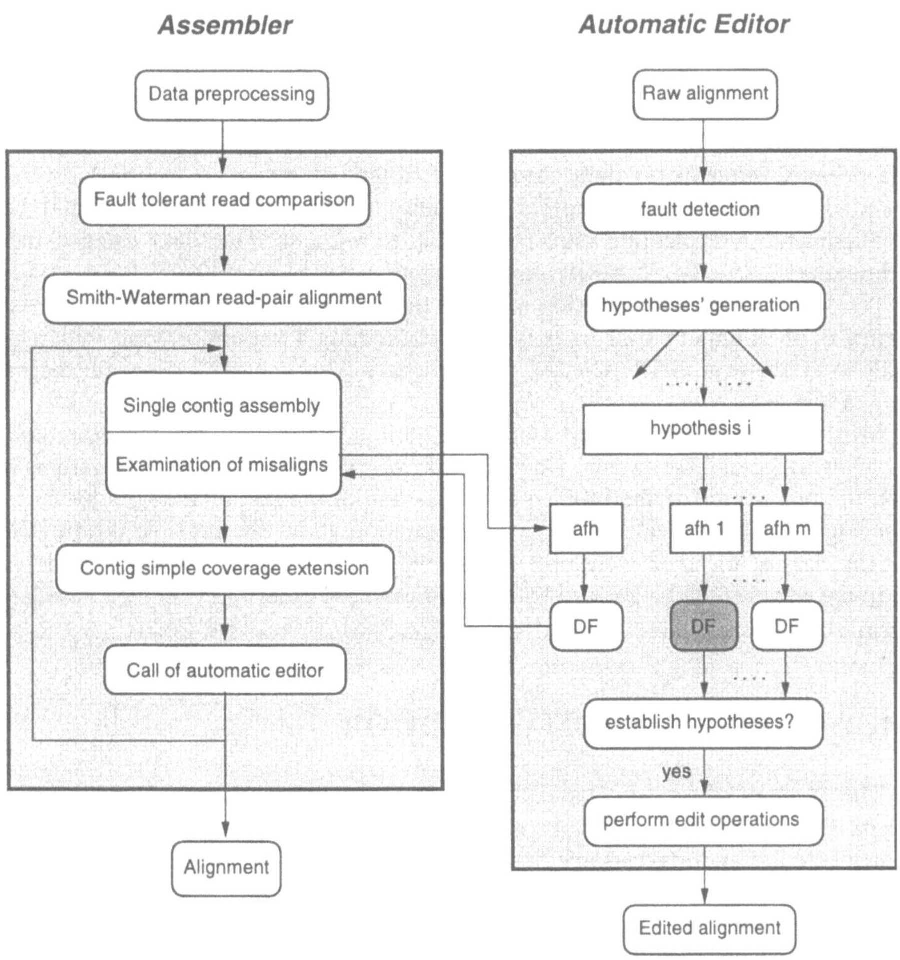

Figure 3. Oversight on the assemblers and editors work-flow and their interaction.

To resolve the conflicting situations given by mismatches at "symbol level", additional information has to be employed to decide whether the discrepancy is due to one or more sequencing or base-calling errors or if it results from a misassembled read. Base quality values and electrophoresis trace data are used in this context.

Base qualities are indicators for the confidence in the called and the non-called bases. The higher the quality value the lower the probability of a wrong call. Quality values can even have a quantitative probabilistic meaning (PHRAP[6,7]). Numerical estimates for base calling accuracy can be used to obviate much of the necessary trace checking activities by overriding weak bases and they can be used to calculate estimates for the probability of the consensus base.[2] This allows for checking only weak consensus bases and ignoring conflicts resulting from low quality bases in the sequences (weak bases have little influence on the quality of the consensus base). The fact that the huge trace files are not obligatory if quality informations are available from other sources is another advantage.

If available, the original electrophoresis trace data can be examined. This is the way how most human editing is performed. Looking at the graphical display of the traces easily reveals most base calling errors or electrophoresis problems.

The decisions of the automatic editor "Auto Edit" from Sanger Centre[4] are based on the original trace data, where simple signal characteristics are calculated and evaluated. Other groups calculate confidence values for sequence readings and use them in the assembly process to determine overlaps and to resolve discrepancies in the consensus sequence,[11] or they use linear discriminant analysis to assign to each position in the primary sequences a data specific probability of being an incorrect, over- or under-predicted nucleotide.[10] Another possibility to use the trace data would be a trace alignment. A nucleotide sequence is aligned with its trace data using dynamic programming.[12]

Beside these automatic decision methods there is also the possibility to improve the efficiency of the human editor or finisher by combining a variety of tools with a user friendly interface (e.g. GAP,[1] Consed,[8] Sequencher). This can go far beyond the mere editing of the sequences.

Most of the errors corrected during the editing of genomic sequences are due to faults at the stage of base calling. One may ask: why do we work on the symptoms and not on the real source of the problems? This is due to the redundancy of the shotgun sequencing data that provides additional information about the possible positions and about the class of the error we expect when looking at the electrophoresis data. Even if most trivial errors could be avoided by an improved base caller this is not the case for all of them.

6. MODEL OF THE AUTOMATIC EDITOR

The human editor has typically a two stage approach to solve an editing problem. On the first stage he looks at the letters of the aligned reads where the discrepancy is found and makes some hypotheses about what could have gone wrong. On the second stage he examines the corresponding traces and tries to verify or refute these hypotheses (of course he can go back and search for a new hypothesis etc.). Results from a first prototype[13] showed us, that it is worth spending some efforts in the first stage of this process.

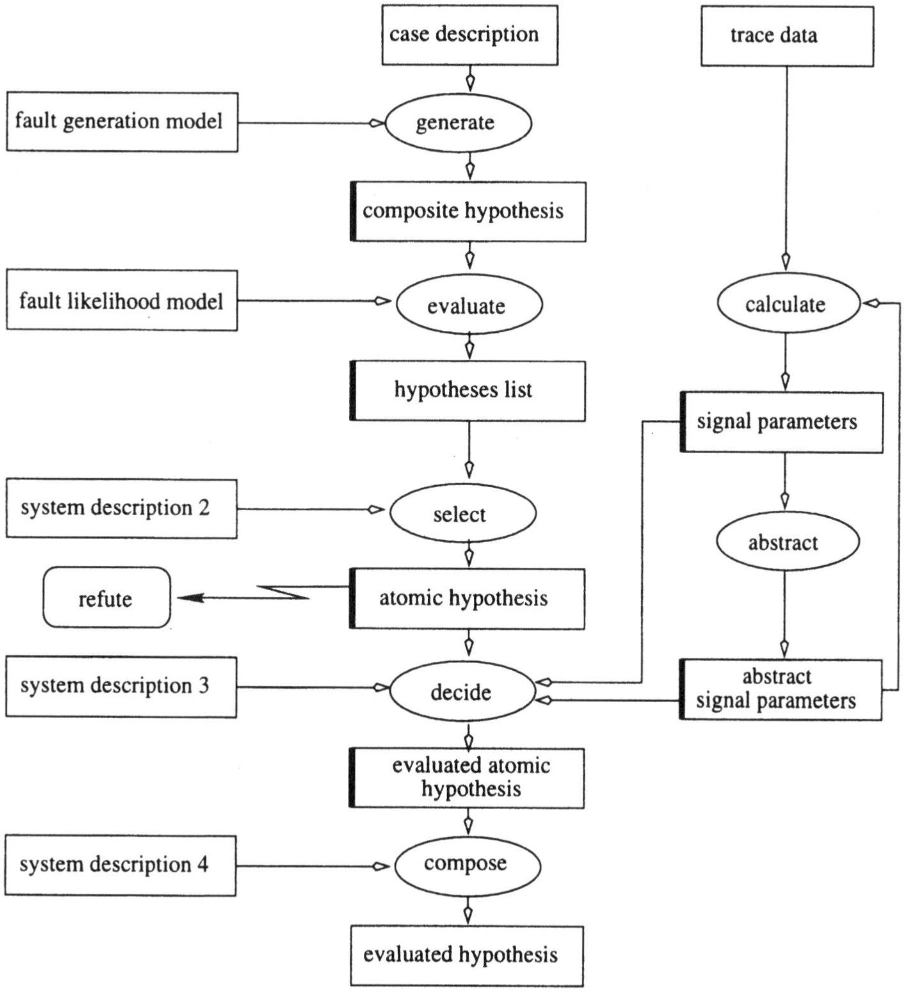

Figure 4. Simplified inference structure of the problem solving methods.

Because the system was intended to support and not to replace the editing it was not necessary for us to solve all problems at once. The system can give problems back to the human editor if it is not able to solve them. Thus it is possible to start with a quite straightforward approach and collect first practical experiences with the prototype. We gained information necessary to select the components which promise the greatest improvement by analysing these experiences.[13]

This situation gave rise to the idea of a scalable design.[16] The inference structure is unchanging whereas distinct components and knowledge sources are replaced by more powerful ones to improve the system's overall performance. This is possible because the main tasks can be described quite independent from each other.

According to the modelling paradigm for knowledge acquisition (following traditional KADS[17]) we built a model of the experts expertise in rating the signal traces, deciding about the editing problems and in performing the necessary operations. Ideally an

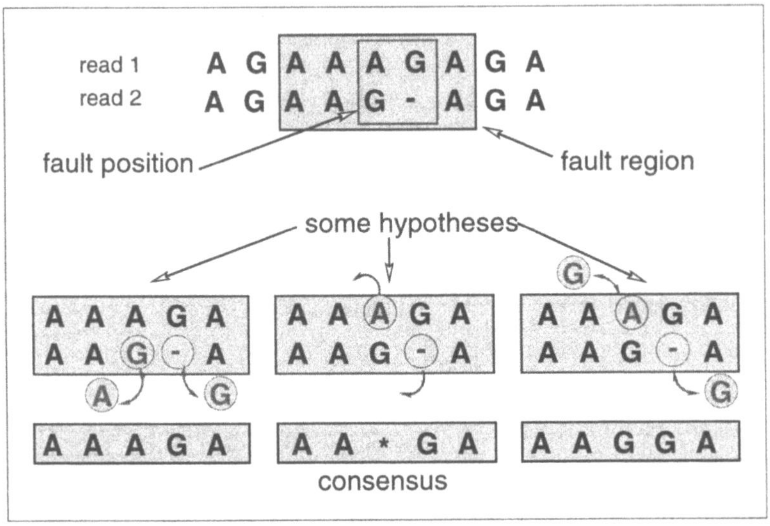

Figure 5. Example of a fault region and some hypotheses each composed of a set of edit operations (atomic fault hypotheses).

interpretation model from a library (e.g., CommonKADS[13]) would be used for the inference layer. But the necessary mixture of diagnostic and repair inferences was not available. The resulting inference structure is shown in Figure 4. Not all knowledge sources in the model can or should be implemented using formal knowledge representation techniques.

Beside decisions of high quality we aim towards reproducible, flexible and modular decision functions for the ubiquitous "atomic" problems (insert, delete, change). With these activities we intend to lay a cornerstone towards noticeably extending the supported editing activities in the future. We already use these decision functions to improve sequence assembly.

6.1. Hypotheses Generation

The mismatches in the alignment can be corrected at "symbol level" by applying a number of edit operations. A sequence of possible operations that would correct the discrepancy is called a hypothesis. Our first goal is to find the most likely hypotheses for an area around the conflicting position (we call it the fault region).

Hypotheses generation is quite simple for most discrepancies. Only a single insert, delete or change has occurred and the alignment is still correct. But for more complex multiple faults, the number of n-step operations to be checked to find a hypothesis is growing exponentially. If m is the width of a fault region and n the depth (the number of aligned sequences) there are about 10*m*n possible operations (5*m*n insert, m*n delete and 4*m*n change operations). Without guidance we would immediately run into a futile combinatorial explosion. But searching too goal oriented for special situations would limit the probability to cope with complex or unexpected fault situations (e.g. at the end of a read or if the alignment was not correct). As a compromise we limited the search space by:

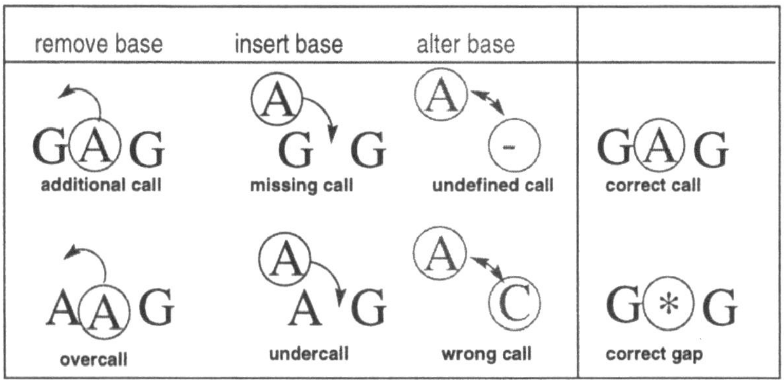

Figure 6. Classes of atomic hypotheses. On the left the different fault hypotheses about expected errors in the sequence and on the right the classes for confirmatory hypotheses.

- identically treating sequences which are identical in the fault region
- avoiding useless combinations of operations (e.g. delete a base we have previously inserted or changed)
- avoiding permutations of operations having the same effect
- doing some preprocessing to find solutions with less operations
- truncating the least promising paths after each step of the breadth-first search-algorithm

Hypotheses are rated by individually penalising the different operations. The best scoring hypotheses are handed over for evaluation, starting with the best scoring until a hypothesis can be confirmed or no hypothesis with a sufficient score is left. The penalty values for the different operations depend on the likelihood of the corresponding fault classes.

Normally we use only the high quality parts of the reads for fault detection and hypotheses' generation. But sometimes there is not enough information in the high quality parts to confirm a hypothesis (e.g. if there are only reads from the same strand available). In these cases we search for suitable reads that can be extended to the fault position (see Figure 7). We align the cutoff part of them against the consensus using the Smith-Waterman algorithm provided by the assembler. If (1) the quality of this alignment has a certain quality and (2) if the trace in a region around the fault region has good signal quality and (3) if the local alignment of the fault region is very good, we also produce editing hypotheses for the hidden data parts (positioning adjustments may be necessary due to the alignment). Thus we can make selective and controlled use of the hidden data if necessary but we do not uncover the hidden data in these cases.

6.2. Hypotheses Evaluation

A hypothesis is confirmed if all necessary operations (which we call atomic hypotheses here) and "some" of the unchanged bases can be verified by examining the trace data.

Each atomic hypothesis is decided by examining a single gel electrophoresis trace. The trace data consists of four signals from the four dyes corresponding to the different nucleotides.

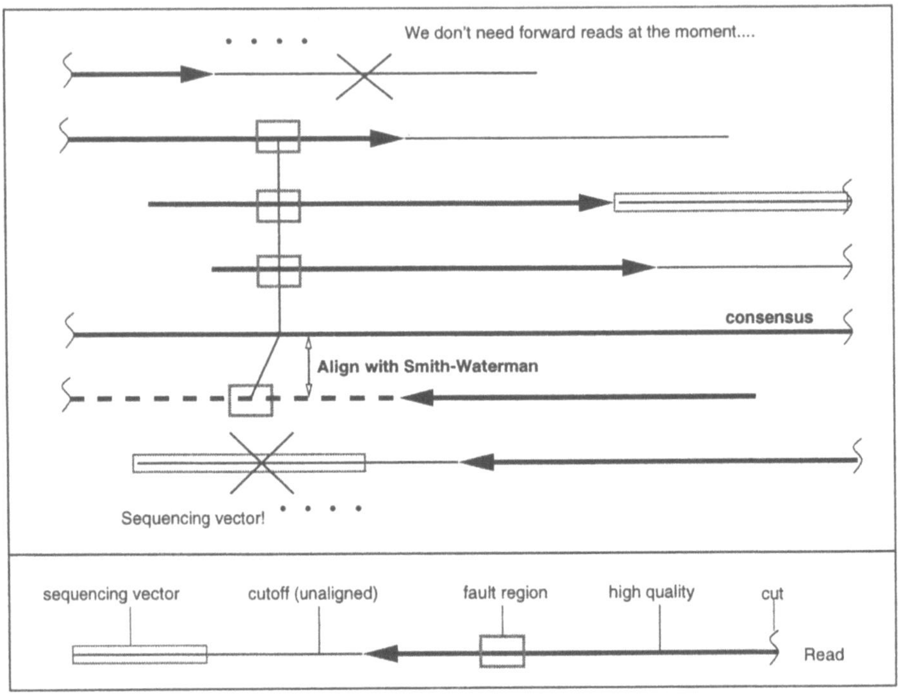

Figure 7. Conditions for generating hypotheses for the hidden data of the sequences if there is not enough information in the high quality parts of the reads.

The relative height of the four traces is somewhat arbitrary because they result from scaling operations of the sequencing software (e.g. ABI/ALF) which bring the traces to a comparable mean height. Thus we have to be careful when comparing their height. We widely ignore the absolute height of a peak or a signal, but we use parameters including a relation like "the height of the peak compared with other peaks from called bases of this nucleotide" or "the height of the peak compared with peaks of non-called bases of this nucleotide" (a kind of signal-to-noise ratio) in a local environment when possible. The same principle of comparing with an attribute in a local environment is also used for parameters concerning the spacing between the bases which is important for deciding about an insert or delete hypothesis.

We have parameters for characterising local signal quality (from 15 bases left to 15 bases right of the problem) and for describing single bases (number of peak, peak position, relative peak height, peak height, peak spacing, peak distinctness, . . .). The traces for all fault hypotheses must have sufficient signal quality to avoid the misinterpretations of very noisy traces.

We implemented a knowledge base using CLIPS (C Language Integrated Production System). CLIPS is an OPS-like forward chaining production system written in ANSI C by the NASA Software Technology Branch (STB).

The knowledge base for hypothesis' evaluation consists of the following main modules or knowledge sources:

- parameter request: control of the calculation of signal parameters depending on the decision problem at hand, values of previously calculated parameters or data abstractions.
- data abstraction: basic data abstractions are made to transform numeric signal parameters, combinations of parameters and dye chemistry properties into qualitative attributes. Abstractions are made to reduce the conceptual gap between the calculated parameters and the concepts used by the expert to describe the patterns. They also weaken the dependency between the decision rules and the individual parameters.
- decision: rules to decide about the truth of an atomic hypothesis based on parameters and data abstractions.
- efficiency: evaluate if—given the previous results—a hypothesis can still become true or if the evaluation of an atomic hypothesis of the classes "correct call" or "correct gap" is necessary.

Because the number of decisions that have to be made is quite high—for a sequencing project of 50 kbp we have to examine about 2000 to 5000 discrepancies (50 kbp with 6 fold coverage and 1% error rate would yield about 3000 errors) and a comparable amount of hypotheses—performance aspects had to be taken into account. The rules for deciding if information about the truth of an atomic hypothesis is necessary or not can speed up system performance notably. Particularly if loading trace data files can be avoided.

At the moment we use crisp rules to decide about an atomic hypothesis, but we intend to try out fuzzy rules as well. Beside of the fuzziness of the decision itself, different purposes (e.g. editing and assembly) require different strengths of confirmation which can simply be achieved by different defuzzification functions.

Using production rules for the knowledge base was a first choice. They are easy to implement, the knowledge base is human readable and decisions can be traced back to the applied rules. It would make sense to try out other knowledge representations to be able to treat the uncertainty implied in the decisions and to be able to learn the knowledge base from examples. We will also examine the stability of the knowledge base for different dye chemistries or for capillary electrophoresis in the future. A hybrid representation where e.g. the fine tuning of parameters for atomic hypotheses can be learned and the requirements for deciding about the overall hypothesis are crisp rules seems suggestive.

7. SYSTEM EVALUATION

In order to carry out a scalable design consistently, it was however necessary to actually perform at a quite early point in time an evaluation in order to determine the components that should be developed top priority in order to attain a comprehensive and efficient system.

The prototype we evaluated could handle only a subset of the fault classes we build (pads in the consensus). This subset was chosen because we needed no explicit hypotheses' generation and only a subset of the decision functions for atomic fault hypotheses.

We wanted to know if the quality of our decisions is high enough to scale up, handle the other fault classes and generate more complex multiple fault hypotheses or if it would be better to improve decision quality first. Our decision policy intended a high positive predictive value (or a low rate of false positive decisions) and we achieved about 99% but at the cost of a low sensitivity of 50%.[13]

The low sensitivity was due to being very careful and ignoring problems when we found other faults in the vicinity or if only information from a single strand was available. This was the case for about 80% of the unsolved problems. We have now implemented decision functions for all fault classes, hypotheses generation and the use of hidden data (see Figure 7) to overcome these problems.

ACKNOWLEDGMENTS

The authors wish to thank Judy Cook who reviewed parts of this article before she died in a car accident, 21 years old, on Christmas Eve 1998.

The research presented in this article was conducted in close cooperation with the Genome Sequencing Centre of the Institute for Molecular Biotechnology, (IMB) Beutenbergstraße 11, 07745 Jena, Germany.

This work is supported by the Bundesministerium für Bildung, Wissenschaft, Forschung und Technologie by grant number 01 KW 9611.

REFERENCES

1. Bonfield, J.K., Smith, K.F., and Staden, R. (1995). A new DNA sequence assembly program. Nucleic Acids Research, 23(24):4992–9.
2. Bonfield, J.K. and Staden, R. (1995). The application of numerical estimates of base calling accuracy to DNA sequencing projects. Nucleic Acids Research, 23(8):1406–10.
3. Breuker, J. and der Velde, W.V., editors (1994). CommonKADS Library for Expertise Modelling. IOS Press Amsterdam.
4. Dear, S., Durbin, R., Hillier, L., Gabor, M., Thierry-Mieg, J., and Mott, R. (1998). Sequence assembly with CAFTOOLS. Genome Research, 8:260–7.
5. Dear, S. and Staden, R. (1992). A standard file format for data from DNA sequencing instruments. DNA Sequence, 3:107–10.
6. Ewing, B. and Green, P. (1998). Base-calling of automated sequencer traces using PHRED. II. error probabilities. Genome Research, 8(3):186–94.
7. Ewing, B., Hillier, L., Wendl, M.C., and Green, P. (1998). Base-calling of automated sequencer traces using phred. I. accuracy assessment. Genome Research, 8(3):175–85.
8. Gordon, D., Abajian, C., and Green, P. (1998). consed: A graphical tool for sequence finishing. Genome Research, 8(3):195–202.
9. Lario, A., Gonzalez, A., and Dorado, G. (1997). Automated laser-induced fluorescence DNA sequencing: Equalizing signal-to-noise ratios significantly enhances overall performance. Analytical Biochemistry, 247:30–3.
10. Lawrence, C.B. and Solovyev, V.V. (1994). Assignment of position-specific error probability to primary DNA sequence data. Nucleic Acids Research, 22(7):1272–80.
11. Lipshutz, R.J., Taverner, F., Hennessy, K., Hartzell, G., and Davis, R. (1994). DNA sequence confidence estimation. Genomics, 19(3):417–24.
12. Mott, R. (1998). Trace alignment and some of its applications. Bioinformatics, 14(1):92–7.
13. Pfisterer, T. and Wetter, T. (1999). Computer assisted editing of genomic sequences—why and how we evaluated a prototype. In: Puppe, F. (editor): XPS-99: Knowledge-Based Systems—Survey and Future Directions, 201–209, Springer.
14. Sedgewick, R. (1994). Algorithmen in C++. Addison-Wesley [Deutschland] GmbH.
15. Wang, L. and Jiang, T. (1994). On the complexity of multiple sequence alignment. Journal of Computational Biology, 1(4):337–48.

16. Wetter, T. and Pfisterer, T. (April 14–18, 1998). Modeling for scalability—ascending into automatic genome sequencing. In Eleventh Workshop on Knowledge Acquisition, Modeling and Management (KAW'98), Banff (Canada).
17. Wielinga, B.J., Schreiber, B.J., and Breuker, J.A. (1992). KADS: a modelling approach to knowledge engineering. Knowledge Acquisition, 4(1):5–54.

6

QUEST

An Iterated Sequence Databank Search Method

William R. Taylor and Nigel P. Brown

Division of Mathematical Biology
National Institute for Medical Research
The Ridgeway, Mill Hill
London NW7 1AA
United Kingdom

1. INTRODUCTION

One of the major aims of bioinformatics in the context of the various genome data, and especially as applied within pharmaceutical companies, is to identify, or at least suggest, new drug targets. A first step in this pursuit requires that some idea of the structure, but more importantly, the function of the protein is established. When there were few sequences available, then predicting the tertiary structure of the protein was the primary goal—as this might give some idea of a function that was unlikely to be discovered by analogy to another protein. Now, however, with so many sequences there is a better chance to find a match to the probe sequence that will shed some light on its structure or function. The emphasis of sequence analysis methods has correspondingly shifted from structure prediction to one of searching for remote relatives. Even if no known function is revealed through searching, the resulting sequence family can help with structure prediction by emphasising structurally conserved regions and motifs associated with secondary structure.

The method described in this chapter was developed in response to a practical need to iteratively search the sequence databanks for distant relatives, starting from a specified probe sequence. In particular, the probes were specified as part of the CASP structure prediction or recognition programme (Critical Assessment of Structure Prediction). The structure prediction (Aszódi et al., 1995; Aszódi and Taylor, 1996) and recognition (Taylor 1997) methods employed by the author and co-workers in CASP (Aszódi et al.,

Genomics and Proteomics, edited by Sándor Suhai.
Kluwer Academic / Plenum Publishers, New York, 2000.

1997) both relied on having a family of diversely related sequences: and the more diversely related this family could be (while still being correctly aligned), then the better was the opportunity to detect important motifs that might give some clue to structure or function.

The initial search protocol that was employed in this pursuit was to start with a BLAST (Altschul et al., 1990) search to extract sequence down to a moderate degree of relatedness. (Ψ-BLAST (Altschul et al., 1997) was not yet available). The extracted sequences were then aligned using a conventional multiple sequence alignment program, specifically, MULTAL (Taylor, 1988). (Although CLUSTAL (Higgins and Sharp, 1988) would have been almost as good). From the resulting sequence alignment, a consensus pattern was determined which formed the basis of a further search using a standard pattern matching tool (the UNIX regex utility). The sequences identified by this search were then realigned and provided the basis for a further pattern search.

This approach suffered from some problems—especially under time constraints (when working to CASP submission deadlines). Although BLAST is fast, analysis of its output was slow. Time was lost in scanning the sometimes very long list of hits to find sequences to include in the family. This was especially problematic when the family of interest contained many close homologues (such as the globins, protein kinases or immunoglobulins). An additional problem with the older BLAST was that it only reported sequence fragments, often requiring work to determine the overall length of the sequence corresponding to the probe. (This is now made easier with gapped BLAST (Altschul et al., 1997)). The multiple alignment stage was generally trouble-free, but the specification of the regular-expression pattern from the alignment was often subjective and arbitrary.

The program described in this chapter (called QUEST) was developed in a attempt to automate these problems. It has its origin in an old (but versatile) pattern matching program (Taylor, 1986b; Taylor, 1989) and was used in conjunction with the multiple alignment program MULTAL (Taylor, 1988). These two programs were applied alternately in successive rounds of searching and alignment (Figure 1).

2. SEARCHING WITH MULTIPLE SEQUENCES

Aligned multiple sequences provide a powerful resource in many areas of structure prediction and recognition through their ability to average-out the noise in calculations. If it can be assumed that each sequence in the multiple alignment adopts the same fold, then inconsistent variations will be averaged-out when the multiple alignment is considered as a whole. This has been applied to good effect in secondary structure prediction (Zvelebil et al., 1987; Rost and Sander, 1993) and in molecular modelling, in particular in fitting the sequence to the structure (threading) (Taylor, 1997). (See Figure 2.)

Most importantly in the current application, multiple sequence alignments reveal conserved features in a family of sequences and when these are identified in two families (or sub-families) then it is easier (less ambiguous) aligning the two alignments than any two individual sequences drawn from each. (See Figure 3 for an illustrative example.) Unfortunately, this power is generally not available when searching a sequence databank. The most commonly used search program, BLAST (Altschul et al., 1990) employs only a one-to-one match in its assessment of relatedness. Its more recent

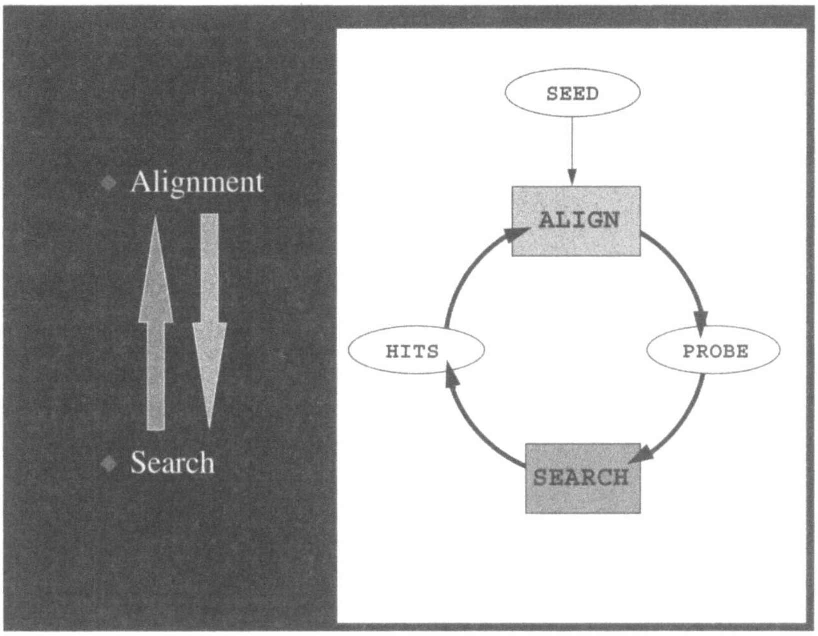

Figure 1. Outline of the approach. The method alternates between search and align phases with the output from each phase forming the data for the next. The alignments were made using MULTAL while the search phase was made using QUEST, a program that was a development of an older template (pattern) matching method.

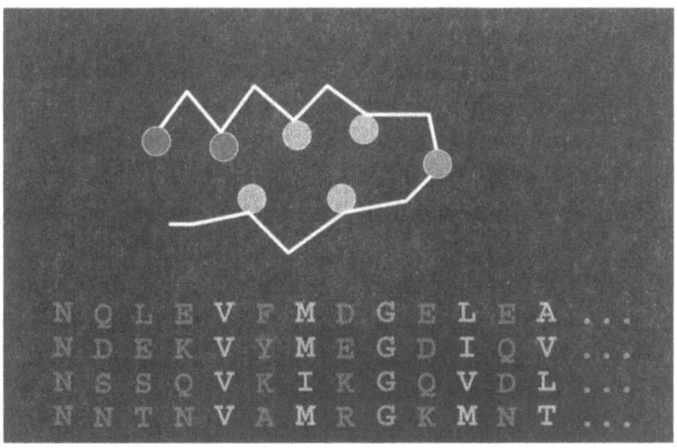

Figure 2. Multiple sequences in threading. The power of multiple sequences in finding an alignment of sequence data onto a structure (threading) is illustrated with a small schematic protein structure in which the residues are represented as circles and coloured green if they are buried. The conserved hydrophobic residues in the multiple alignment have also been coloured green and the unique register of these onto the structure can be seen more easily than if any single sequence in the alignment had been considered. Other features, such as the conserved glycine also align well.

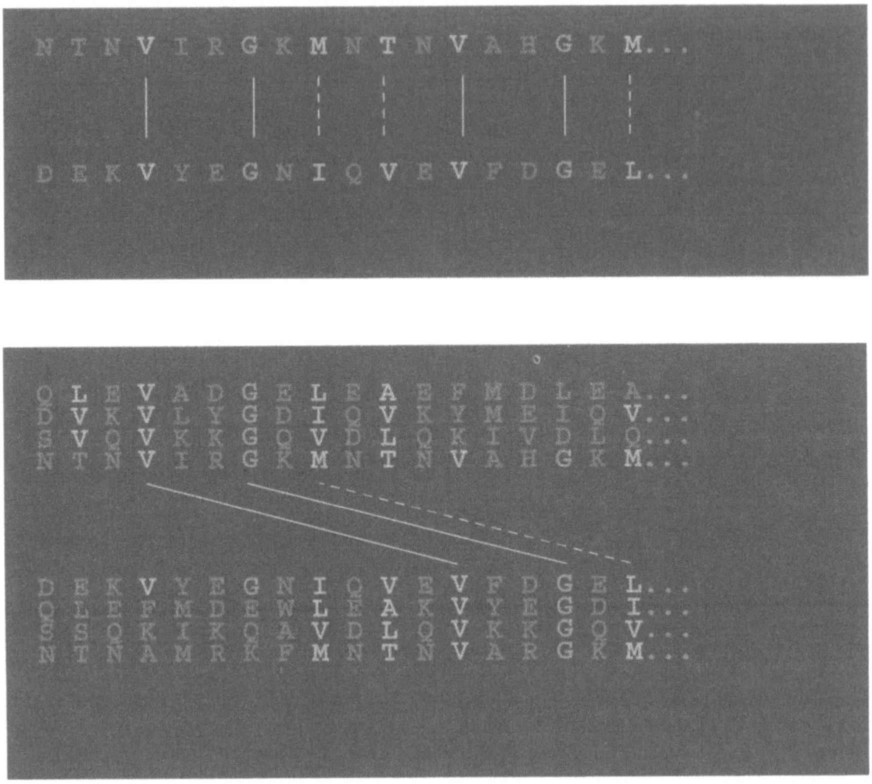

a

b

Figure 3. Multiple sequences in alignment. (*a*) pairwise alignment of two sequences with a reasonable degree of similarity. (*b*) the same sequences now in multiple sequence families with the true alignment revealed by conservation.

relative, Ψ-BLAST (Altschul et al., 1997) goes a stage further and allows an alignment (sometimes called a profile) to be matched against a databank—but this is still only a many-to-one match. To properly achieve a many-to-many match would require a databank that had been pre-aligned into families. While this has been attempted (Smith and Smith, 1990), such databanks are necessarily less complete than those currently available and are complicated by the level at which the similarities should be combined.

The QUEST program described here attempts to move towards a many-to-many match in a less direct way: not by pre-aligning the whole databank, but only a small fraction of it that has been identified as being potentially similar to the probe family. It is then hoped that the defining comparison that decides whether a family drawn from the databank is similar to the probe family will be made on the basis of a profile/profile (many-to-many) alignment rather than a one-to-many. This approach was made possible by the speed of MULTAL which is able to align a large number of sequences (hundreds) in a reasonable time (minutes). This makes it less critical that the initial search is perfect, and as long as the search overestimates the similarity then the onus is placed on the multiple alignment phase to sort the true from the spurious relationships.

In comparison to Ψ-BLAST, the current method (QUEST) places more emphasis (and computing time) on the alignment phase—Ψ-BLAST simply "stacks-up" the individual segment matches to create a multiple alignment. By contrast, the PROBE program (Neuwald et al., 1997), which also has a BLAST based search phase, spends considerable effort on the alignment phase (using a complex Gibbs sampling technique), almost to the point of rendering its use impractical. In broad outline, the current method can be considered as lying somewhere between Ψ-BLAST and PROBE.

3. NEW TEMPLATE METHOD: QUEST

3.1. Origins

As stated above, the current method derives from an old pattern matching method in which the patterns were called "templates" (Taylor, 1986b; Taylor, 1989)—which, in turn derived from an even older secondary structure prediction program (Taylor and Thornton, 1983; Taylor and Thornton, 1984). The template-matching program was rather slow as it matched each pattern from the probe alignment at each position in the sequence being scanned (Figure 4), and with the current databank sizes, this would make its use impractical. To rectify this problem, the current formulation of the program incorporates a BLAST-like pre-filter to focus each template only onto positions in the sequence where a match is likely. This method was based on a fast look-up table in which the location of each tri-peptide in the probe was recorded.

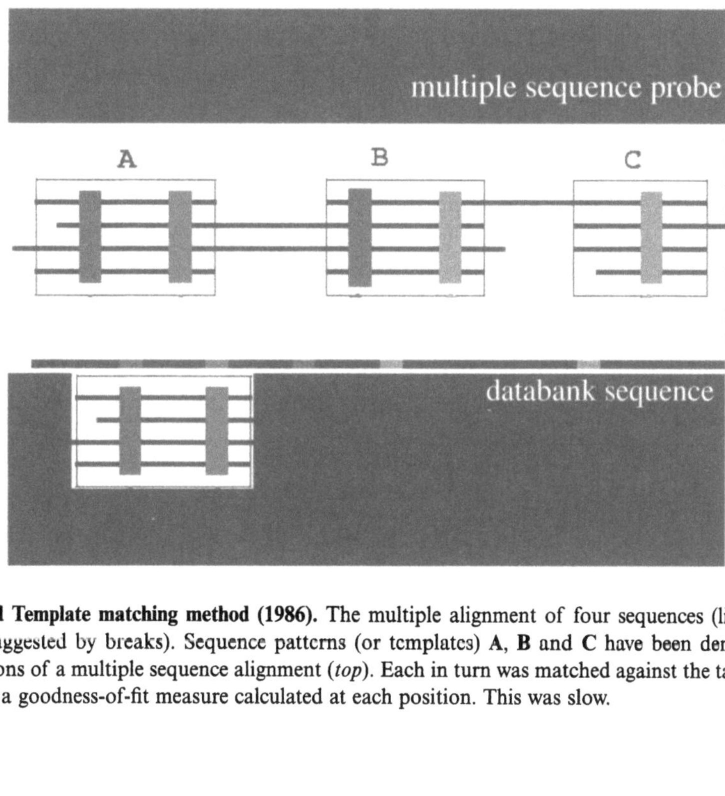

Figure 4. Old Template matching method (1986). The multiple alignment of four sequences (lines) is shown (with gaps suggested by breaks). Sequence patterns (or templates) **A**, **B** and **C** have been derived from the aligned portions of a multiple sequence alignment (*top*). Each in turn was matched against the target sequence (*bottom*) and a goodness-of-fit measure calculated at each position. This was slow.

3.2. Amino Acid Match-Sets

As in the previous template-matching program (Taylor, 1986b), the possibility was allowed to extend the amino acid matches beyond what was observed in the profile. This approach incorporates a predictive element by including other acids that form related groups (or sets) (Taylor, 1986a). By using simple set operations on the predefined groups, the smallest subset that includes all acids occurring at the profile position can be taken as the match-set. While useful, this approach depends on the predefinition of groups which may be ambiguous. A more severe limitation, however, is the sensitivity of the approach to degradation through error. A single misaligned sequence (or sequencing error) can reduce the specificity of a position to a point where the smallest subset includes all 20 amino acids.

The problems and ambiguities of match-set assignment (including the problem of the all-or-nothing match) were avoided by using a weighting scheme for the amino acids, similar to the use of pseudo-counts (Henikoff and Henikoff, 1992) or Dirichlet mixtures (Karplus, 1995). For a given position in a profile, a vector of amino acid weights (${}^{a}W$) was defined as; 1 for observed acid types and otherwise 0. A modified vector (${}^{b}W$) was then built-up such that each component was the sum of the original vector components distributed according to a matrix of amino acid similarity (M):

$$ {}^{b}w_i = {}^{a}w_i + \frac{\alpha}{100}\sum_{j=1}^{20}({}^{a}w_j + 1)M_{ij}, \tag{1} $$

where M_{ij} is a measure of the similarity of amino acids i and j (stored in the matrix $\mathbf{M}$) and α is a constant that controls the extent of the redistribution of weight. (The PAM_{120} matrix of Dayhoff et al. (1978) was used). To attain some independence over the numeric range of the values of the components in any given matrix $\mathbf{M}$, the modified weights were normalised to have unit variance and zero mean, then shifted back to the original mean of ${}^{a}W$. The procedure was then repeated with ${}^{b}W$ becoming ${}^{a}W$ in the second round. Figure 5 shows some match-sets extended by this method.

3.3. Gaining Speed

As the current method is intended to be used on problems similar to those currently tackled by Ψ-BLAST, then its speed should be at least comparable. Clearly, the "brute-force" approach of the old template method (Figure 4) is impractical and, following the approach used in BLAST itself, a faster algorithm was employed. This was based on using tri-peptides to "seed" templates on potentially matching locations. For detecting remote relationships, it is not possible to rely on finding exact tri-peptide matches. (For example; the FASTa program recommends dipeptides for proteins.) BLAST gets round this problem by "softening" its tri-peptide matches using a score based on a matrix of amino acid relationships (such as the PAM_{120} matrix of Dayhoff et al. (1978)). In the current method, the range of variation seen in the multiple alignment was used instead. This was implemented simply by allowing the match to be made with any combination of acids made-up from those seen in the sequences at the three positions considered. These were then encoded into a look-up table that allowed "immediate" location of any potential matches in the probe to a tri-peptide in the target sequence (Figure 6).

```
GDAEAAAKTS    *..**.*...*..7..2.**.....     ADEGKST ---> ADEGKNQST
EEDEEEDDDQ    ...**........1..*........         DEQ ---> DENQ
WWFKKRKKKA    *....*....*......*0...*..       AFKRW ---> AFKRSW
QNDSSATSAA    *..*4.0......*..*.**.....      ADNQST ---> ADEGNQST
QHAATANVRL    *......*0..*0*..**1*.*...    AHLNQRTV ---> AHILMNQRSTV
VVVVIIVVVV    ........*............*...          IV ---> IV
LLLTSKKKKK    ..........**1.....**.....        KLST ---> KLMST
NGKAAAAAAS    *..2..*...*..*....*2.....       AGKNS ---> ADGKNST
VICLVLAFLS    *.*..*..*..*4.....*3.*...     ACFILSV ---> ACFILMSTV
WWWWWWWWWW    ......................*..           W ---> W
GAGGGGGGDE    *..**.*......2....1......        ADEG ---> ADEGNS
KKPKKKKKKE    ....*.....*....*2........         EKP ---> EKPQ
VVVVVIVIIF    .....*..*..20........*...         FIV ---> FILMV
EEENNDGSEN    2..**.*......*....*......       DEGNS ---> ADEGNS
APAVIVAGGA    *.....*.*......*..22.*...       AGIPV ---> AGIPSTV
```

Figure 5. Amino acid match-set extension. A sequence alignment of ten globins (over a short region near the amino terminus) runs vertically on the left. This is recoded as a matrix showing the weight on each amino acid in alphabetical order (A . . . Y). Amino acids present at the aligned position are marked by an asterisk (*) and have weight 1, while a weight less than zero is represented by a dot. Intermediate weights are indicated as the integral part of ten times their value. The new match-sets were defined as all acids having positive weight. (See Equ^n. 1.)

3.4. Template Combinations

In the template approach, combinations of template matches were selected to form a match over the expected extent of the probe. Previously, and in the current method, this selection was made by combinatoric enumeration of the matched templates subject to the simple constraint that they occur in the correct order along the target sequence and that their spacing does not exceed limits derived from the gaps observed in the probe alignment. However, to avoid the time-consuming enumeration of trivially different low scoring template combinations, the combinatorial search was restricted to combinations that contain at least one of the top ten scoring matches. Furthermore, if the letters ABCD represent a valid template combination, then the trivial variants such as: A-CD, AB-D, etc., were not accepted. The template matches are considered in decreasing order of score, so the first solutions are also likely to be formed from the highest scoring template matches (see Figure 7 and Taylor (1989) for further details).

3.5. Treatment of Gaps

As previously, the current method does not have a gap-penalty but allows free gap formation within limits derived from the probe alignment. (This approach is more closely related to the specification of mis-match ranges in regular-expression pattern matching (Smith and Smith, 1990).) Gap ranges were derived from the probe alignment by recording, for all pairs of positions in the alignment, the maximum and minimum number of intervening residues found in the individual sequences in the alignment. For an alignment of reasonably divergent sequences, the observed gaps provide a good guide for the location or size of gaps in the target sequence being matched. However, for

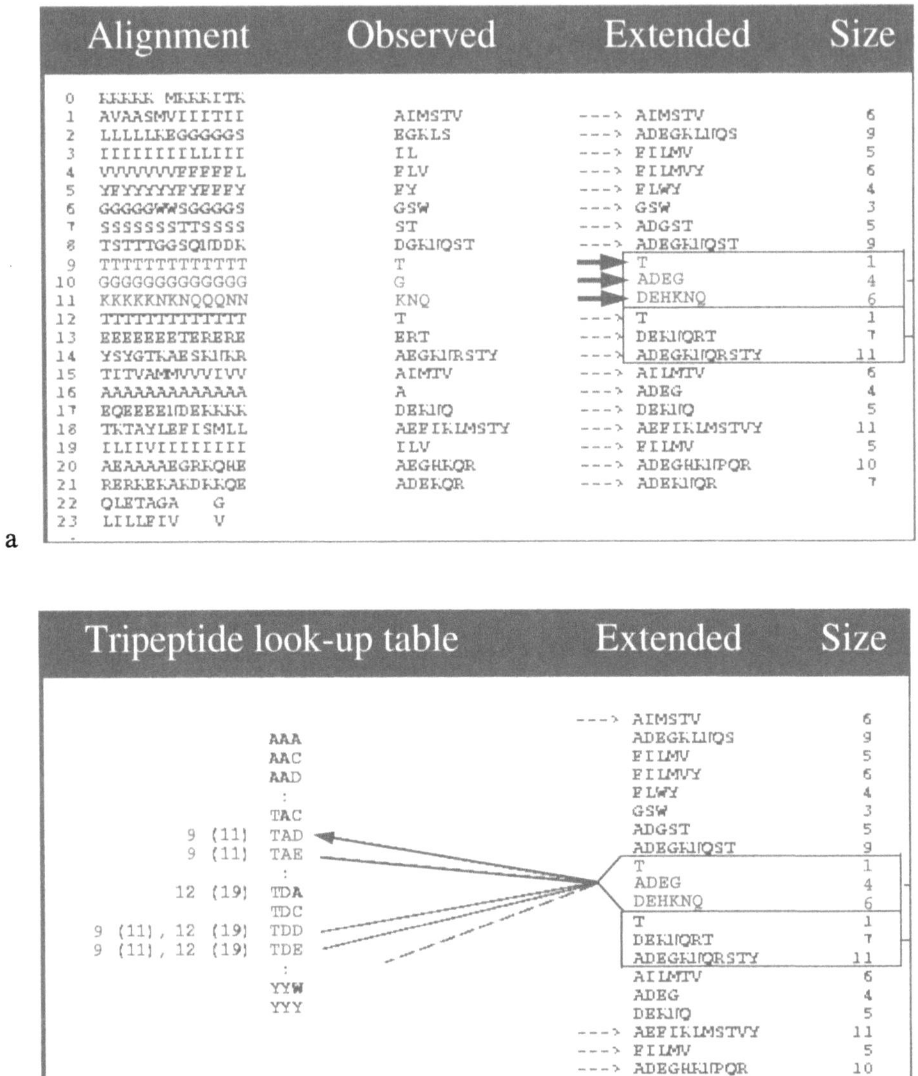

Figure 6. Tri-peptide look-up table. (*a*) a tri-peptide in a multiple sequence alignments gives rise to amino acid match-sets. (*b*) all possible peptides are recorded in the look-up table with their associated location.

more closely related sequences, the observed gaps may not adequately represent the likely location of new gaps and in this situation, an additional tolerance was allowed in the constraints.

The gap constraints were applied between each pair of templates encountered in the combinatoric search. Retaining the simple example used above of four template matches, A, B, C and D; if A is the starting selection to which B is added (forming AB), the number of residues between the carboxy terminus of A and the amino terminus of B is compared to the observed maximum and minimum range in the alignment (allowing any tolerance). If the gap between A and B is within this range, then B is added to

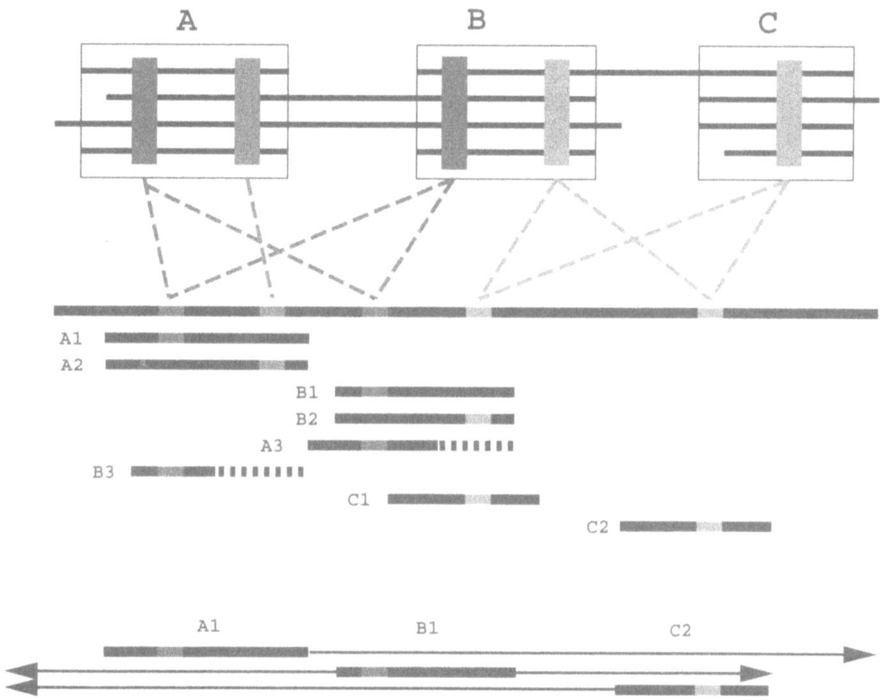

Figure 7. Template fitting and selection. *Top*: multiple alignment of four sequences (lines) (gaps suggested by breaks). In each region where three-quarters of the sequences align, templates (A B C) are defined from the amino-acid match-sets. The three highest scoring tri-peptides are shown in different colours. *Middle*: a databank sequence (long thick line) is scanned and each peptide corresponding to those identified from the alignment provides a site at which the match of a template is tested. The first occurrence of the red peptide allowed a match of the template A to be located (A1) and also a partial match to template B (B3) in which the unmatched portion is shown dashed. Each template match is shown in decreasing order of score below the sequence. When identical template fits have been seeded from different tri-peptides (e.g.; A1, A2 and B1, B2) only one is passed to the following stage. *Lower*: combinations of templates are selected that have a spacing that does not exceed those observed in the alignment. For example, template fit A1 and B1 have no gap between them, which is allowed as two of the sequences in the alignment run straight from template A into B. Under these constraints, the only valid complete selection is A1-B1-C2 with B3-C1 forming a partial valid selection. The selected templates are then used to make a consensus prediction of the sequence segment termini corresponding to the alignment probe. The arrows show the termini predicted by each template (for no arrow take the template terminus) and a weighted mean is taken of these.

the selection; otherwise it is discarded and the next template match is evaluated with A. If the next template to be selected is C then the gaps between both C and A and C and B are checked before deciding if C is selected. (See Figure 7.)

3.6. Score Cutoffs

In the method outlined above, there were three distinct levels at which the match was assessed:

1. **peptide** level; at which each peptide has a score reflecting its degree of conservation.
2. **template** level; at which each matched template has a score.

3. **domain** level; which has a score made up from the sum of the combined template scores.

Three cutoffs can be associated with the scores at each of these levels to restrict the number of matches that must be considered on each search. Although the choice of these cut-off values is central to the sensitivity and speed of the method, it is not possible to predefine a fixed value for of them, as the scores on which they operate will vary depending on the number and quality of the sequences in the probe alignment. It is equally undesirable to leave them all as free parameters to be set manually with each search as this would present too many possibilities. To circumvent these difficulties, the following method was developed to set the cut-offs automatically with each search.

3.6.1. Tri-Peptide Score Cut-Off. The first, at the peptide level, will restrict the number of templates that will be located. If this is high (allowing only the consideration of conserved positions) then time will be saved as less template scores will be evaluated. However, if set too high, then weak matches (that might be "seeded" by a less conserved position) might be missed. A compromise between these competing drives was found by allowing (roughly) the top quarter of the peptides to pass the cutoff.

3.6.2. Template Score Cut-Off. At the template level a measure was introduced to provide an estimate of the scores that can be achieved by an unrelated (or "random") sequence. "Random" sequences are often generated by shuffling a set of sequences while keeping constant composition and length or by using a "background" of scores of unrelated sequences. The former is rather time-consuming, while the latter has the danger that some of these "unrelated" sequences may be unrecognised relatives.

To avoid these problems, a simple approach, previously used in template matching (Taylor, 1986b), was employed that considers the scores obtained from matching against the probe sequences when these are reversed. This device has the advantage that the length and composition of the 'random' sequences are identical to those in the probe while, in addition, non-specific (direction symmetric) features associated with secondary structure are retained. A cutoff was chosen that excluded 95% of the matched template scores generated from matching the probe against all the sequences (in reverse) that composed the probe profile itself (Figure 8).

3.6.3. Domain Score Cut-Off. The cut-off on the total score (the sum of template scores in the domain) was kept as an adjustable parameter to allow the number of hits found in a search to be controlled either interactively or automatically. A range for this cut-off was set with its low-end at the score of the highest scoring "random" (reversed) sequence and at its upper-end at the mean score of the native sequences. (The mean native score was taken rather than the minimum score to give robustness to the transient misalignment of unrelated sequences in the probe.) The domain cutoff was then specified as a percentage between these two values. (0% = best random score; 100% = mean native score.)

4. DYNAMIC CONTROL

The full search and align cycle is shown in Figure 9, summarising the steps outlined above. Into this cycle, two points of control are possible one in the multiple alignment phase and another in the search phase.

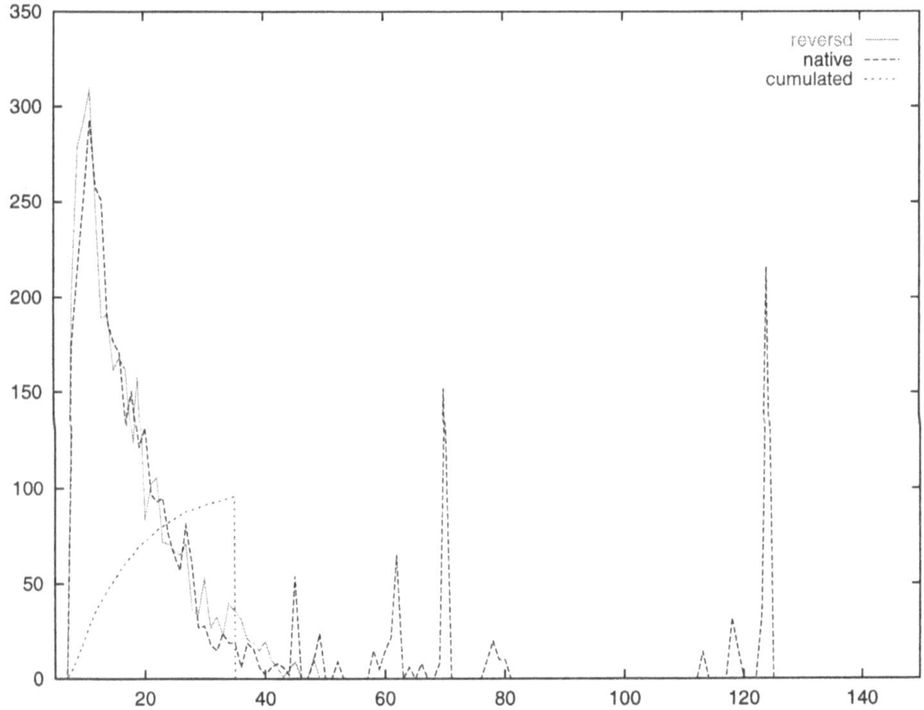

Figure 8. Template score cutoff. The templates derived from a globin probe were matched against the individual sequences that made up the probe and the frequency of scores plotted (*native*). A large broad peak can be seen at low score (false matches) followed by sharp "spikes" at higher scores (true matches). A cutoff on the score should lie between these. This was found by reversing the individual sequences and re-matching the templates giving the the score distribution (*reverse*) that matches the native distribution at low score but does not have any high scoring matches. A cutoff was determined by excluding 95% of the reversed scores as shown by the cumulative distribution of the reversed scores (*cumulated*) plotted up to the 95% level.

4.1. Multiple Alignment Control

The multiple alignment stage will not be described in detail as it used the program MULTAL effectively unmodified from its original implementation (Taylor, 1988; Taylor, 1990) (see Taylor 1998, Appendix II for details of the parameters used in the current application).

Besides its speed, an aspect of the alignment strategy used in MULTAL that is important for the current method is the way in which MULTAL does not "force" all its input sequences to align—unlike tree-based methods such as CLUSTAL (Higgins and Sharp, 1988). This allows sub-families to be identified and only those that relate to the probe need be retained in the subsequent cycle. These were identified by adding the original probe sequences into the pool of sequence domains identified on the search phase. After alignment, the only relevant family for the next search will be that which contains the added probe sequences (which can be identified by the pre-addition of a tag to their code-name).

This strategy advantageously incorporates some additional checks. One serious danger that is avoided is that the search probe cannot be 'usurped' by another family. This is sometimes seen in other iterated search strategies (such as Ψ-BLAST) when there is an erroneous incorporation of a member of a large sequence family into the probe.

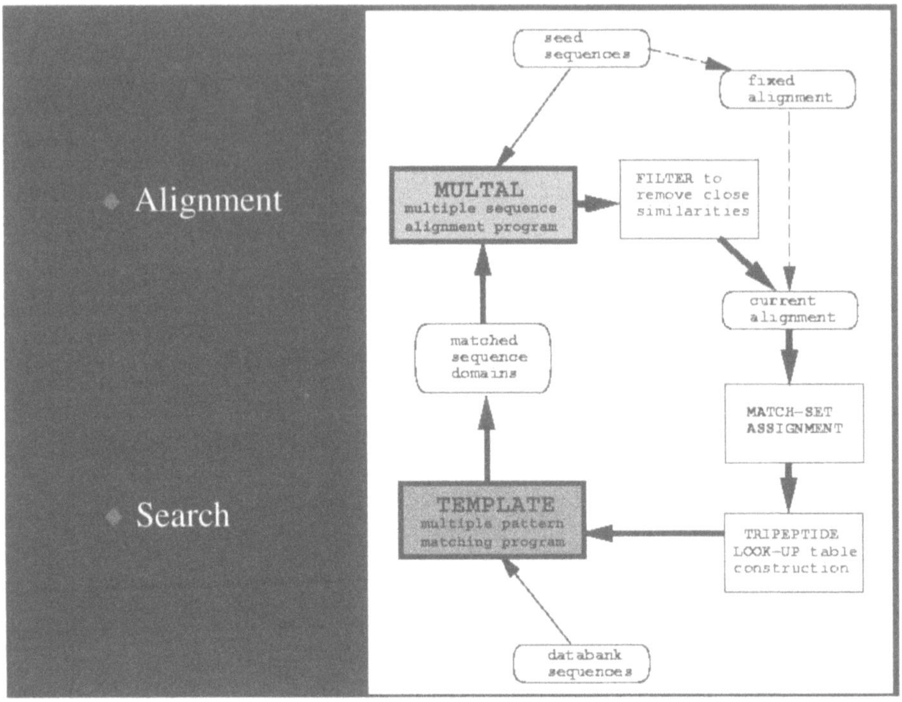

Figure 9. Full iteration cycle. The cycle can begin with one or more unaligned probe sequences. These are filtered to remove close homologues and those remaining aligned. The resulting multiple alignment is processed to generate the fast look-up tables which guide the template matches. The matched sequence fragments (domains) are collected (and combined with the original probe sequences) and together this collection is presented to the multiple alignment program. This has the capacity to expel sequences if they have not aligned with the original probe sequences. On each round of the cycle, if more sequences are found then the cutoff on the domain score is increased; otherwise it is decreased. This negative feedback allows the iteration to evolve to a stable state.

This "encourages" the recruitment of others from the wrong family, resulting in a shift of probe specificity towards the spurious family.

A further internal check is also available on the alignment quality: the probe squences are added and if these are found in different subfamilies, then either MULTAL is not properly set to align sequences of sufficient divergence (correctable by a parameter adjustment) or the original probe sequences are were too divergent. The added sequences also provide some resilience against the failure of the search phase: if, at one extreme, this finds nothing then the next probe will simply consist of the original probe sequences (back to "square-one") or, at the other extreme, if many spurious sequences are found through a loss of probe specificity, then the original probe sequences will still be represented in the next cycle.

4.2. Score Cutoff Control

While it is possible to intuitively anticipate a suitable value of the cut-off for the next cycle from the results of the previous, a simple automatic scheme was adopted to eliminate any subjective bias from the searches. This was based on the principle that if novel sequences were found, the (domain score) cutoff would rise by 5% whereas if nothing new was added, then the cutoff would fall by the same amount. An upper limit was placed at 50% and below 5% the decreasing step size was 1%. This embodies a negative-feedback element which should allow the profile to develop to a certain level but not continue to expand to such an extent that specificity is lost (Figure 10).

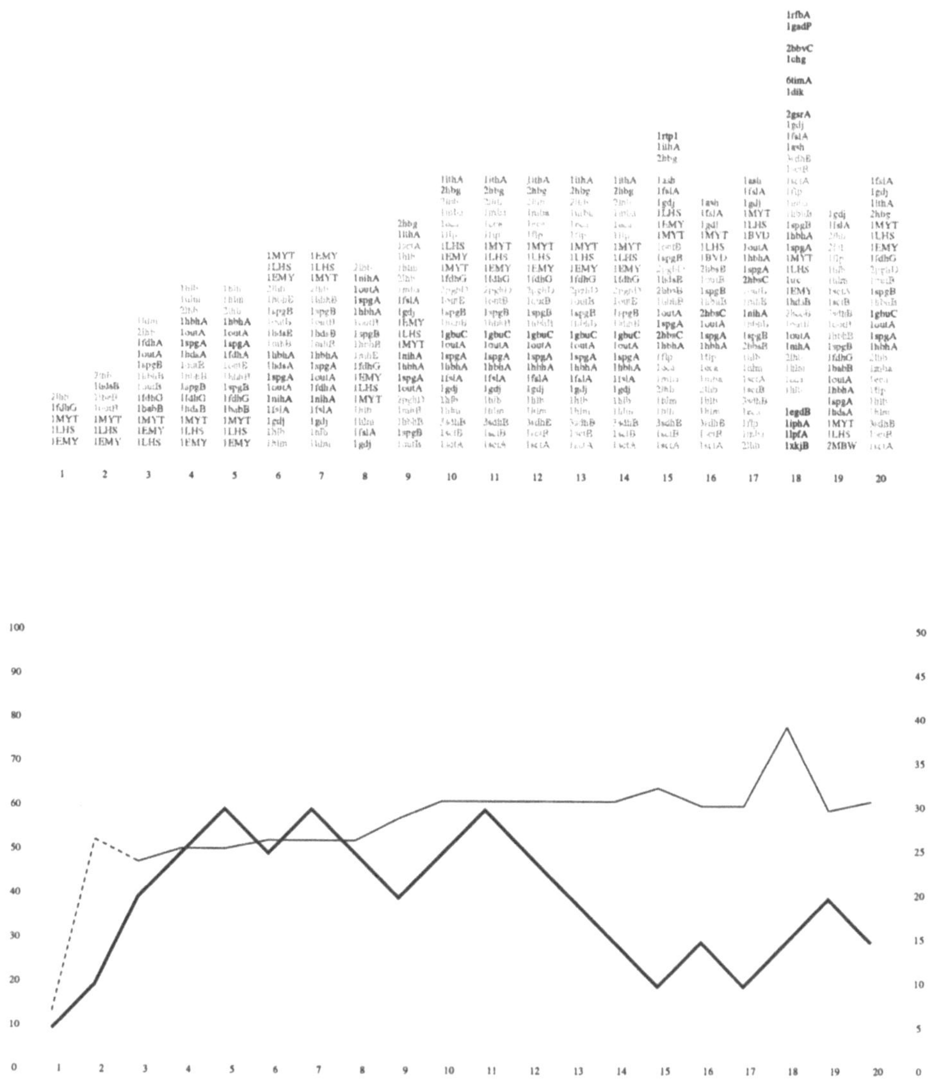

Figure 10. Probe evolution over 20 cycles. *Top*: Plot of profile development of the proteins identified starting with a three myoglobin sequences (upper case PDB codes). Grey PDB codes are members of the globin family while solid black codes are not. Spaces in each column of codes separate aligned sub-families. (e.g. cycles 15 and 18). *Lower*: graph of the domain score cut-off (bold line and right scale) with the total number of sequences before removal of homologues (fine line and left scale). The probe family expands to a stable profile (cycles 10–15) which contains all but one globin (1ash). As the cutoff falls, this is eventually identified but s non-globin (1rtp1) is also found which misaligns with two other globins.

Occasionally, this method resulted in too many sequences scoring over the cutoff. This situation was defined when the number of sequences in the current list was more than three times the previous number. To reduce these, only those scoring over their mean score value were passed to MULTAL for alignment and the cut-off value was increased by a further 5% to reduce the likelihood of recurrence.

5. COMPARISON WITH OTHER METHODS

5.1. Quality Assessment

Some iterated sequence databank search methods were assessed from the viewpoint of someone with the sequence of a novel gene product wishing to find distant relatives to their protein and, with the specific searches against the PDB, also hoping to find a relative of known structure. Three methods were compared, regular-expression matching, Ψ-BLAST and SAM, spanning a range from simple pattern-matching to sophisticated weighted profiles. Rather than apply these methods "blindly" (with default parameters) to a large number of test queries, tests concentrated on the globins, so allowing a more detailed investigation of each method on different data subsets with different parameter settings. (See Taylor and Brown 1999 for full details.)

Despite their widespread use (for example with PROSITE patterns), regular-expression matching proved to be very limited—seldom extending beyond the sub-family from which the pattern was derived. To attain any generality, the patterns had to be "stripped-down" to include only the most highly conserved parts.

The QUEST program avoided these problems by introducing a more flexible (weighted) matching. On the PDB sequences this was highly effective, missing only a few globins with probes based on each sub-family or even a single representative from each sub-family. In addition, very few false positives were encountered, and those that did match, often only did so for a few cycles before being lost again. On the larger sequence collection, however, QUEST encountered problems with maintaining (or achieving) the alignment of the full globin family.

Ψ-BLAST also recognised almost all the globins when matching against the PDB sequences, typically, missing three or four of the most distantly related sequences while picking-up a few false positives. In contrast to QUEST, Ψ-BLAST performed very well on the larger databank, getting almost a full collection of globins although still retaining the same proportion of false positives.

The Hidden-Markov Model (HMM) method, SAM (Krogh et al., 1994), when applied to the PDB sequences performed reasonably well with the myoglobin and hemoglobin families as probes, missing, typically several of the more difficult proteins but performed poorly with the globin probe that was most distantly removed from the main family (leghemoglobins). Only with the full family range as a probe did it produce results comparable to Ψ-BLAST and QUEST. With the larger databank, it produced a good result but, again, this was only achieved using the full range of sequence variation with the default regulariser and use of Dirichlet mixtures completely failed in this situation.

5.2. Speed Assessment

Search time is an important discriminating factor, however, all the above methods are sufficiently fast that anyone with an important single query would be prepared to wait: execution times only become important when a single resource is shared by many users or a single user has a large number of queries (such as a genome).

Searching the PDB+SWISS-PROT databank (73,427 sequences) on a single Pentium processor (333MHz) Ψ-BLAST took 100 seconds for 6 cycles while QUEST took 10–100 sec. However, for each cycle, SAM took over 2000 seconds for one search step. (In principle, HMM times are equivalent to dynamic programming). In our experiments with SAM, the search phase was the rate-limiting step, however, with manual iteration, it is likely that the development of the model would consume a much greater proportion of the time. Although not tested on a comparable platform, the PROBE program was also expected to take a time more comparable to SAM than Ψ-BLAST or QUEST.

6. CONCLUSIONS

In the construction of a multiple alignment, Ψ-BLAST takes a "lazy" approach and simply "piles-up" the matched segments on to the original query sequence. This saves the independent calculation of a multiple sequence alignment (which is expensive) but has the disadvantage that the resulting alignment (which is not explicitly reported by the program) is biased by the original query sequence. The convergence of the results for the searches over the larger databanks, however, suggest that this is not a problem when the features of the original queries become sufficiently "diluted". By contrast, QUEST spends a large fraction of its effort constructing a multiple alignment afresh with each cycle. This effort is justified, however, by the resulting ability to discard sequences from the probe between cycles, giving QUEST a very low rate of false-positive hits. As was seen with the larger databank searches, however, simple multiple alignment (without sequence weighting) becomes difficult when the family grows large and one of the weaker aspects of QUEST is its reliance on a simple (and old) multiple sequence alignment method. This aspect is currently being investigated.

REFERENCES

Altschul, S.F., Gish, W., Miller, W., Myers, E.W., and Lipman, D.J. (1990). Basic local alignment search tool. *J. Molec. Biol.*, 214:403–410.

Altschul, S.F., Madden, T.L., Schäffer, A.A., Zhang, J.H., Zhang, Z., Miller, W., and Lipman, D.J. (1997). Gapped BLAST and PSI-BLAST: a new generation of protein database search programs. *Nuc. Acid Res.*, 25:3389–3402.

Aszódi, A. and Taylor, W.R. (1996). Homology modelling by distance geometry. *Folding & Design*, 1:325–334.

Aszódi, A., Gradwell, M.J., and Taylor, W.R. (1995). Global fold determination from a small number of distance restraints. *J. Molec. Biol.*, 251:308–326.

Aszódi, A., Munro, R.E.J., and Taylor, W.R. (1997). Protein modelling by multiple sequence threading and distance geometry. *Proteins*, pages 38–42.

Dayhoff, M.O., Schwartz, R.M., and Orcutt, B.C. (1978). A model of evolutionary change in proteins. In Dayhoff, M.O., editor, *Atlas of Protein Sequence and Structure*, pages 345–352. Nat. Biomed. Res. Foundation, Washington DC, USA. Volume 5, Supplement 3.

Henikoff, S. and Henikoff, J.G. (1992). Amino acid substitution matrices from protein blocks. *Proc. Natl. Acad. Sci. USA*, 89:10915–10919.

Higgins, D.G. and Sharp, P.M. (1988). Clustal: a package for performing multiple sequence alignment on a microcomputer. *Gene*, 73:237–244.

Karplus, K. (1995). Evaluating regularizers for estimating distributions of amino acids. In Rawlings, C., Clark, D., Altman, R., Hunter, L., Lengauer, T., and Wodak, S., editors, *The third international conference on*

Intelligent Systems for Molecular Biology (ISMB), pages 188–196. AAAI Press, Menlo Park, CA, USA. Cambridge, U.K., Jul 16–19.

Krogh, A., Brown, M., Mian, I.S., Sjölander, K., and Haussler, D. (1994). Hidden Markov models in computational biology: Applications to protein modeling. *J. Mol. Biol.*, 235:1501–1531.

Neuwald, A.F., Liu, J.S., Lipman, D.J., and Lawrence, C.E. (1997). Extracting protein alignment models from the sequence database. *Nucleic Acid Res.*, 25:1665–1677.

Rost, B. and Sander, C. (1993). Prediction of protein secondary structure at better than 70-percent accuracy. *J. Molec. Biol.*, 232:584–599.

Smith, R.F. and Smith, T.F. (1990). Automatic generation of primary sequence patterns from sets of related protein sequences. *Proc. Natl. Acad. Sci. (U.S.A.)*, 87:118–122.

Taylor, W.R. and Brown, N.P. (1999). Iterated sequence databank search methods. *Comp. chem.* in press.

Taylor, W.R. and Thornton, J.M. (1983). Prediction of super-secondary structure in proteins. *Nature*, 301:540–542.

Taylor, W.R. and Thornton, J.M. (1984). Recognition of super-secondary structure in proteins. *J. Molec. Biol.*, 173:487–514.

Taylor, W.R. (1986a). The classification of amino acid conservation. *J. Theor. Biol.*, 119:205–218.

Taylor, W.R. (1986b). Identification of protein sequence homology by consensus template alignment. *J. Molec. Biol.*, 188:233–258.

Taylor, W.R. (1988). A flexible method to align large numbers of biological sequences. *J. Molec. Evol.*, 28:161–169.

Taylor, W.R. (1989). A template based method of pattern matching in protein sequences. *Prog. Biophys. Molec. Biol.*, 54:159–252.

Taylor, W.R. (1990). Hierarchical method to align large numbers of biological sequences. In Doolittle, R.F., editor, *Molecular Evolution: computer analysis of protein and nucleic acid sequences*, volume 183 of *Meth. Enzymol.*, chapter 29, pages 456–474. Academic Press, San Diego, CA, USA.

Taylor, W.R. (1997). Multiple sequence threading: an analysis of alignment quality and stability. *J. Molec. Biol.*, 269:902–943.

Taylor, W.R. (1998). Dynamic databank searching with templates and multiple alignment. *J. Molec. Biol.*, 280:375–406.

Zvelebil, M.J., Barton, G.J., Taylor, W.R., and Sternberg, M.J.E. (1987). Prediction of protein secondary structure and active sites using the alignment of homolgous sequences. *J. Molec. Biol.*, 195:957–961.

7

AN ESSAY ON INDIVIDUAL SEQUENCE VARIATION IN EXPRESSED SEQUENCE TAGS (ESTs)

Jens Reich, David Brett, and Jens Hanke

Max Delbrück Center of Molecular Medicine
Berlin Buch
Humboldt University of Berlin
Charite Medical Faculty

Expressed sequence tags (ESTs) are short sequence segments (usually up to 500 nt long) obtained by reverse-transcription into cDNA clones from m-RNA preparations of a cell or tissue in a specified functional or developmental stage. They are produced by automatic procedures and released by their producers (after a certain time lag) into public databases (Boguski, 1995; you may visit the dbEST database, 1999). At present EST collections (about two thirds of them of human origin) grow much faster than any other genomic sequence information. The main applications of EST analysis are sketched in Figure 1:

- The main application is analysis of gene expression in cells. Which cell type expresses which gene? This question is urgent, as for example in the human genome there are about 75,000 genes, but only about 10,000 proteins appear even in very protein-diverse cells such as neurons
- Gene expression in a given cell type can be studied in different functional and developmental stages. Such analysis becomes indispensable when the overwhelming wealth of data of cell biology is to be integrated into a comprehensible picture
- A special application to gene expression is of great importance: study of pathological conditions of a cell. The expression pattern of tumor cells, for instance, may be compared to the pattern in pertinent normal tissue, thus possible revealing clues to origin and metabolism of tumors (Strausberg et al., 1997)
- ESTs can be helpful in the search for candidates genes responsible for certain traits, in particular for disease genes. The partial sequence may lead, by homology search in genomic databases of other species (mouse, rat, fruit fly, worm,

Genomics and Proteomics, edited by Sándor Suhai.
Kluwer Academic / Plenum Publishers, New York, 2000.

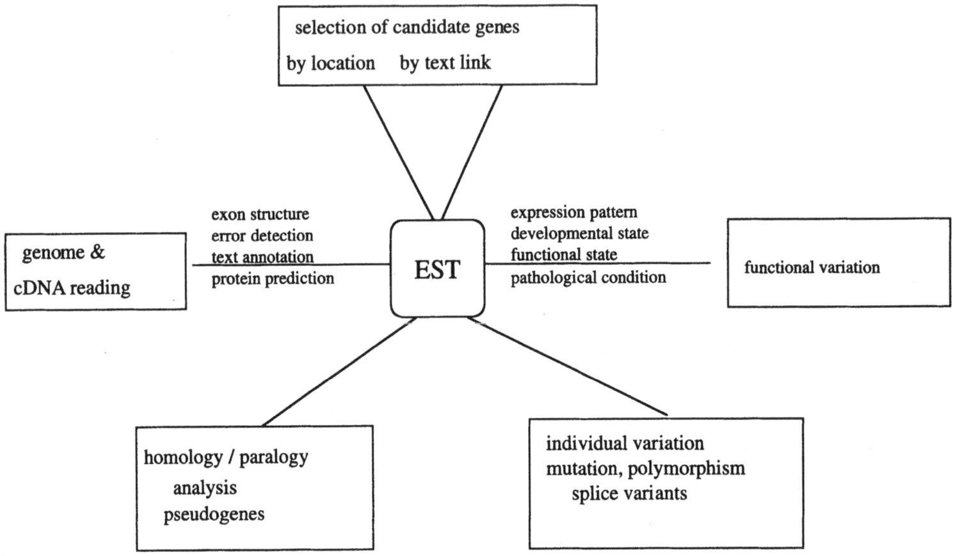

Figure 1. ESTs at the cross-road of pathways.

fugu fish, zebra fish, yeast, E. coli and others; for web adresses, see: mdc-pointers, 1999) to valuable hints as to the nature of a gene looked for. The mouse genome (see Note Added in Proof) has been mapped in much more detail than the similar human genome. Fly genome data contain a wealth of information on mutations and pertinent traits. The fully sequenced genome of the worm Caenorhabditis is a rich source of information on developmental regulation and neural differentiation. Fugu fish has a more compact genome than man and is therefore a tool for studying genome organisation. Zebra fish is at present intensively screened for genes of vertebrate development. Yeast, also fully sequenced, is a valuable source in particular for the regulation of cell division. And E. coli and other microorganisms can yield information about the basic metabolism of any living cell.

- Many of the ESTs in the databases are by now mapped to their chromosomal location (they are then called STSs, sequence tagged sites). These sites may serve as markers of that region. Such markers are particularly valuable, because they reside within a coding part of the genome and are therefore more liable than extragenic markers to cosegregate (with only very rare recombination) with the pertinent gene. A STS mapped to the chromosomal locus may lead to genes or markers in its neighborhood and may, by text or homology links into other data bases, suggest a possible candidate gene, perhaps even with a pertinent mutation.
- ESTs are a short stretch of information that may be translated into protein sequence. In genome sequencing this may help to reveal the genomic organization of an obtained sequence segment. After all, the numerous tools (useful web-adresses, see e.g. MDC pointers (1999)) of gene identification, intron-exon separation, gene head and tail identification can predict only with limited certainty, so that a piece of definitely expressed genomic region simplifies the search for genes in the ocean of non-coding information

- ESTs may be used for the search of homologous region in other species as well as of paralogous regions in the same species, and also of pseudogenes (i.e. transcribed and spliced, but because of certain defects not functionally active parts of the genome). This usage is possible because such regions retain some sequence similarity or identity which may be visible in ESTs
- Each EST stems from a unique specimen of that species, or from a human individual, and may, when aligned in clusters, reveal individual variation of the expressed genomic information: gene polymorphism, gene defects, mutations, splice variants etc.
- Mining for SNPs in EST databases requires only computer resources and does not incur experimental cost (as do the various techniques of large-scale DNA chip analysis). There are now databases that integrate the numerous ESTs into a human genome map (Schuler, 1996; Schuler, 1997). The dbEST database currently (mid-1999) contains about 2.8 million entries of EST text.

Clusters of aligned ESTs and assembled to gene segments (Figure 2) are available in the public domain (see Unigene Web server, 1999). They are a valuable source of genomic information:

- The assembly covers contiguous stretches of the expressed genome, due to overlap much longer than the relatively short individual ESTs
- The number of EST hits at a certain site is an indication (though not a proportional one) of the intensity of expression in the pertinent cell
- The EST alignment permits identification of single nucleotide polymorphisms (SNPs), deletions, insertions and splicing variants.

USE OF ESTs FOR STUDYING INDIVIDUAL VARIATION

An EST ideally suited for this purpose should fulfill the following criteria (see Figure 3):

- It is a short contiguous reverse-transcribed segment excised from a spliced mRNA. It should contain either the 5′ untranslated region (5′ UTR) and/or spliced exonic sequence, and/or 3′ untranslated region (3′ UTR)
- The ensemble of ESTs in the available databases cover all genes of the expressed genome and all parts of each gene. At present, there are about 1.5 million ESTs covering the greater part of the about 75,000 human mRNAs. Thus one mRNA is hit on the average by 11 ESTs, but one EST can cover only a fraction of mRNA sites (about 500 nt per about 2000 sites)
- It neither contains intergenic material away from the coding region nor intronic sequences
- Its abundance is approximately proportional to the equilibrium between synthesis and hydrolysis of mRNA
- To avoid heavy overrepresentation of mRNA species typical for the respective tissue (like globin in red blood cells) normalization procedures reduce the redundant population. This increases coverage by rarely expressed genomic sections, but at the cost of losing proportionality to cellular concentration
- Alignment of autologous EST stretches from different donors reflect individual genomic variation in the coding region (missense and silent), and/or the adjacent expressed regulatory parts (promoter region, terminator region etc.).

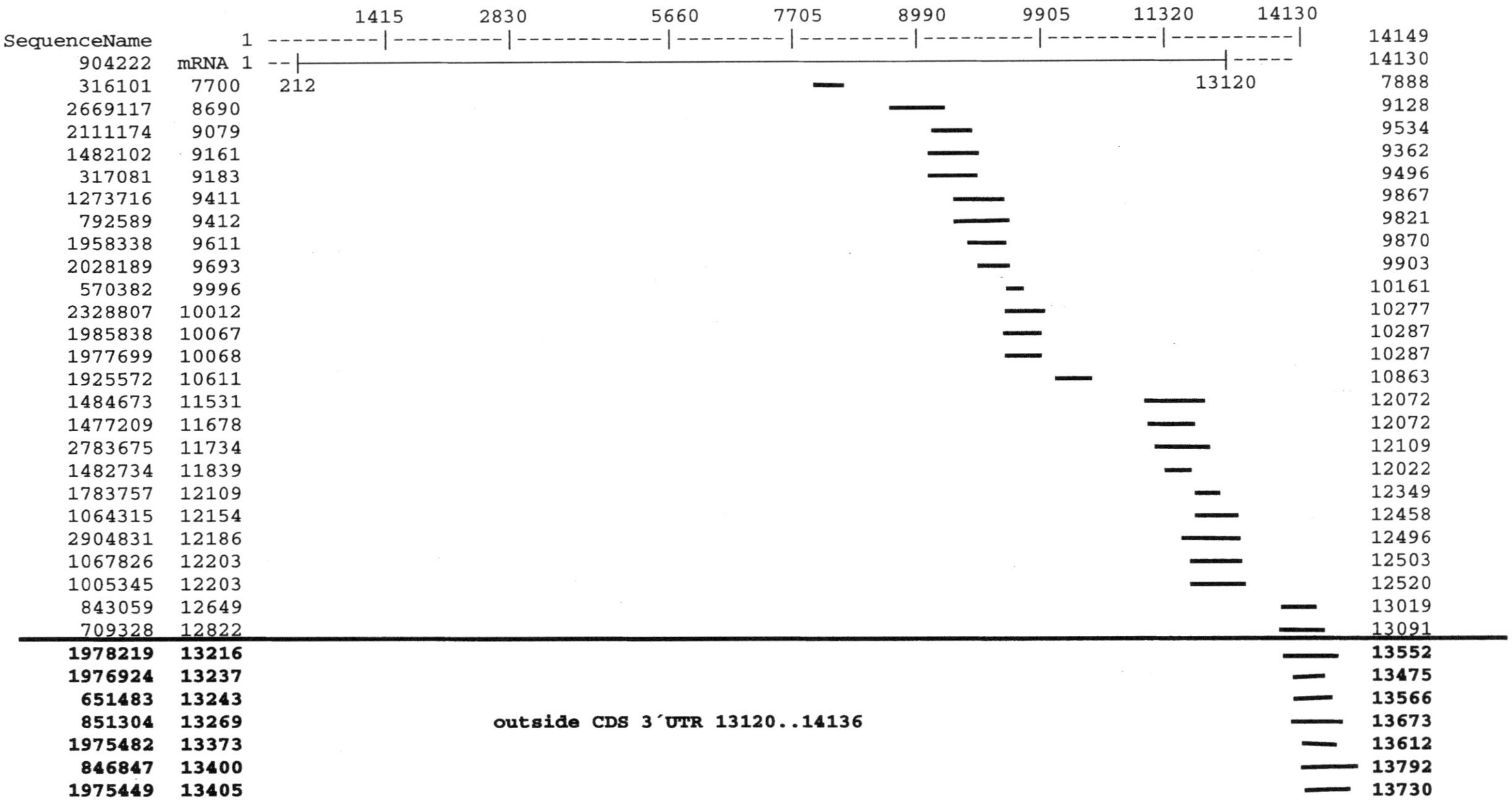

Figure 2. EST-Sequences against positionally-cloned gene of the polycystic kidney disease. The scale on top numbers the genomic region of the cloned gene (polycystic kidney disease gene, GI accession code 904222) from 1 to 14131. Partial regions:

1-211: 5′ UTR (untranslated region) of mRNA
212-13120: coding region of mRNA (coding for 4303 amino acids)
13121-14131: 3′ UTR of mRNA

Schematically shown in the first line (long solid horizontal line) is the coding region of the mRNA of the gene, and of all ESTs from dbEST (with accession code) that align to this sequence with sequence similarity >95% and >100 nucleotides overlap. There are 25 ESTs that cover, at least in part, the coding region, and further 35 (of which only 7 shown: under the horizontal line below EST 709328) that cover only the 3′ UTR region. The figure exemplifies the rule that an EST alignment covers a considerable part of the mRNA, but with preference of the 3′ region, in particular the 3′ UTR.

- **An EST is a scrap of a m essenger caught as c DNA**
- true transcribed genome region (unless cloning artifact)
- available as library clone for probing
- amenable to PCR technology (and its pitfalls)
- 300-400 bases of expressed gene text
- automatic single-pass reading (error prone, 2-5%)
- from either end of the clone
- mostly 3′ and 5′ end of m-RNA
- from "normalized" cDNA libraries

Figure 3. Expressed sequence tags (ESTs).

- It displays part of the correct amino acid sequence of the gene product when read in the correct complementarity and reading frame. This can be helpful if one is searching for a gene whose approximate location, but not its sequence, has been located by genetic studies (genes swim in an ocean of non-coding genomic text and are difficult to find).
- It reveals splice variants, if parts of the genome text are present in the interior of one EST but are not present in the interior of another one, while the remainder shows close sequence identity

In practice, the EST collection does not live up to these ideal demands:

- It can cover only a fraction of the expressed part of the genome, because some genes are read off at a very low level or not at all, others are difficult to clone
- Coverage of expressed information is far from being uniform. Figure 2 is a typical example of a gene whose mRNA sequence is known so that one can calculate the coverage rate of the ESTs aligning to it. It is seen that the coverage is heavily skewed towards the 3′ end of the mRNA. Only about 30% of all mRNA sites are at the present time covered by more than one EST library. This reduces the chance of finding all of the existing SNPs. As a result any large scale in-silico analysis of polymorphic variations will be biased towards the tail region of strongly expressed genes
- The error in the sequences (in the percent range) is no problem for the usual whole-sequence-based approach to expression analysis, but it is a drawback when individual positions are studied. In particular, the automatic base-calling by a computer may increase the error
- There is also a small error (said to be about 1/10,000) due to reverse transcription and synthesis involved in the generation of cDNA clones from mRNA
- Incompletely spliced primary messenger as well as unprocessed genomic material may be present as impurities in a mRNA preparation and may obscure the alignment of autologous ESTs necessary for finding variants
- SNP candidates derived from ESTs refer to one allele of the donor person, so the zygosity of the carrier may remain obscure
- Some EST preparations come from pooled material rather than from one person, which may make statistical calculation dubious

- It happens that several EST libraries are from one person, which also confuses statistical calculation
- An error source is that a variation in an alignment of EST sequences may not come from the same gene but rather from a highly similar paralogous copy elsewhere on the genome or from a pseudogene ("dead" gene: transcribed, processed, but not translated into protein). This necessitates restriction to high sequence identity as criterion of inclusion of an EST into the aligned cluster. This does not fully rule out the paralalogy error and nevertheless risks to exclude some true variants that do not pass such a strict threshold.

In spite of all these problems EST databases are a valuable source of large scalc analysis of the genome and its expression. They will become even more valuable when the data grow at the present rate. An algorithm for computer-aided SNP mining should contain filters to eliminate the potential sequence errors. Such filters can be based on the probabilistic analysis of sequence features. It can also take into account that multiple occurences of a variant are more trustworthy, and it may furthermore focus on improving the quality of base-calling if the fluorescent traces are available for closer srcutiny. I describe now the results of a systematic search, by our group and by others, of the variation present in the whole expressed genome and visible by EST-driven data base search.

GENERATION AND EVALUATION OF cSNP CANDIDATES FROM EST ALIGNMENTS

ESTs were obtained from dbEST (Boguski et al., 1993) as regularly expanding database as division of GenBank (Benson et al., 1999) that contains sequence data and other information on "single-pass" cDNA sequences and/or expressed sequence tags from a number of organisms including homo sapiens. A brief account on the temporal development of that collection is given by Boguski et al. (1995).

mRNA was obtained from GenBank entries identified by the appropriate annotation ("mRNA", "complete cDNA"). GenBank is the genetic sequence database maintained at the NCBI of NIH in Bethesda, Md. There are approx. 3 billion nt in 4 mio sequence records in this database as of June, 1999. About 9000 mRNA or cDNA entries may be used as fully sequenced master template for studies of variation.

- usually a gene scrap
- sometimes without coding text (long polyA-tail)
- sometimes with remnant intron
- many technical insertion/deletion errors
- occasional chimeric constructs
- mutation, similarity of duplicate or plain error ?
- operative reading frame often dubious (although alignment tools help to establish)
- not all expressed genes covered (some mRNA segements difficult to retro-transcribe into cDNA)

Figure 4. Expressed sequence tags—caveats!

Several groups focused on hunting SNPs from assembled EST clusters such as collected by UNIGENE data base (Schuler, 1997, 1998; UNIGENE, 1999; for such applications see Miller et al., 1998; Buetow et al., 1999; Picoult-Newberg et al., 1999). The two latter groups used the Phred base calling program together with the phrap sequence assembling tool (Ewing et al., 1998; Ewing and Green, 1998; Green, 1998; Gordon et al., 1998). This procedure yields a quality score for each base as called, which expresses its statistical trustworthiness on a logarithmic scale (e.g. phred value >20 is already a reliable call, while values below 20 are increasingly doubtful).

The approach of Buetow et al. excludes possible paralogous hits and applies algorithmic filters in order to avoid erroneous reading of fluorescence traces. Picoult-Newberg et al. also apply filters to avoid sequencing errors, in particular indels or further mismatches nearby in the sequence. They neglect variants suggested in the first 100 EST positions and discard variants seen only once in the EST collection.

Our group (Sunyaev et al., 1999) performed a benchmark analysis on a set of genes for which the full mRNA and/or the pertinent protein sequence was documented in the literature. Instead of clustering ESTs without any template we aligned them by a BLAST search (Altshul et al., 1977; see BLAST server) to this set of master mRNA sequences and looked in these alignments for variant letters. We applied a set of filters as follows:

- Only subalignments of length >100 nt above 99% sequence identity and with >15 exact matches at both ends were considered. This is a hard criterion for excluding paralogues and other unreliable candidates
- We also excluded positions when there were closely located further mismatches
- We furthermore excluded sequence patterns liable to cause gel compressions or homopolymer stretches, which often lead to base miscalls
- We excluded ESTs that aligned to >1 mRNA of the panel
- A significant improvement of the prediction reliability is achieved by considering only variants that occur more than once. The prize to be paid is a strong sampling bias towards frequent variants.
- About 60% of the data collections offer also the pertinent EST chromatograms. In these cases we applied a filter based on Phred quality.

SNPs IN EST CLUSTERS

Buetow et al. (1999) report more than 3000 candidates of a score >0.99 from the set of more than 8000 UNIGENE clusters. A subset of nearly 200 candidates was directly checked in a pooled preparation from 10 individuals (20 chromosomes). More than 80% of these candidates were indeed confirmed in this pool.

Picoult-Newberg et al. (1999) analysed more than 21,000 5′ ESTs and more than 19,000 3′ ESTs. More than 6000 candidates were localized, but only 850 passed the filters applied. They inspected the fluorescence traces of 100 randomly selected specimens. 88 verified candidates were then validated as common variants by sequencing from a panel of individuals. 55 out of 88 sites were confirmed to be polymorphic.

Our own study focused on a subset of 500 disease-associated genes from the OMIM data base of Mendelian traits (McKusick, 1999). All ESTs were aligned to the mRNA sequences of these genes. In order to test also doubtful candidates we applied here a less strict sequence identity threshold of 95% at the amino acid level for inclusion into the alignment. We selected 100 predicted non-synonymous SNP candidates from this alignment and subjected them to direct re-sequencing of the cDNA clone. In 61 cases we obtained also the fluorescence traces. Thus we could evaluate the reliability of the phred-scores as predictors of nucleotide variants. It turned out that variants with phred value >20 are fairly confident candidates of a true SNP. The results of this benchmark test allowed us also to cross-validate what fraction of known polymorphisms were found and what the fraction of false positives and false negative was (details see Sunyaev et al., 1999).

These test results encouraged us to do a SNP search in all presently available mRNA sequences. The EMBL database contains approx. 9000 fully sequenced mRNAs. 50% of mRNA nucleotide sites were hit by at least one EST; whearas 32% were covered by more than one EST. Perigenic 3′ regions are more intensively covered. A small fraction (but still ten thousands) of mRNA sites were represented by between 10 up to a maximum of 87 different libraries.

About 29,000 mismatches (SNP candidates) were identified, but only 1535 were present in more than one library. About half of all these passed algorithmic plausibility filters. Traces were available for 55% of the candidates, and the algorithmic filter based on phred >20 confirmed 74% of the SNP candidates represented more than once. 5464 confirmed candidates are characterized in Table 1.

POPULATION DIVERSITY ESTIMATED FROM EST DATA

Our EST studies cover about 9000 mRNAs. About 6.3 million positions were found to be aligned to more than one EST. There were about 5500 reliably reported occurences of SNPs. As the range of different genes probed is greater than in the previous studies by other authors (which focused on certain stretches of the genome), it is interesting to calculate the population genetic parameters estimated from such data.

Omitting details, we state that about 5 to 6 SNPs are present per 10,000 perigenic sites. About 10 per 10,000 were found in coding sites as silent polymorphisms (not changing the amino acid), while only 4 SNPs per 10,000 coding non-synonymous sites were found. This can be interpreted as selection pressure on replacement sites as compared to neutral selective value at silent sites, the sequences of perigenic sites being in between. The results are very similar to those of Cargill et al. and Halushka et al. which concentrated on disease-relevant genes while we studied every cloned gene structure available.

Table 1. Number of candidate SNPs with phred values > 20

Untranslated		Coding			
2461		3003			
3'UTR	5'UTR	Synonymous		Nonsynonymous	
2280	181	1254		1749	
		4 fold degenerate sites	2 fold degenerate sites	Nondegenerate sites	2 fold degenerate sites
		713	480	1514	227

ALTERNATIVE SPLICE FORMS IN EST ALIGNMENTS

ESTs can be used to identify expressed paralog gene members with in the same family and/or ortholog genes expressed in other species. When the tissue type is reported a simple expression profile can also be generated. Processed pseudogenes (lacking introns) are also identifiable from within the EST data base. ESTs also represent a valuable source of structural information within a gene. Alternative splicing occurs within genes and provides a mechanism by which a specific cell or tissue type can generate a variant protein product by changing the sequence of exons normally expressed. In practice the splicing mechanism is able to choose alternative donor and acceptor sites in the DNA sequence from which to splice out introns. This alternative splicing leads to the gain of an additional exon or the loss of an exon or part of an exon. These inframe alternative splice forms evidently lead to a change in expressed peptide sequence and can radically alter a protein's function and or location (for examples, see Klamt et al., 1998; Qi et al., 1998). ESTs have been derived from a wide variety of tissue types including normal tissues, diseased tissues and immortalized cell lines. There is also a wide degree of time points represented ranging from the 2 week old embryos to old age (75 years old). This inherent variability within the EST databases can give rise to a number of alternative splice forms of a gene occurring as single hit or multiple EST hits to a gene.

HOW DO WE IDENTIFY SPLICE VARIANTS?

We have developed a method of identifying splice variants in EST databases if cDNA libraries are available from different tissues. The method comprises:

- selection of master sequences (DNA, protein) from genomic databases
- BLAST alignment of ESTs to these templates (chosing appropriate search parameters for the task
- applying algorithmic filters to exclude implausible variants (artifacts)
- validation by sequencing the splice variants on independent preparations

The method is described in more detail elsewhere (Hanke et al., 1999; Sunyaev et al., 1999).

ALTERNATIVE SPLICING IS FREQUENT IN DISEASE-ASSOCIATED GENES

An interesting question arising from the production of alternative splice forms is that of disease association. Are alternative splice forms of a gene associated with the development of a specific disease type? Another possibility is that a specific splice form might present as a strong risk factor in the development of more complex disease types like heart disease or diabetes. A number of such examples have been reported. These range from the drastic reduction of a specific alternative splice form leading to a distinct form of disease (WT1gene/Frasier syndrome: Klamt et al., 1998; Menkes gene/occipital horn syndrome: Qi et al., 1998) to specific alternative splice forms exclusively expressed or over expressed in diseased tissue (G-protein beta 3 subunit/hypertension: Siffert et al.,

1998; presenilin gene/Alzheimer's disease: Sato et al., 1999; CD44 gene/esophageal carcinomas: Koyama et al., 1999).

The discovery of new alternative splice forms of genes associated with disease has the exciting potential to lead to new rapid PCR based diagnostic markers. The ability to extract such alternative splice forms together with as yet unknown new disease-associated genes from the EST data bases has made private EST collections a valuable commercial resource.

HOW FREQUENT IS ALTERNATIVE SPLICING?

Given that ESTs are derived from a wide variety of human tissues and individuals, the number of possible alternative splice forms extracted from an EST data base can be argued to give a reasonable estimate of the general level of alternative splicing occurring in human genes. We studied a sample of 475 proteins annotated in the SWISSPROT data base (Bairoch and Apweiler, 1999) as disease associated were searched against the EST data base for the presence of possible alternative splice forms (Hanke et al., 1999). After filtering the data to remove possible premature mRNAs or pseudogenes 204 candidate sites were predicted from 162 of the proteins in the set. 34% of the proteins studied had a candidate alternative splice site. This initial study was extended to cover 8500 full length mRNAs and confirmed the figure in the first study with an initial value of approximately 30% gene products that carry a splice variant (work in progress). We found about 5000 splice variants in these EST clusters, mainly (about two thirds) exon skipping events, about one third with inserted sequence. The coverage of matching ESTs in the set of 475 proteins was approximately 50% of all positions only, and the average report was from about 2 different tissues per position. A coverage of only 50% of all position by at least one EST suggests that 30% is an underestimate of the true value. Previous estimates had lower incidence of splice variants (of around 5%: Sharp, 1994; Wolfsberg and Landsmann, 1997, but see Mironov et al., 1999). To what degree this represents reality in terms of alternative protein forms finally expressed at any one time in a given tissue type remains to be experimentally verified. In many cases different alternative splices forms coexist at a given ratio within the same cell. Whether or not the existence of a particular alternative splice form represents a functional protein is open to question. It is quite possible that cells could tolerate quite high levels of incorrect alternative splicing if the half lives of the mRNAs or peptides produced were relatively short and/or if the variants do not impair function.

ENVOI

In conclusion, the present state of knowledge indicates that protein sequences are subject to genetic variation, in the range of a few single nucleotide polymorphisms per thousand sites (being silent or sense-changing on the protein level). The extent of influence of those individual variants on the physiology and pathology of the organism is to be elucidated in future.

Our experience shows that EST databases are a convenient source of genetic variation, currently in man, and only with limited statistical representation, but with the exponential growth of EST databases one may expect full coverage of all genes of an organism with a considerable number of "hits". This will be a rich source for detection of genetic variants and their role in disease pathogenesis.

REFERENCES

Bairoch A. and Apweiler R. (1999) The SWISS-PROT protein sequence data bank and its supplement TrEMBL in 1999. Nucl. Acids. Res. 27, 49–54.

Benson D.A., Boguski M.S., Lipman D.J., Ostell J., Ouellette B.F., Rapp B.A., and Wheeler D.L. (1999) Genbank. Nucl. Acids Res. 27, 12–17.

BLAST server (1999): http://www.ncbi.nlm.nih.gov/blast.

Boguski M.S. (1995) The turning point of the history of human ESTs in Genome Research. Trends Biochem. Sci. 20, 295–296.

Boguski M.S., Lowe T.M.J., and Tolstoshev C.M. (1993) dbEST—data base for expressed sequence tags. Nature Genet. 4, 332–333.

Buetow K.H., Edmonson M.N., and Cassidy A.B. (1999) Reliable identification of large numbers of candidate SNPs from public EST data. Nat. Genet. 21, 323–325.

Cargill M., Altschuler D., Ireland J., Sklar P., Ardlie K., Patil N., Lane C.R., Lim E.P., Kalyanaraman N., Nemesh J., Ziaugra L., Friedland L., Rolfe A., Warrington J., Lipshutz R., Daley G.Q., and Lander E.S. (1999) Characterization of single-nucleotide polymorphisms in coding regions of human genes. Nature Genet. 22, 231–238.

dbEST database (1999): http://www.ncbi.nlm.nih.gov/dbEST/index.html).

Ewing B., Hillier L., Wendl M.C., and Green P. (1998) Base-calling of automated sequencer traces using Phred. I. Accuracy assessment. Genome Res. 8, 175–185.

Ewing B. and Green P. (1998) Base-calling of automated sequencer traces using Phred. II. Error probabilities. Genome Res. 8, 186–194.

Gordon D., Abajian C., and Green P. (1998) Consed: a grapical tool for sequence finishing. Genome Res. 8, 195–202.

Green P. (1998) Phrap, sequence alignment and contig assembly program. http://genome.washington.edu.

Halushka M.K., Fan J.B., Bentley K., Hsie L., Shen N., Weder A., Cooper R., Lipshutz R., and Chakravarti A. (1999) Patterns of single-nucleotide polymorphisms in candidate genes for blood-pressure homeostasis. Nature Genet. 22, 239–247.

Hanke J., Brett D., Zastrow I., Aydin A., Delbrück S., Lehmann G., Luft F., Reich J., and Bork P. (1999) Alternative splicing, more the rule than the exception? Trends Genet. 15, 389–390.

Klamt B., Koziell A., Poulat F., Wieacker P., Scambler P., Berta P., and Gessler M. (1998) Fraier syndrome is caused by defective alternative splicing of WT1 leading to an altered ratio of WT1 +/– KTS splice isoforms. Hum. Mol. Genet. 4, 709–14.

Koyama S., Maruyama T., and Adachi S. (1999) Expression of epidermal growth factor receptor and CD44 splicing variant sharing exons 6 and 9 on gastric and esophageal carcinomas: a two flow-cytometric analysis. J. Cancer Res. Clin. Oncol. 125, 47–54.

McKusick V. (1999), available as "NCBI-distributed Online Mendelian Inheritance in Man"; http://www.ncbi.nlm.nih.gov/omim.

MDC-pointers (1999): http://www.bioinf.mdc-berlin.de/pub/pointers.html.

Mironov A.H., Fickett J.W., and Gelfand M.S. (1999) Frequent alternative splicing of human genes. Genome Res. 9, 1288–1293.

Picoult-Newberg L., Ideker T.E., Pohl M.G., Taylor S.L., Donaldson M.A., Nickerson D.A., and Boyce-Jacino M. (1999) Mining SNPs from EST databases. Genome Res. 9, 167–174.

Qi M. and Byers P.H. (1998) Constitutive skipping of alternatively spliced exon 10 in the ATP7A gene abolishes Golgi localization of the the Menkes protein and produces the occipital horn syndrome. Hum. Mol. Genet. 7, 465–9.

Sato N., Hori O., Yamaguchi A., Lambert J.C., Chartier-Harlin M.C., Robinson P.A., Delacourte A., Schmidt A.M., Furuyama T., Tohyama M., and Takagi T. (1999) A novel presenilin-2 splice variant in the human Alzheimer's disease brain tissue. J Neurochem. 72, 2498–505.

Schuler G.D. (1996) A gene map of the human gene. Science 274, 540–546.

Schuler G.D. (1997) Pieces of the puzzle: ecpressed sequence tags and the vatalog of human genes. J. Mol. Med. 75, 694–698.

Sharp P.A. (1994) Split genes and RNA splicing. Cell 77, 805–815.
Siffert W., Rosskopf D., Siffert G., Busch S., Moritz A., Erbel R., Sharma A.M., Ritz E., Wichmann H.E., Jakobs K.H., and Horsthemke B. (1998) Association of a human G-protein beta3 subunit variant with hypertension. Nature Genet. 18, 45–48.
Strausberg R.L., Dahl C.A., and Klausner R.D. (1997) New opportunities for uncovering the molecular basis of cancer. Nat Genet 15, S415–416.
Sunyaev S., Hanke J., Aydin A., Wirkner U., Zastrow I., Reich J., and Bork P. (1999) Prediction of single nucleotide polymorphisms (cSNPs) in human disease-associated genes. J. Mol. Med. 77, 754–760.
Taillon-Miller P., Gu Y., Li Q., Hillier L., and Kwok P.Y. (1998) Overlapping genomic sequences: a treasure trove of single-nucleotide polymorphisms. Genome Res. 8, 748–784.
Unigene web server (1999): http://www.ncbi.nlm.nih.gov/UniGene/index.html.
Wolfsberg. T.G. and Landsman D. (1997) A comparison of expressed sequence tags (ESTs) two human genomic sequences. Nucl. Acids Res. 25, 1626–1632.

NOTE ADDED IN PROOF

The full genome of a mouse is expected to be released in April, 2000.

8

SEQUENCE SIMILARITY BASED GENE PREDICTION

Roderic Guigó,[1*] Moisés Burset,[1*] Pankaj Agarwal,[2†] Josep F. Abril,[1*] Randall F. Smith,[2†] and James W. Fickett[2†]

[1]Grup de Recerca en Informàtica Mèdica
Institut Municipal d'Investigació Mèdica
Universitat Pompeu Fabra
C/ Dr. Aiguader 80, E-08003 Barcelona, Spain
[2]Department of Bioinformatics
SmithKline Beecham Pharmaceuticals
UW2230, 709 Swedeland Road
P.O. Box 1539
King of Prussia, Pennsylvania 19406

1. INTRODUCTION

The Human Genome Project is entering the large scale sequencing phase. During the next few years, millions of bases will be sequenced daily in the genome centers worldwide, and, in order to analyze them, methods to reliably predict the genes encoded in genomic sequences are becoming essential. As the databases of known coding sequences increase in size, gene prediction methods based on sequence similarity to coding sequences—mainly, proteins and ESTs—are becoming increasingly useful, and they are routinely used to identify putative genes in anonymous genomic sequences (see, for instance, The *C. Elegans* Sequencing Consortium, 1998). There is little systematic knowledge, however, on the accuracy of sequence similarity based gene predictions, in particular of the ability of these methods to correctly infer the exonic structure of the genes in higher eukariotic organisms. In this chapter, we will address this shortcoming, by evaluating the accuracy of gene predictions derived exclusively from sequence similarity database searches. In practice, we will use two programs from the popular BLAST suite (Altschul et al., 1990; Altschul and Gish, 1996): BLASTX (Gish and States, 1993), using a

* {rguigo,mburset,jabril}©imim.es
† {Pankaj_Agarwal,Randall_F_Smith,James_W_Fickett}©sbphrd.com

Genomics and Proteomics, edited by Sándor Suhai.
Kluwer Academic / Plenum Publishers, New York, 2000.

non-redundant amino acid sequence database as the target database, and BLASTN using dbEST as the target database (Boguski et al., 1993) (dbEST includes all the publicly available EST sequences). BLASTX performs a translation of the query sequence into six frames, and searches for similarities between each of these translations, and the amino acid sequences in the database. BLASTN searches for similarities at the nucleotide level (both the query and the database sequences are nucleic acids and are compared as such). None of these programs have been explicitly developed to predict complete gene models in genomic sequences, and some post-processing of their output is required to infer gene predictions from the search results. We will delineate here a procedure to infer gene models from similarity searches of genomic sequences against databases of coding sequences, and we will evaluate the accuracy of the procedure in a set of "well-annotated" human genomic sequences.

2. BENCHMARK SET

A relatively "well-annotated" set of sequences, extracted from the EMBL database release 50 (1997), has been used to evaluate the accuracy of sequence similarity based gene predictions. These are human genomic sequences coding for single complete genes for which both the mRNA and the coding exons are known. The procedure used to extract the sequences is described in Burset and Guigó (1996) and Guigó (1997). The procedure resulted in 178 human gene sequences for which one can assume the annotation is mostly correct. The characteristics of these sequences are given in Table 1.

3. EVALUATING ACCURACY

Gene predictions obtained after running sequence similarity searches on the sequences in the benchmark set were compared with the actual gene annotations. A number of accuracy measures may be used to compare gene predictions with the annotated genes. The measures of accuracy used here are extensively discussed in Burset and Guigó (1996). But for completeness, these are defined briefly. Accuracy is measured at the nucleotide and exon level. At the nucleotide level, the proportion of actual coding nucleotides that have been correctly predicted is called SENSITIVITY, and the proportion of predicted coding nucleotides that are actually coding is termed SPECIFICITY. At the exon level, the MISSING EXONS measure the proportion of actual exons that overlap no predicted exon, and the WRONG EXONS measure the proportion of predicted exons that overlap no actual exon.

Table 1. Characteristics of the benchmark sequence seta[a]

		Length sequence			Genes (average)			CDS (average)		
#	G + C	Average	Min	Max	Length	Density		#Exons	Length	Density
178	50%	7169	622	86,640	3657	53%	7169	5.1	968	21%

[a]The columns "genes (average)" and "CDS (average)" provide values averaged over all sequences. "genes #" is the number of genes. "genes length" is the number of nucleotides that occur in genes. "gene density" provides the % nucleotides that occur in genic regions (exons, introns, and UTRs), and the number of kilobases per gene. "CDS # exons" is the number of coding exons. "CDS length" is the number of nucleotides that occur in coding regions, and "CDS density" is the % of coding nucleotides.

4. FINDING GENES BY SEARCHING THE PROTEIN SEQUENCE DATABASE

The genomic sequences in the benchmark set were compared with the sequences in the two databases of known protein sequences using BLASTX, and gene models were inferred from the results of the searches. In all cases, sequences were previously masked for repeat regions using REPEATMASKER (Smit and Green, unpublished, http://ftp.genome.washington.edu/RM/RepeatMasker.html). The set of non-redundant protein sequences at the NCBI, as in july 1998, and the set of proteins predicted in the genome of *C. elegans*, as on the same date were used as the two databases of known protein sequences. While typical computational gene finders predict genes, that is a list of positions along the query genomic sequence corresponding to the exons, database searches produce just lists of sequence database hits along the query sequence. Each hit above a given similarity threshold is assumed to be a coding exon. For different database entries, however the set of hits may be different (see Figure 1). The problem is then to infer a gene model from the set of database hits. In this section, we first delineate a

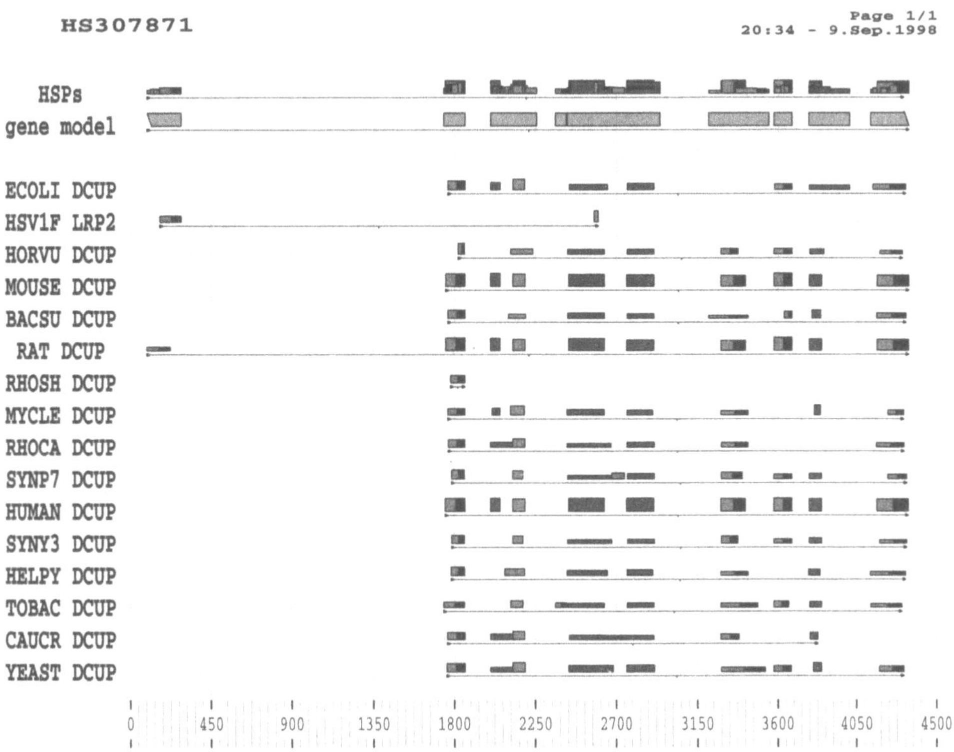

Figure 1. Database hits found when the EMBL sequence HS307871 is compared against the non-redundant amino acid sequence database using BLASTX. Database hits (High scoring Segment Pairs, HSPs) are plotted as boxes along the query sequence. The high of the boxes is proportional to the % similarity. Colors indicate frame and remainder, but are mostly irrelevant in this context. Hits corresponding to different database sequences are plotted in different lanes. The top lane (labeled HSPs) is the projection into a single axis of HSPs corresponding to different database entries. Below, in the lane labeled gene model, the gene directly inferred from this projection. The display above has been obtained with the program GFF2PS (Abril and Guigó, 2000).

procedure to infer gene models from similarity searches of genomic sequences against databases of coding sequences, and we then evaluate the accuracy of the procedure in our benchmark set.

4.1. Inferring Gene Models from Protein Searches

A simple solution is to project the hits into a single axis along the genomic sequence, and to assume the projections to be the coding exons. However, we need to take into account that in large genomic regions genes may occur on both strands. It would also be useful to link predicted exons sharing hits to the same database entry, as these exons are likely to correspond to the same gene. Thus the procedure to infer gene models from database searches needs to be a little more complex. We proceed in the following way:

1. Database hits (the high scoring segment pairs, HSPs, in BLAST) are "projected" along the query DNA sequence, and categorized by the minimum probability obtained in each sequence segment after this projection. Then, the DNA sequence is segmented into non-overlapping regions having the same minimum probability. These are call "similarity regions" (SRS). Each SR is assigned the set of locus names of database sequences matching it. SRS are constructed within frame and strand. That is, HSPs are actually independently projected into the six different frames.
2. Adjacent SR'S are clustered together into "contiguous similarity regions" (CSRS). Each CSR is assigned the union of the sets of locus names in the clustered SRS. CSRS could be interpreted as exons. CSRS are constructed within frame and strand. That is, only contiguous SRS in the same frame are clustered together.
3. CSRS sharing locus names (that is, database hits) are connected into "connected contiguous similarity regions" (CCSRS). Each CCSR could be interpreted as a gene. CCSR'S are constructed within strand. That is, only CSRS in the same strand are connected.

This process results in a set of potential "genes" predicted in both strands—that is, an output similar to the one produced by gene finders.

4.2. Accuracy of Predicitions

4.2.1. BLASTX-NR Gene Models. BLASTX searches against the non-redundant protein sequence database were perfomed with default parameters. Gene models were inferred from the results of database searches as described above. Probability, score, and similarity cutoff values used to consider HSPs for further processing are given in Table 2. Predicted gene models were compared with the annotated genes, and the measures of accuracy described in Section 3 were computed. Accuracy values are given in Table 2. These values were averaged only over the set of sequences in which predictions were obtained (175 out of 178). Predictions in the wrong strand, even when overlaping actual exons, were considered incorrect.

The proteins encoded by the sequences in the Benchmark set are all included in the non-redundant database of amino acid sequences (nr). Thus, one would expect near perfect predictions from BLASTX searches of these genic sequences against the nr database. However, the accuracy achieved is substantially lower. There is one intrinsic

Table 2. Accuracy of BLASTX based gene prediction in the benchmark set of single gene sequences[a]

		Nucleotide			Exon	
	#	Sn	Sp	CC	ME	WE
BLASTX nr	175	0.90	0.84	0.85	0.10	0.02

cutoffs *probability* = 10^{-10} *score* = 60 *identity* = 40

[a]BLASTX was run with default parameters. Values are averaged only over the sequences in which predictions have been produced.

reason why gene-modeling based on amino acid sequence database searches—as those performed by BLASTX—can not produce perfect predictions. BLASTX alignments are performed at amino acid level, and therefore the coordinates of the HSPs—and of the inferred coding exons—are given on the amino acid translation of the genomic sequence; however, exon boundaries often occur within codons. These boundaries could never be reported by a BLASTX search. This reason can explain, however, only an small decrease in accuracy at the nucleotide level, but certainly not in the amount observed. Two other factors may explain this additional decrease in sensitivity. The first factor is the existence of very short (mostly initial) coding exons—as short as 10 nucleotides or even less—which may not be detected by BLASTX searches at our choice of parameters. The second reason is the use of filters to mask low complexity regions before performing database sequence similarity searches. BLASTX searches are masked with the XNU (Claverie and States, 1993) and SEG (Wootton and Federhen, 1993) programs. Low complexity regions often give spuriously high scores that reflect compositional bias rather than significant alignments. However, a few actual coding exons do occur in low complexity regions (or overlap with low complexity regions), and they go undetected when the query sequence is masked (for an example, see Figure 2). The decrease in specificity, on the other hand, can partially be explained by the tendency of BLASTX alignments to expand into non-coding regions. However, it is also partially artifactual. Despite the procedure used to guarantee database annotation correctness during the process of extraction of the benchmark sequences, we have detected annotation errors in a few of these sequences, resulting in actual exons being incorrectly annotated as occurring in non-coding regions (for an example, see Figure 2).

5. FINDING GENES BY SEARCHING THE EST DATABASE

Large fractions of genes from many organisms have been discovered by sequencing ESTs from random cDNA libraries (see, for instance, Hillier et al., 1996). However, little has been done to investigate the actual accuracy of EST-based gene predictions. Wolfsberg and Landsman (1997) compared ESTs to genomic sequences for a few human genes, but their main goal was not to evaluate ESTs as tools for gene prediction. In this section, we develop an strategy to use ESTs for gene prediction. Our goal is to balance sensitivity and specificity.

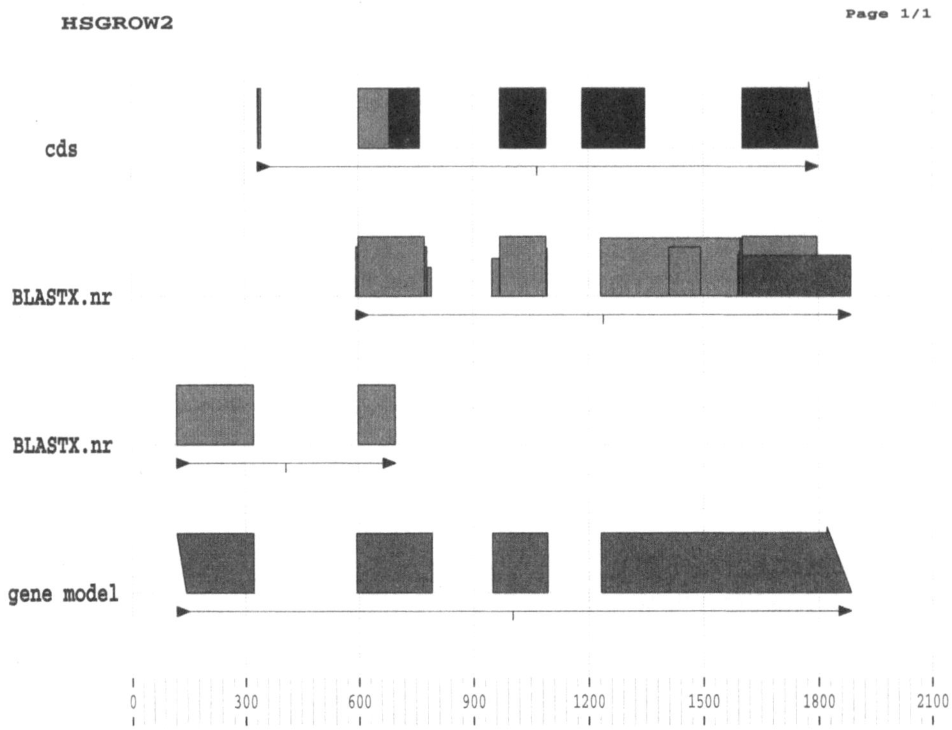
HSGROW2
Page 1/1
cds
BLASTX.nr
BLASTX.nr
gene model
0
300
600
900
1200
1500
1800
2100

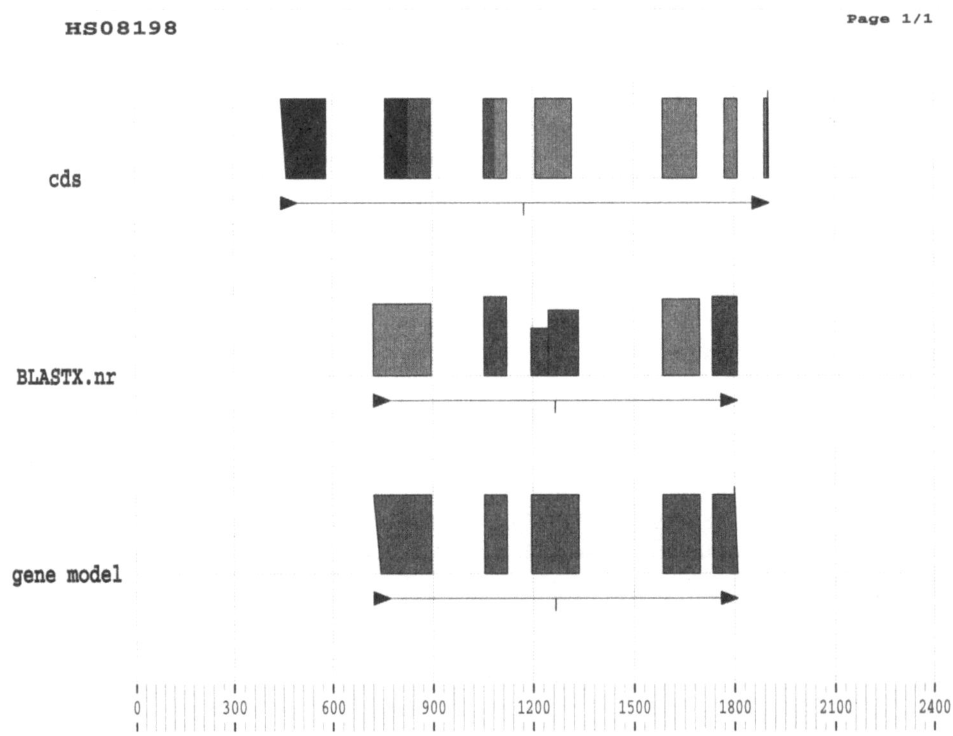
HS08198
Page 1/1
cds
BLASTX.nr
gene model
0
300
600
900
1200
1500
1800
2100
2400

5.1. Inferring Gene Models from EST Searches

5.1.1. Simple Parsing. Initially we considered a simple parsing, proceeding as follows:

1. ignore strand and orientation and project all HSPs into a single axis.
2. cluster together overlapping HSPs.
3. link clusters sharing EST id and/or clone id. These links of clusters is what we have called before "connected contiguous similarity regions" (CCSRs), and they could be considered as sets of connected "exons" defining a gene.

This procedure was applied to the matches found when the single gene benchmark sequences were compared to the sequences in dbEST using BLASTN. BLASTN was run with stringent criteria, the specific parameters are provided in Table 3. Probability, score, and similarity cutoff values used to consider HSPs for further processing are also given in Table 3. 21,657 HSPs corresponding to 7,380 ESTs were found to match the genomic sequences in the bechmark sets. 10 sequences in the benchmark set lack any EST matches after applying the cutoffs. The inferred "genes" were compared with the annotated mRNA exons, and the resulting accuracy is shown in Table 3. Accuracy values are averaged only over the set of sequences in which prediction were obtained. Although in this particular case, it is irrelevant because all predictions are assumed to occur in the forward strand, in all cases through this section, accuracy has been computed ignoring predictions in the reverse strand. Both these choices overestimate slightly the accuracy of gene prediction using ESTs.

5.1.2. More Complex Parsings. The main problem with this simple procedure is that it does not provide information on the strand in which the gene occurs. However, from each random cDNA clone, two ESTs are usually obtained: one when the cDNA is sequenced from the 5′ end, and another when the cDNA is sequenced from the 3′ end. Often, the orientation on the cDNA in which the EST has been sequenced is given in dbEST. In theory, 5′ ESTs should match their original genomic sequence in the forward strand, while 3′ ESTs should match the genomic sequence in the reverse strand. Table 4 shows the results actually obtained in the benchmark sequences—which encode genes only in the forward strand. As it is possible to see, there is a clear association between orientation of the match and EST orientation: 95% of the 5′ ESTs match in the forward strand, while 85% of the 3′ ESTs match in the reverse strand. However, there is a substantial fraction of the ESTs matching the benchmark sequences (about 20%) whose orientation is not annotated in dbEST.

Figure 2. Problems encountered when deriving gene models from BLASTX searches against the non-redundant amino acid sequence databases. Top. Over prediction of exons. BLASTX finds strong similarity between a region of a genomic sequence and a region of a known protein (BLASTX.nr lanes). The region is, then, included in the predicted gene model (gene model lane). This region, however, is not annotated as coding in the EMBL database (cds lane). In this case, however, the problem is in the database annotation, and the predicted exon is indeed an actual exon—corresponding, however, to a gene other than the annotated in EMBL entry for HSGROW2). Bottom. Missing exons. The first and last actual coding exons in EMBL sequence HS08198 are not detected by a BLASTX search. The coding fraction of the last coding exon is too short to be detected, while the first exon has a low complexity composition and is masked when BLASTX is run with the SEG program. Note that, often, the BLASTX matches expand into the non-coding introns.

Table 3. Accuracy at the nucleotide level of T-based gene structure reconstruction in the 178 single gene human sequences[a]

	Sn	Sp	CC	ME	WE	noESTs	CCSR
All ESTs into one axis	0.73	0.78	0.62	0.16	0.12	10 (6%)	343 (2.0)
Use EST and match orientation	0.72	0.80	0.63	0.16	0.11	10 (6%)	330 (2.0)
No single-HSP ESTs	0.72	0.87	0.69	0.16	0.07	25 (14%)	260 (1.7)
No single-HSP ESTs + UniGene id	0.72	0.87	0.69	0.16	0.07	25 (14%)	215 (1.4)

BLASTN parameters B = 100 V = 100 E = 0.01 P = 8 M = 2 N = –9 hspmax = 50 progress = 0 S2 = 40

cutoffs *probability* = 10^{-10} *score* = 60 *identity* = 40

[a]noESTs is the absolute number (and the percentage) of sequences without ESTs. CCSR is the absolute number (and the number per sequence with ESTs) of linked clusters of ESTs (ideally, one per sequence). Values are averaged only over the sequences in which predictions have been produced.

Table 4. Number of ESTs in each orientation, 5′, 3′ or unknown (?), matching the genic sequences in each orientation, forward (+) or reverse (–)

	5′	3′	?	
+	8756	1149	1245	11,150
–	496	6911	3100	10,507
	9252	8060	4345	21,657

Because of the strong association between orientation of the EST, orientation of the match and orientation of the gene, EST matches can be used to indicate the orientation of the gene:

- 3′ ESTs matching in the reverse strand (3′–) and 5′ ESTs matching in the forward strand (5′+) indicate genes in the forward strand.
- 5′ ESTs matching in the reverse strand (5′–) and 3′ ESTs matching in the forward strand (3′+) indicate genes in the reverse strand.

This suggest the following more complex strategy to derive gene models from EST hits in genomic sequences:

Figure 3. Problems encountered when deriving gene models from BLASTN searches against the dbEST database. Top. Over prediction of exons. All actual exons are predicted by EST searches, however an extra exon is predicted at towards the 3′ end of the gene, which is not annotated in EMBL. This is not necessarily an error, but could reflect an instance of alternative splicing. Similarly, coding exons 3 and 4 are glued together by EST matches. Again, the possibility of alternative splicing could not be discarded, but in this case aberrant incomplete splice products in the cDNA libraries from which ESTs are obtained could be to blame. Note that, often, BLASTN matches expand into the non-coding introns. Bottom. Missing exons. EST matches are able to reproduce the exonic structure of the genes only partially. Some actual exon are missing. Note also that in this case, the procedure used to derive gene models from sequence similarity EST searches is unable to derive a single continuous gene model expanding the whole actual gene. Note, finally, that in both cases it is impossible to unquestionable call the strand in which the gene occurs, because EST matches are assigned into the wrong strand.

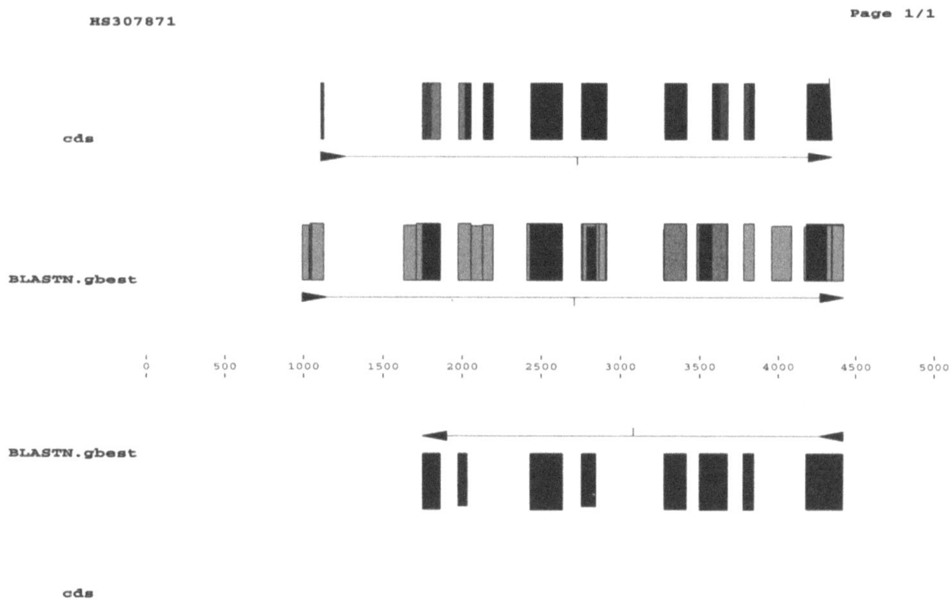
HS307871
Page 1/1
cds
BLASTN.gbest
0
500
1000
1500
2000
2500
3000
3500
4000
4500
5000
BLASTN.gbest
cds

HSATP1A2
Page 1/1
cds
BLASTN.gbest
BLASTN.gbest
0
2700
5400
8100
10800
13500
16200
18900
21600
24300
27000
BLASTN.gbest
BLASTN.gbest
cds

1. (a) assign 5′+ and 3′− ESTs to the forward strand
   (b) assign 5′− and 3′+ ESTs to the reverse strand
   (c) assign ESTs of unknown orientation to both strands
2. cluster overlapping HSPs in the same orientation and strand (similarity regions, SRS)
3. join adjacent clusters in the same strand (continuous similarity regions, CSRS)
4. link CSRS sharing EST id and/or clone id's in the same strand (connected continuous similarity regions, CCSRS)

The results when the inferred "genes" using this procedure are compared with the annotated mRNA exons are shown in Table 3. Apparently, there is no increase in accuracy using this parsing when compared with the more simple parsing. Note however that, now, we are calling the strand in which the gene occurs, thus making the predictions effectively more precise.

Table 3 also shows the results when single-HSP ESTs (which are a large subset of unspliced ESTs: ESTs that align to the genomic sequence with no gaps in the EST) are ignored, and when the UniGene (Boguski and Schuler, 1995) id, in addition to EST and clone id, is used to link the EST clusters (CSRS). Unspliced ESTs are more likely to be the result of artifacts during the construction of the cDNA libraries. Indeed, a substantial increase in specificity is observed when single-HSP ESTs are ignored during EST-based gene construction: the percentage of wrong exons (WE) drops from 11% to 7%. The drawback, however, is that the number of sequences without usable EST matches also increases from 6% to 14%. Using the UniGene id, on the other hand, we obtain a substantial reduction in the number of linked clusters of ESTs (predicted genes) per sequence: from 2.0 to 1.4. The ideal number would be 1.0, if each CCSR corresponded to a gene (Table 3). In summary, using this more complex parsing, ESTs were detected only for 153 out of the 178 sequences in h178 (86%). In those cases, ESTs matches cover on average 75% of the gene. This is about what one expects given that ESTs do not necessarily cover the 5′ end of the genes. Specificity of EST-based gene predictions, on the other hand, is high. Incompletely spliced RNAs included into the EST libraries could explain part of the loss of specificity observed, but some loss of specificity is likely to have an artifactual component, reflecting database miss-annotation, or instances of unknown alternative splicing. Indeed, alternative splicing could be more prevalent than previously expected, and it could affect more than 30% of the genes (Mironov et al., 1999). Figure 3 illustrates some of the problems encountered when inferring gene models from EST database searches.

6. CONCLUSION

The Human Genome Project is in the large scale sequencing phase with a draft of 90% of the human genomic sequence expected by spring 2000. Computational methods are, unfortunately, still not powerful enough to accurately annotate the genes. Computational genefinders produce acceptable predictions of the exonic structure of the genes when analyzing single gene sequences, but are unable to correctly infer the exonic structure of multigene genomic sequences. Sequence similarity searches on databases of known protein sequences may help on deciphering the exonic structure for the genes that have known close homologs. In such a case, however, sophisticated DNA to amino acid splicing alignment tools are required to get the correct exonic structure of the genes. EST database searches, on the other hand, may help to identify a larger proportion of the human genes, but in that case, their exonic structure is often only partially predicted.

REFERENCES

Abril, J.F. and Guigó, R. (2000). gff2ps: visvalizing genomic annotations, *Bioinformatics*, in press.
Altschul, S. and Gish, W. (1996). Local alignment statistics. *Methods Enzymol.*, 266:460–480.
Altschul, S.F., Gish, W., Miller, W., Myers, E., and Lipman, D. (1990). Basic local alignment search tool. *Journal of Molecular Biology*, 215:403–410.
Boguski, M., Lowe, T., and Tolstoshev, C. (1993). dbest, database of "expressed sequence tags". *Nature Genetics*, 4:332–333.
Boguski, M. and Schuler, G. (1995). Establishing a human transcript map. *Nature Genetics*, 10:369–371.
Burset, M. and Guigó, R. (1996). Evaluation of gene structure prediction programs. *Genomics*, 34:353–357.
The *C. Elegans* Sequencing Consortium (1998). Genome sequence of the nematode *c. elegans*: A platform for investigating biology. *Science*, 282:2012–2018.
Claverie, J.-M. and States, D. (1993). Information enhancement methods for large scale sequence analysis. *Comput. Chem.*, 17:191–201.
Gish, W. and States, D. (1993). Identification of protein coding regions by database similarity search. *Nature Genetics*, 3:266–272.
Guigó, R. (1997). Computational gene identification: An open problem. *Computers and Chemistry*, 21:215–222.
Hillier, L. et al. (1996). Generation and analysis of 280,000 human expressed sequence tags. *Genome Research*, 6:807–828.
Mironov, A.A., Fickett, J.W., and Gelfand, M.S. (1999). Frequent alternative splicing of human genes. *Genome Research*, 9:1288–1293.
Wolfsberg, T.G. and Landsman, D. (1997). A comparison of expressed sequence tags (ests) to human genomic sequences. *Nucleic Acid Research*, 25:1626–1632.
Wootton, J. and Federhen, S. (1993). Statistics of local complexity in amino acid sequences and sequence database. *Computers in Chemistry*, 17:149–163.

9

FUNCTIONAL PROTEOMICS

Joachim Klose

Humboldt-Universität
Charité, Campus Virchow-Klinikum
Institut für Humangenetik
13353 Berlin

1. GENOTYPES AND PHENOTYPES

It is well known what a gene is, but what is a genotype? We always obtain one half of our genes from the mother, the other half from the father. Consequently, we have each gene twice—the two alleles of every gene, and this is brought about by the two homologous chromosomes. Frequently, however, the two alleles of a gene differ slightly from one another, due to mutations. These mutations are usually spontaneously arisen point mutations which lead to amino acid substitutions. In principle, a point mutation may occur in the maternal allele, in the paternal allele, or in both alleles of a gene. These three possibilities are described by the term "genotype": a gene *a* and its mutant allele *a′* may create the genotypes *aa* (homozygous, wild type), *aa′* (heterozygous), and *a′a′* (homozygous, mutant type). Considering two different natural populations, the chance that the two alleles of a gene differ is the higher the greater the genetic distance between the two populations is. For comprehensive genetic investigations in an organism it is an advantage, if many genes show allelic variation in this organism. Working with mice, instead with human populations, one can take the mother from one strain and the father from another strain, or even from an other species, to reach a maximum in the genetic distance between the parents. Moreover, taking inbred strains, all the genes show the homozygous genotype, which facilitates genetic studies considerably. In our investigations, for example, we use the two mouse species *mus musculus* (strain C57BL/6; B6) and *mus spretus* (SPR). Because these two strains belong to different species, the genetic distance is relatively large. On the other side, the genetic distance in this case is within a range where cross breeding (at least in the direction B6 ♀ × SPR ♂) is still possible.

A gene creates a phene, i.e. a "visible" character, and because a mutation in a gene may alter its phene, the three genotypes of a gene usually create three different

Genomics and Proteomics, edited by Sándor Suhai.
Kluwer Academic / Plenum Publishers, New York, 2000.

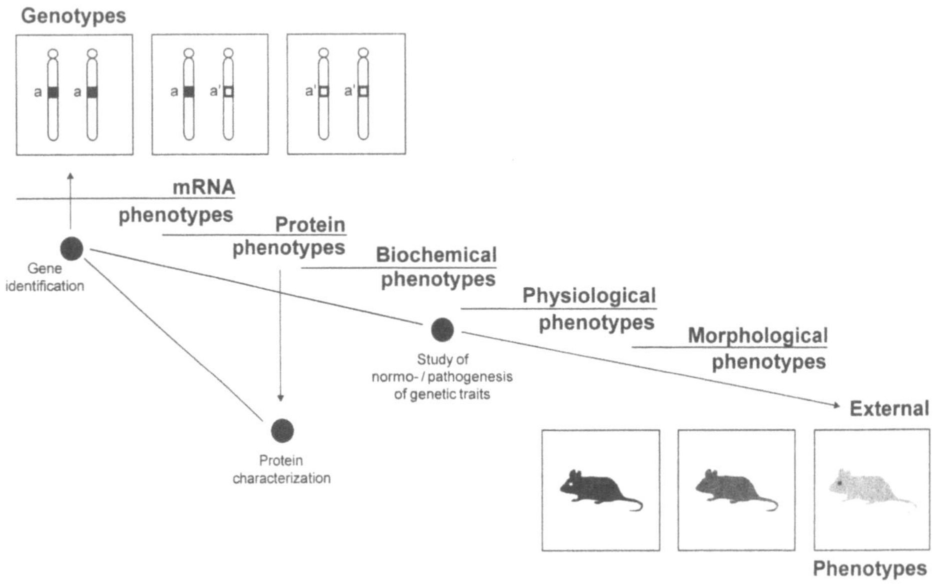

Figure 1. Genotype-phenotype relationships and a strategy to analyse normal genetic traits and genetic diseases.

phenotypes. According to King and Stansfield (1990a), the term phenotype is defined as the observable properties of an organism produced by the genotype (in conjunction with the environment, see below). In the classical sense of genetics, observable properties are external traits of an organism, such as the hair color of the mouse, morphological characteristics of an animal, or clinical symptoms in humans. Nowadays, however, many different instruments and techniques are available (e.g. microscopes, physiological tests, electrophoresis, molecular analytical techniques) that allow us to observe properties of an organism on many different levels of gene expression. Therefore, one may distinguish between morphological, physiological, biochemical, and molecular phenotypes, the latter including phenotypes of proteins and mRNAs (Figure 1).

Figure 1 suggests that the relationship between genotype and phenotype is a linear one. This, however, is not the case. On the contrary, any phenotype may be the result of the genotype of more than one gene, most likely of many genes, which may, however, contribute to a phenotype to a different extent (major and minor genes). Moreover, environmental factors may modify a phenotype. This leads us into a dilemma, if we try to determine the precise and specific functions of a particular gene. The function of a gene is reflected by its phene. But where is the place on the long road from the genes to the external traits of an organism that most directly and specifically reveals the function of a gene? This question is considered in Figure 2. The DNA sequence of a gene tells us nothing about its function. The mRNA is somewhat more informative in this respect. If a distinct mRNA species occurs, for example, in the brain yet in no other tissue, we may conclude that the function of this mRNA species has something to do with brain functions. The cellular concentration of the different mRNA species reflects the degree of activity of the corresponding genes, but does not necessarily correlate with the concentration of the proteins translated from these mRNAs. Therefore, quantitatively, mRNAs

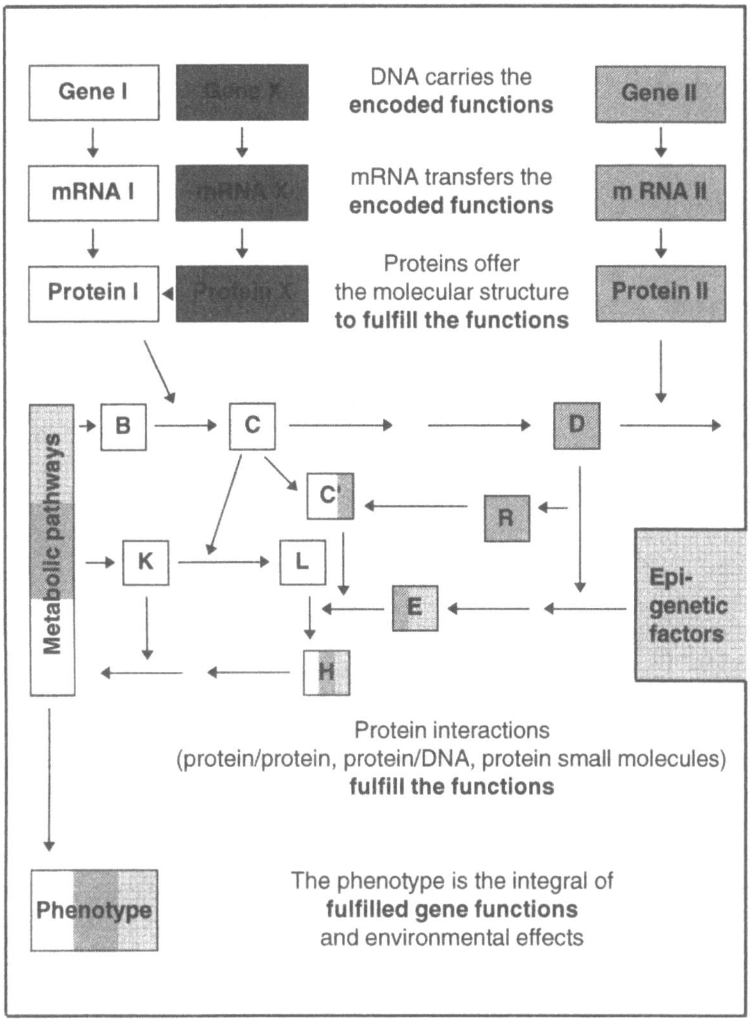

Figure 2. The way from genotypes to phenotypes is shown in some detail to illustrate the problem of determining the specific function of a gene. On this way the proteins are in a particular position. On one side they are yet directly related to the individual genes, and, on the other side, they offer all the molecular properties necessary to interact with other molecules to fulfill the functions of the individual genes. At higher levels of gene expression other genes and epigenetic factors become involved in creating distinct phenotypes so that the specific function of genes, i.e. their specific contribution to a distinct phenotype, becomes more and more obscured. In the figure the special case is shown in which even the function of a single protein depends on two genes: the molecular reaction from B to C needs the presence of protein I, but protein I can fulfill its function only in connection with protein X (see text).

are not very informative with regard to gene function. The next level in gene expression, the protein level, reflects gene function to a much higher degree. The protein of a gene offers all the molecular structures and properties needed to fulfill the functions of a gene. For example, the protein X of a gene X (Figure 2) may occur specifically in the cell nuclei and show a sequence motive for DNA binding. We would assume that the function of this gene concerns the regulation of the transcription of a particular structural gene I. This would be the most direct and specific information about function obtainable from gene X. This information, however, is soon obscured, if other proteins (transcription factors XI, XII) would be necessary to activate by interacting with protein X the target gene I. "Activation of gene I" is then no longer the function of gene X, but the combined function of gene X + XI + XII. Here, the way from genotypes to phenotypes enters the network of gene regulation, and, in a broader sense, the network of metabolic pathways. The metabolic pathways further obscure the specific function of a gene. Many genes (II, III, . . .) contribute to the cascades of metabolic reactions which lead to phenotypes of higher levels, and finally to the external genetic traits of an organism. In this complex process of gene expression, the proteins offer the most suitable target for gaining information about specific functions of individual genes. Elucidating gene functions means therefore, determining the chemical, biochemical and biological characteristics of proteins. These characteristics include the molecular structure of the individual proteins, the co- and post-translational modifications, the binding properties of the various protein species, the quantitative properties, such as synthesis rate, cellular concentration and degradation rate, and all the biological characteristics of proteins: tissue specificity, cell structure and organelle specificity, sex specificity, specificity to the various stages of embryonic and postnatal development, and specificity to the stages of ageing.

Two-dimensional electrophoresis (2-DE) is a unique method for large scale protein characterization. By comparing 2-DE protein patterns from different tissues, cell fractions and developmental stages, proteins can be characterized according to different biological parameters. Western blotting followed by immunological based procedures for glyco- or phospho-staining allows the detection of proteins which are post-translational modified. The structure of proteins can be investigated by extracting protein spots from 2-DE gels and employing analytical techniques such as mass spectrometry and partial sequencing. Using such a global strategy, individual proteins, whether known or unknown, become characterized according to many different parameters. Taking the features attributed to a distinct protein spot altogether, conclusions about the function of that protein—and, consequently, of its gene—can be drawn. One may learn, for example, that protein spot no. xy is brain-specific, occurs in the membrane fraction of neural cells late in life, shows increasing phosphorylation in the course of ageing, and reaches higher levels in cellular concentration in males than in females. One may conclude that this protein plays a role in the process of ageing.

When proteins have been characterized in several respects, the genes of these proteins must be identified, if discovering functions for individual genes is the aim. There are, in principle, two ways to detect the gene of a particular protein: (1) genetic linkage studies and gene mapping on the basis of protein pholymorphisms, and (2) mapping genes on a physical map of chromosomes on the basis of the sequence homologies between proteins and their corresponding genes. Protein polymorphisms indicate that the gene of this protein exists in different alleles. Protein polymorphisms represent different phenotypes of a gene existing in different genotypes. Two-DE protein patterns offer a unique opportunity to detect protein polymorphisms on a large scale, and to observe various protein phenotypes (Figure 3). Working with distantly related mice, many proteins can be genetically mapped. However, thinking in terms of total genomes, one has to realize that the vast majority of proteins does not reveal polymorphisms in 2-DE patterns. Additional strategies are necessary for gene-protein identification, such as mentioned above in item (2) and further explained elsewhere (Nock et al., 1999).

2. FUNCTIONAL GENOMICS AND FUNCTIONAL PROTEOMICS

The term "genomics" covers the whole genome of a single organism, and "genome analysis" means sequencing of the total DNA and mapping of all genes of a genome (structural genomics, see Editorial, Genomics 1997, 45, 244–249). At present, genome analysis is world-wide performed in both human and model organisms. Genomes of several microorganisms (URL: http//www.tigr.org/tdb/mdb/mdb.html) and the first genome of a multi-cell organism (C. elegans; The C. elegans Sequencing Consortium, 1998) have already been completely sequenced. In consequence of the rapidly proceeding genome projects, subject and aim of the post-genome (post-sequence) era are problems of present interest. It is, however, already commonly agreed that the topic of the era to follow will be what is called "functional genomics". "Functional genomics is the attachment of information about function to knowledge of DNA sequence" (Goodfellow, 1997). But what should be attached to the sequences that offers this information? Following considerations mentioned in Chapter I, genome-wide analysis of the proteins of an organism, and genome-wide gene/protein identification would be the most basical, i.e. the most single-gene-related approach towards discovering gene functions.

Genome-wide analysis of the proteins of an organism is an idea put forward already 20 years ago, shortly after 2-D electrophoresis has been published. In particular Leigh and Norman Anderson presented the idea to separate and catalogue all the human proteins (Anderson, 1979; Anderson and Anderson, 1982; Wade, 1981), a concept called today proteome analysis. The term proteome[1] was introduced to describe the entire protein complement of an organism (Swinbanks, 1995). According to the terminology used in genomics, one should distinguish between structural and functional proteome analysis. Structural proteome analysis would mean isolation and sequencing of all the proteins encoded in the genome of an organism (the "primary proteins"), and functional proteome analysis would mean determining of all the chemical, biochemical and biological characteristics of the different primary proteins. With other words, identification of functionally significant sequence motives in primary proteins would be a matter of functional proteome analysis. However, functional proteome analysis would not be restricted to the amino acid sequence of proteins, but would include the broad spectrum of structural modifications and quantitative changes, the proteins are subjected in different tissues, cell organelles and developmental stages, i.e. in the various spacial and temporal dimensions of an organism. The structural and quantitative heterogeneity, the proteins create to fulfill their functions, is the central subject of functional proteomics. Studies, as performed at present in many laboratories, use 2-DE protein patterns to detect proteins, which may be involved in a biological or pathological process of particular interest. This, however, is not what has been called proteome analysis, as well as genome analysis does not mean searching for a distinct gene. Another interest of many studies in this field is the detection of known proteins in complex 2-DE patterns of tissue proteins. Protein spots from the gels are analysed by mass spectrometry, and the data obtained are

[1]The term genome, first used by H. Winkler in 1920, was created by elision of the words GENes and chromosOMEs. Therefore, the word GENOME is purely artificial, but signifies: the complete set of chromosomes and their genes (see Editorial, Genomics 1997, 45, 244–249). The word proteome, consequently, is artificial as well, and signifies according to the term genome: the complete set of chromosomes and their encoded proteins

used to screen sequence databases to find matches with known proteins. Studies of this kind cover an important part of the work that has to be done in analysing proteomes structurally and functionally. In this way, proteins known with respect to their amino acid sequences and some functional properties are sorted out from the bulk of unknown proteins. However, proteome analysis, in its real sense, aims at the analysis of all proteins of a cell type, tissue or organism, and this includes also the yet unknown proteins which may, at present, be considered to be the vast majority of proteins of an organism. In conclusion, proteome analysis should include the following features: (1) The use of techniques (protein extraction, 2-DE) which offer the chance to detect the vast majority, if not all of the proteins of a tissue. (2) The inclusion of the unknown as well as the known proteins in structural proteome analysis. (3) Characterization of the separated proteins (both the known and the unknown proteins) on the basis of a broad spectrum of biochemical and biological parameters, i.e. performing functional proteome analysis. Finally, (4) the genes corresponding to the separated and characterized proteins should be identified and mapped on the chromosomes. Proteome analysis done that way results in functional genomics.

3. ANALYSIS OF THE MOUSE PROTEOME

We started a systematic analysis of the mouse proteome. The whole procedure consists of four steps: (1) Extraction of proteins from selected tissues and cell fractions, (2) separation of proteins by 2-D electrophoresis, (3) image analysis of protein patterns, and establishing protein standard patterns as the basis for a mouse protein database, and (4) spot identification or, in case of unknown spots, spot characterization by mass spectrometry or partial sequencing. This procedure is followed by mapping genes of polymorphic proteins on the mouse chromosomes. At the same time, the proteins registered in our protein database are characterized on the basis of a broad spectrum of biochemical and biological parameters.

For the analysis of the mouse proteome we selected an inbred strain, the strain C57BL/6, which is one of the most commonly used mouse strain in research. In a first approach we analyse the proteins of the three organs brain, liver and heart, which represent the three germ layers ectoderm, endoderm and mesoderm, respectively. These organs were collected from males and females, from different developmental stages, from post-natal and adult stages, latter including the final stages of ageing. In order to reveal as many proteins as possible from a particular tissue, we fractionate the total tissue proteins into three fractions: (1) the buffer-soluble proteins (supernatant I + II), which may represent the cytoplasmic proteins, (2) the urea/CHAPS-soluble proteins (pellet extract), which may consist of proteins normally bound to the cell structures, and (3) a DNase digested rest pellet suspension that reveals chromosomal proteins such as histones. The fractionation procedure has been described in detail elsewhere (Klose, 1999a). The fractionation procedure was based on a concept, that avoids any loss of particular groups or classes of proteins. The 2-DE patterns of these three fractions may represent the vast majority of the total proteins of a tissue. In addition to these basic fractions, we prepare highly concentrated protein extracts from purified cell organelles, primarily from cell nuclei. Protein patterns from these extracts reveal many minor proteins, not detectable in the three basic patterns.

In order to reach a maximum of resolution of the proteins extracted, we developed a 2-DE technique for large gels (Klose, 1999b), a modification of our original 2-DE

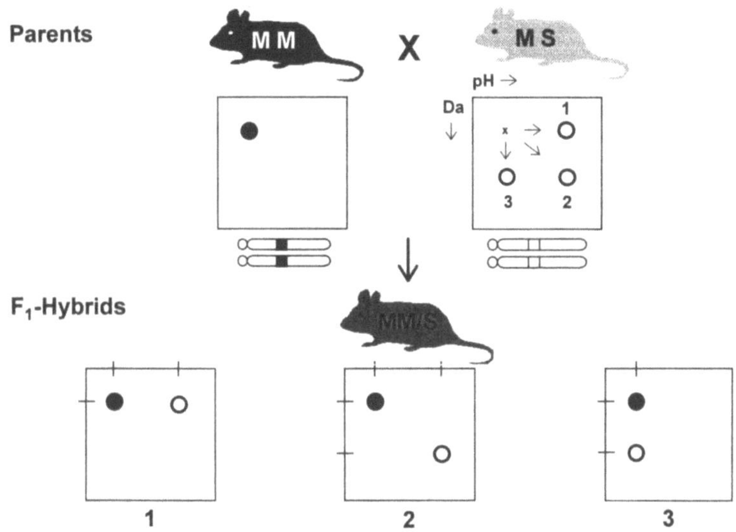

Figure 3. The three protein phenotypes are shown electrophoretic mobility variants may reveal in two-dimensional protein patterns. The two parental mouse species *mus musculus* (MM) and *mus spretus* (MS) differ in the electrophoretic position of a protein spot. The difference can be caused by changes in the isoelectric point, the molecular weight, or in both parameters of a protein. Consequently, among the hybrids (MM/S) three different protein phenotypes may occur. The homozygues genotypes of the parental strains are shown by schematic chromosomes.

technique (Klose, 1975). Isoelectric focusing is performed in capillary tube gels, 40 cm in length (tubes 46 cm). The separation distance in the second dimension, the SDS flat gel, is 30 cm. Carrier ampholytes were used for isoelectric focusing. Immobilines compared to carrier ampholytes were found to have a lower resolving power in large distance gels (Klose and Kobalz, 1995). Protein detection in 2-DE gels was performed by silver staining (Klose, 1999b).

Protein patterns from liver and brain were shown elsewhere (Klose, 1999a; Klose and Kobalz, 1995). A protein pattern from the mouse heart muscle is shown in Figure 4. A rough estimation of the total number of protein spots detectable in the 2-DE patterns of the three organs and the three fractions mentioned, is shown in Table 1. In comparing 2-DE patterns from different tissues and cell fractions, it is actually impossible to avoid that in some cases a distinct protein registered in one pattern is registered again in an other pattern, that also reveals this protein. The same protein may occur in the patterns of different tissues at different places, if differently modified. On the other hand, a protein spot that occurs in different patterns in the same position is not necessarily the same protein. In Table 1, the figures given for the supernatant fractions indicate the total number of spots of corresponding 2-DE patterns. For the other two fractions the attempt was made to count only spots which were not registered already in one of the other two patterns. The three organs, however, were not compared, i.e. redundant spots were not subtracted from the total number of spots found per pattern and per organ. The highest number of spots revealed in one pattern was found so far for the supernatant fraction of mouse testis. Here, more than 10,000 protein spots were detected per pattern (Klose and Kobalz, 1995). In spite of the high resolution we reach with our technique, and even if protein patterns from cell organelles would be taken into account, we cannot say that all the different proteins of a tissue could be presented by our patterns. Certain protein

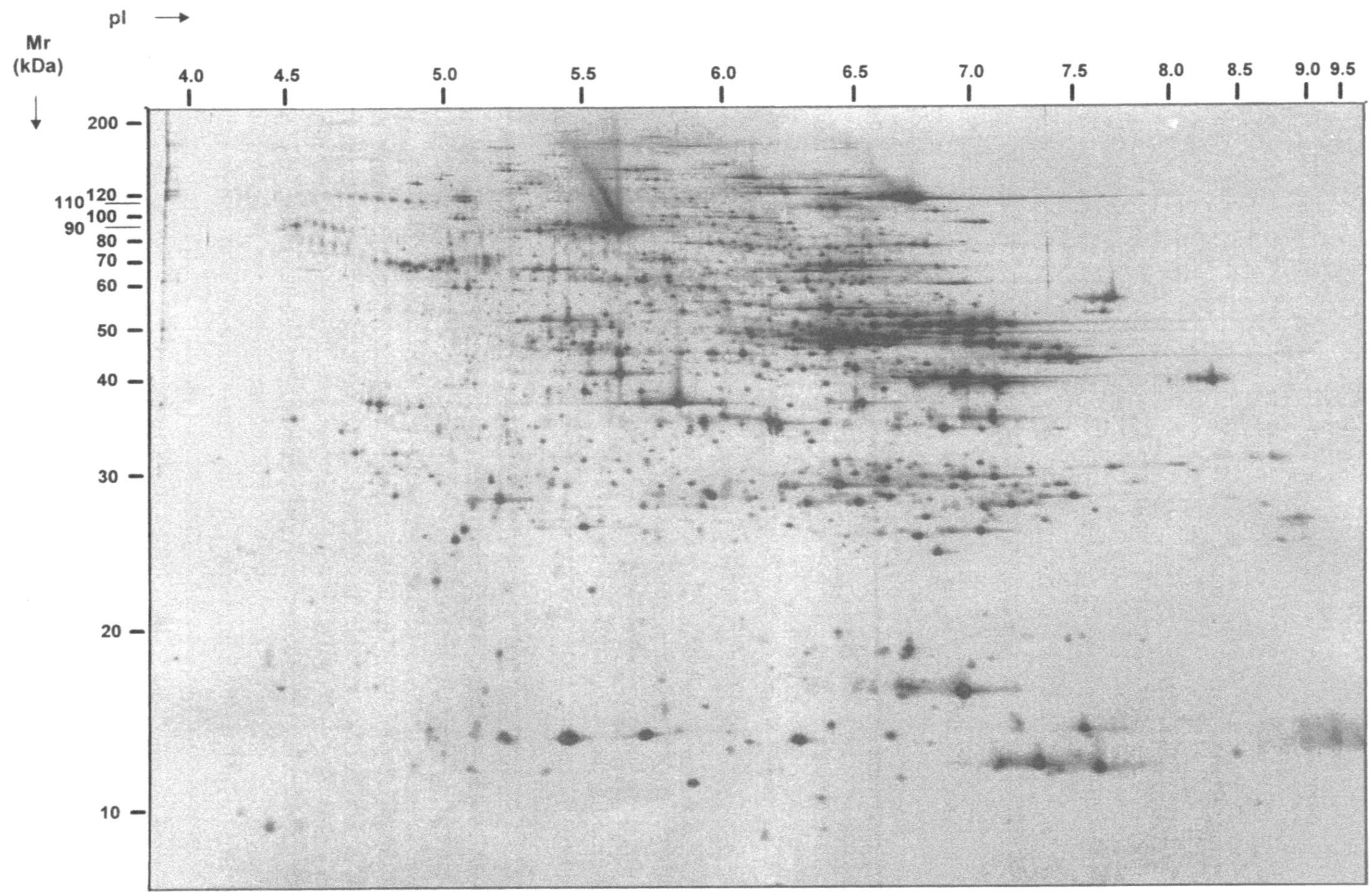

Figure 4. Two-dimensional protein pattern from the mouse heart supernatant fraction. The proteins were extracted from the heart in three fractions, the supernatant, the pellet extract and the rest pellet suspension (Klose, 1999a). The proteins were separated by large-gel two-dimensional electrophoresis and revealed in the gel by silver staining (Klose, 1999b).

Table 1. Number of protein spots as revealed by large-gel two-dimensional electrophresis of proteins from three different organs of the mouse

	Number of protein spots		
Tissue fractions	Liver	Brain	Heart
A) Supernatant[1] (buffer)	9,204	8,458	4,790
B) Pellet extract[2] (urea, CHAPS)	1,975	1,692	1,470
C) Pellet suspension[3] (DNA digestion)	73	50	40
Total No. of spots/Organ[4]	11,252	10,200	6,300
Total No. of spots/Mouse		**27,752 protein spots**	

[1]Spots/pattern.
[2]Spots not present in A or C.
[3]Spots not present in A or B.
[4]Spots which may occur in two or three organs were not identified to bring the total number of spots completely down to the level of unique spots.

species may exist only in a few copies per cell, or not even in all cells of a tissue. These proteins would not be detected in a 2-DE pattern. Three important questions cannot be answered at the present stage of our studies: How many proteins are expressed in a certain tissue (celltype)? How many of these are specific for this tissue? How many proteins arise by modifications of the primary proteins?

From each tissue and protein fraction we establish a 2-DE standard pattern. This is a synthetic pattern produced from a stained 2-DE gel by scanning, digitizing and analysing the image with a computer program for spot detection (Prehm et al., 1987). The pattern generated by the computer is carefully compared, spot by spot, with the original gel pattern and with several other patterns produced from the same kind of tissue. The computer pattern is then interactively corrected on the screen by searching for spots which were not recognized perfectly by the program. The final pattern is divided into 40 sections, and in each section the spots are provided with numbers. The standard patterns constructed from the different mouse tissues constitute the basis for our mouse protein database. Any information we obtain for a distinct protein spot is stored in the database with reference to the corresponding spot number.

The standard pattern of the mouse brain supernatant fraction together with a first set of data, concerning the protein spots identified so far, was recently presented by our homepage http://www.charite.de/humangenetik (Gauss et al., 1999). We analysed 560 protein spots by using mass spectrometry in combination with a genetic approach (Gauss et al., 1999). From these spots, 331 have been identified, and out of these, 90 indicated different proteins.

In the past several years, 2-DE protein patterns from many different cell types and tissues of various organisms, including man, have been published (WORLD-2DPAGE URL: http//www.expasy.ch/ch2d/2d-index.html; 2DWG metadatabase URL: http//www.lecb.ncifcrf.gov/2dwgDB). Federated 2-DE databases were established in the Internet network allowing laboratories world-wide to share 2-DE data (Sanchez et al., 1995). In practice, however, matching 2-DE patterns from different laboratories turned out to be difficult or impossible, due to the different techniques used (carrier ampholytes, IPGs, gel format, staining procedure, sample preparation). In future, this problem will be overcome by the increasing improvements in analysing 2-DE patterns by mass spectrometry. This will allow the laboratories to compare 2-DE spots on the level of mass spectrometry data rather than by matching 2-DE patterns. Consequently, establishing a 2-DE technique which has to be used precisely in all laboratories to allow sharing of data will be no longer an indispensable aim.

Following protein extraction of selected mouse tissues, 2-D electrophoresis, image analysis of 2-DE patterns and chemical analysis of protein spots, detecting the genes of the separated proteins is the next step in our pilot study on the mouse proteome. Gene-protein identification was started by genetic linkage analysis of genes revealing protein polymorphisms between the two mouse species *mus musculus* (B6) and *mus spretus* (SPR). Among the ~8700 protein spots revealed in 2-DE patterns of brain supernatant proteins, more than 1000 genetically variant spots were found by comparing B6 and SPR. About one half of these variants showed electrophoretic mobility changes, the other half showed changes in spot volume (protein amount). By an European collaborative project a comprehensive mouse backcross (the European Collaborative Interspecific Backcross, EUCIB) has been produced using B6 and SPR as the parental strains. About 1000 animals were generated in the backcross generation. We used 64 of these animals to study the segregation patterns of polymorphic proteins. By genetic linkage studies and gene mapping procedures, we mapped the genes of several hundreds of protein spots on the mouse chromosomes (publication in preparation).

4. PROTEIN PHENOTYPES

Two-dimensional electrophoresis is a unique tool to study the effect of gene mutations on properties—or in terms of genetics: on phenes—of proteins. Applying large-gel 2-DE to a genetic mouse system that reveals in this way more than one thousand polymorphic proteins solely in one organ (brain), protein phenotypes can be investigated on a large scale and all under the same conditions. The protein phenes visible in 2-DE gels include the electrophoretic position, the spot volume (spot area x optical density) and the heterogeneity of proteins (spot series, spot families). The investigation of genetic changes in proteins allow us to ask interesting questions; for example: Does a variant protein that occurs in several tissues, show the variation in each of these tissues, and, if so, is this variation then always of the same type? Does the occurrence of a certain protein alteration in an individual depend on its age? Is a quantitative deviation in the early developmental profile of a protein stable throughout the whole embryonic development and even in the post-natal life? Does an amino acid substitution in a protein, due to a point mutation, affect the post-translational modification, the conformation, the turn-over rate or some of the binding properties of this protein? All these questions point toward problems which are of fundamental significance for human genetic diseases. Investigations of these questions may explain, why genetic diseases usually show tissue specificity, why diseases often set in at a certain age of the persons affected, or why the expression of a particular disease depends on certain environmental factors (food, drugs). Moreover, with respect to the heterogeneity frequently observed in genetic diseases, it is of interest to search for genes which act on a protein apart from the structural gene. Findings like this would explain, why the same disease, i.e. the same clinical symptoms, may result from mutations in different genes.

Mutations alter the position of protein spots in 2-DE gels by affecting the charge, the molecular weight or (and) the conformation of proteins (positional variants = electrophoretic mobility variants, mV) (Figure 3). Mutations may also have consequences on the synthesis rate or degradation rate of proteins. In 2-DE patterns, this is revealed by changes of protein spots in size and intensity (variation in spot volume = variants of protein amount, aV). The three phenotypes of a quantitatively variant protein are composed of the two homozygous parental spots, one with a high, the other one with a low spot volume, and the heterozygous spot of the F_1 generation with a spot volume in between. In an extreme situation, a protein may completely disappear in a mouse strain or in a human individual (presence/absence variants, paV). The amount of a protein, i.e. its cellular concentration, is regulated by transcription factors and factors involved in the process of translation and protein processing. Therefore, quantitative protein variants most frequently may reflect mutations not in the structural gene, but in genes or DNA sequences which are components of the regulatory system of proteins.

When we compared hundreds of polymorphic proteins from B6 and SPR mice in the hybrid patterns, we frequently observed that the two spots of the heterozygous positional variants differed not only in the horizontal position, but also, or only, in the vertical position (Figure 3). This indicates differences in the molecular weight between the two variants of a protein. The maximum effect of an amino acid substitution on the molecular weight of a protein would be given, if tryptophan (204 Da) is replaced by glycine (75 Da); the difference is then 129 Da. In many cases, however, we observed differences much higher than that, the maximum ranging at 2500 Da (unpublished results). This indicates that mutations may frequently affect the structure of a protein much more extensively than just by amino acid substitution. These alterations may include changes in co- or post-translational modifications, truncations or altered conformations of protein molecules. While this observation is currently investigated in more detail, other findings support this assumption. We observed in 2-DE patterns frequently spot families, another indication for protein modifications. We detected the spot families by mass spectrometry in combination with genetic criteria: Variant protein spots which showed in the hybrid pattern exactly the same distance (mm), the same relative position, the same positional orientation with regard to the parental positions, and, moreover, which mapped to the same locus on the mouse chromosomes, were considered as spots which originate from the same protein. Usually, we identified or characterized the most prominent spots of a spot family by mass spectrometry. In this way we confirmed to some extent the family character of these spots, and, at the same time, identified tentatively many minor spots of the pattern which may be difficult to analyse by mass spectrometry directly. As a result, we found in the brain protein patterns, for example, that gamma enolase revealed 23 spots, synapsin 38 spots and L-lactate dehydrogenase H chain (LDH-H) 23 spots (Sanchez et al., 1995). The protein tau, a protein involved in Alzheimer disease, showed more than 100 spots in 2-DE patterns from human brain proteins. By analyzing the complexity of these spots, we found that alternative splicing and phosphorylation were one of the protein modifying mechanisms (Janke et al., 1996). Other proteins, e.g. LDH-H, formed spot family patterns interpretable as protein degradation pattern (Sanchez et al., 1995). Some of these spot families were found to be extremely reproducible with regard to spot composition. Apparently, the degradation products were rather stable in the cells and seemed to be the result of an ordered cleavage process rather than of random degradation. Limited and ordered degradation is known to be a mechanism which is of significance for certain cell functions (Glotzer et al., 1991; Stuart and Jones, 1997).

Positional variants as shown in Figure 3 and quantitative variants as mentioned above were the most frequently occurring protein phenotypes in 2-DE patterns obtained from B6 and SPR mice. Moreover, however, we made the interesting observation that the size of a spot family, i.e. the number of spots found to belong to a certain family, may also vary between B6 and SPR. Out of the 14 degradation spots found for the LDH-H family in the SPR pattern, 5 did not occur in B6. Additionally, the degradation spots showed higher intensities in SPR than in B6. This can be interpreted as a higher degradation rate occurring in the LDH-H of SPR than in the LDH-H of B6. Synapsin, as an other example, revealed in 2-DE patterns two extended horizontal spot series, due to a protein modification not clarified so far. The number of spots of the series differed between B6 and SPR. In both cases, LDH-H and synapsin, the variation in the phenotype "size of spot families" segregated in the backcross progeny of B6 and SPR. Preliminary results show, that these phenotypes mapped to the locus of the structural gene of these proteins. Apparently, a mutation in the structural gene led in one case to an altered degradation rate of the protein, in the other case to an alteration in the degree of modification of the protein.

Genetic variation of the complexity of spot families was found also in "one-spot-families". Proteins were observed which create in one mouse species one spot, but in the other species, by splitting, two spots. This suggests, that a protein can be modified in one species, but not in the other one. Protein variants of this type were found to be even more interesting, when compared in different tissues. A protein was detected that split into two spots in one organ (liver), but not in other organs (brain, heart) [Kaindl et al., in preparation]. In this case protein modification was not only genetically determined, but also tissue specific regulated. This may be an example to explain, why genetic diseases in some cases (e.g. Huntington's chorea) affect one organ, but not others.

5. THE PROTEIN, A POLYGENIC TRAIT

As mentioned, proteins in 2-DE patterns show different phenes and phenotypes, and the different phenotypes may result from changes in molecular weight or charge of proteins, from variations in the amount of proteins, from the degree of degradation, from the degree of post-translational modifications, or may result from alterations in tissue specificity or developmental stage specificity of proteins. For understanding multifactorial diseases, and more basically, for understanding the network of gene activity, it is a question of fundamental significance whether the various phenes of a protein depend on different genes. A mutation in the structural gene of a protein certainly may affect several phenes of this protein at the same time; for example: charge, molecular weight and prosthetic groups attached to the amino acids substituted by the mutation. However, quantitative changes of proteins most likely may result from mutations in regulatory sequences. Furthermore, the degree to which a protein is modified by phosphorylation or glycosylation, for example, may depend on the concentration and structure of certain enzymes, and, therefore, on other genes than the structural gene of the protein. If different phenes of a protein would be affected by different genes, this could be detected by genetic linkage studies. In this case the different phenes of the protein would segregate differently in the progeny and map to different loci on the chromosomes. A protein present in a 2-DE pattern in a high amount but in another mouse species in a low amount, frequently shows an intermediary level in the hybrid pattern. However, in the most cases, quantitative protein variants show various levels of concentrations in the progeny. This indicates that the cellular concentration of such a protein depends on several genes. Segregation studies of these loci (quantitative trait loci, QTL) requires, however, very precise measuring spot volumes from a large number of animals. A genetic analysis of QTL of proteins, quantitatively variant in 2-DE patterns, has been performed by Damerval et al. (1994) in maize. In this interesting and important investigation it was shown that the cellular concentration of a single protein species can depend on several chromosomal loci. Up to five, or even up to 12 loci were mapped for single proteins. At least some of these loci were located on different chromosomes. This finding demonstrates that proteins represent polygenic traits. In our studies on mice, we found that protein modifications, as revealed by protein spots which split into two spots in an other mouse species, can be caused by genes which do not map to the locus of the structural gene of the protein. If confirmed, observations like this would show that not only the amount of a protein but also its structure can depend on several genes.

Another genetic phenomenon worth to consider under the aspect of proteins is pleiotropy. Contrary to polygeny, the situation in which several genes act on the same genetic trait, pleiotopy is the phenomenon where a single gene is responsible for a number of distinct and seemingly unrelated phenotypic effects (King and Stansfield, 1990b). We described recently a pleiotropic effect observed in the crystallins of the mouse eye lens (Jungblut et al., 1998). A mouse strain carrying a cataract mutation in the gene for γB-crystallin was investigated by 2-DE. First, the lens proteins of normal mice were separated and analysed by mass spectrometry and partial sequencing. All the various crystallins of the crystallin family, encoded by different genes, were identified. Then, analysing the proteins of the mutant strain, the unexpected observation was made that not only the amount of the γB-crystallin was drastically reduced, but also all the other γ-crystallins (subfamily γA–γE). In principle, one may assume, that the gene of γB-crystallin had a pleiotropic effect (e.g. via gene regulation, frameshift) on the other γ-crystallin genes, which form a cluster on chromosome no. 1, or that the protein γB-crystallin had a pleiotropic effect on the other γ-crystallins by affecting the normal development of the whole lens. In any case, this is an example that leads to a general conclusion in the analysis of genetic diseases. Trying to elucidate a genetic defect, a useful strategy might be, at least in model organisms, to start from the protein level instead from the DNA level. Looking at an enormous number and at a broad spectrum of proteins offers the chance to detect not only the primarily defect protein, or several proteins of this type, but also co-affected proteins. The next step would be identifying the genes of

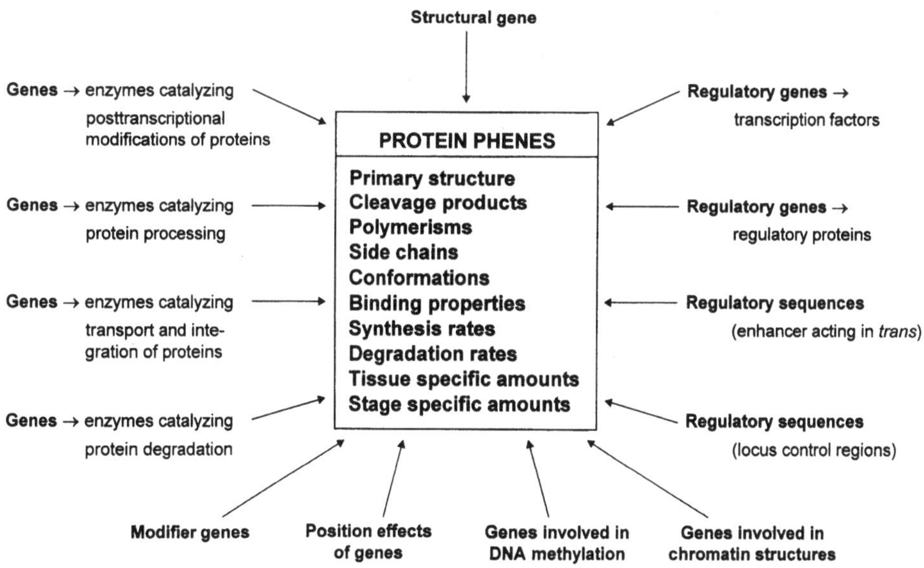

Figure 5. The polygenic nature of proteins.

the abnormal proteins. Then the way leads back to the proteins and to higher levels of abnormal phenotypes. This strategy starts with a survey of the compelex level of gene expression and does not postulate, the rather unrealistic situation, that only one gene is responsible for a genetic disease. Moreover, the proteins found to be affected may give us some hints towards the pathogenesis induced by the genetic defect.

Considering multifactorial diseases, which include polygenic factors, findings mentioned in this article suggest that even a single protein, involved in such a disease, might be a polygenetic trait (Figure 5). Human diseases are usually defined in terms of clinical symptoms such as high blood pressure, heart malformations or mental retardation. There is no doubt that several genes are involved in regulating blood pressure, and that many genes contribute to the normal development of heart and brain. Consequently, one can assume that any genetically based disease can be caused by a defect in one of the numerous genes involved in a particular disease. Many different genes, if mutated, may be able to induce, for example, microcephaly. A distinct patient, however, showing the symptoms of microcephaly, usually carries a mutation only in one of these genes. This means, whereas any genetic disease, as defined by clinical symptoms, might be polygenic in nature, with respect to an individual patient, each genetic disease is monogenic. This implies that the phenotype of a certain disease, the symptoms, may differ more or less among patients as far as different genes are involved. If far in the future the specific function of each of the genes will be known, as well as their role in the regulatory and metabolic network, one will realize, that each patient has its own disease. Genetic diseases will then be defined by the special genes affected rather than by the abnormal phenotypes described by the physicians.

REFERENCES

Anderson, N., *Nature* 1979, 278, 122–123.

Anderson, N. and Anderson, L., *Clin. Chem.* 1982, 28, 739–748.

Damerval, C., Maurice, A., Josse, J.M., and deVienne, D., *Genetics* 1994, 137, 289–301.

Gauss, C., Kalkum, M., Löwe, M., Lehrach, H., and Klose, J., *Electrophoresis* 1999, 20, in press.

Glotzer, M., Murray, A.W., and Kirschner, M.W., *Nature* 1991, 349, 132–138.

Goodfellow, P., *Nature Genet.* 1997, 16, 209–210.

Janke, C., Holzer, M., Klose, J., and Arendt, T., *FEBS Letters* 1996, 379, 222–226.

Jungblut, P., Otto, A., Favor, J., Löwe, M., Müller, E.-C., Kastner, M., Sperling, K., and Klose, J., *FEBS Letters*, 1998, 435, 131–137.

King, R.C. and Stansfield, W.D. (a), *A Dictionary of Genetics*, Oxford University Press, New York, Oxford 1990, p 239.

King, R.C. and Stansfield, W.D. (b), *A Dictionary of Genetics*, Oxford University Press, New York, Oxford 1990, p 244.

Klose, J., *Humangenetik* 1975, 26, 211–243.

Klose, J. (a), *Methods in Molec. Biol.* 1999, 112, 67–86.

Klose, J. (b), *Methods in Molec. Biol.* 1999, 112, 147–172.

Klose, J. and Kobalz, U., *Electrophoresis* 1995, 16, 1034–1059.

Nock, Ch., Gauss, Ch., Schalkwyk, L.C., Klose, J., Lehrach, H., and Himmelbauer, H., *Electrophorsis* 1999, in press.

Prehm., J., Jungblut, P., and Klose, J., *Electrophoresis* 1987, 8, 562–572.

Sanchez, J.C., Appel, R.D., Golaz, O., Pasquali, C., Ravier, F., Bairoch, A., and Hochstrasser, D.F., *Electrophoresis* 1995, 16, 1131–1151.

Stuart, D.I. and Jones, E.Y., Nature 1997, 386, 437–438.

Swinbanks, D., *Nature*, 1995, 378, 653.

The C. elegans Sequencing Consortium, *Science* 1998, 282, 2012–2018.

Wade, N., Science 1981, 211, 33–35.

10

THE GENOME AS A FLEXIBLE POLYMER CHAIN

Recent Results from Simulations and Experiments

Jörg Langowski, Carsten Mehring, Markus Hammermann, Konstantin Klenin, Christian Münkel, Katalin Tóth, and Gero Wedemann

Division Biophysics of Macromolecules (H0500)
Deutsches Krebsforschungszentrum
Im Neuenheimer Feld 280, D-69120 Heidelberg, Germany
and
Interdisciplinary Center for Scientific Computing (IWR)
Im Neuenheimer Feld 368, D-69120 Heidelberg, Germany

ABSTRACT

For most of its large-scale structural features, DNA can be regarded as a string-like flexible polymer. Its higher folding units such as chromatin and chromosomes can in turn be regarded as flexible linear chains. The structure and dynamics of such chain molecules are determined by intramolecular elastic, electrostatic and hydrodynamic interactions. We have developed Monte-Carlo (MC) and Brownian dynamics (BD) models which are used in connection with solution scattering, optical and scanning microscopy and hydrodynamic techniques to describe the structure and dynamics of DNA, chromatin and chromosomes.

Here we show a study where we determined the diameter of superhelical DNA directly by neutron scattering in comparison with results from a Monte-Carlo model, and a simulation of the condensation and decondensation of chromosome territories.

1. INTRODUCTION

The three-dimensional folding of the DNA in the cell plays a prominent role in determining gene activity. First, gene regulation by transcription factors is often

Genomics and Proteomics, edited by Sándor Suhai.
Kluwer Academic / Plenum Publishers, New York, 2000.

mediated by DNA looping between enhancer and promoter, and the strength of this interaction is to a great deal determined by the structure of the intervening DNA (as reviewed in[1]). Second, many examples exist where the expression of a gene is determined by the compaction state of chromatin. Finally, it is now well established that interphase chromosomes occupy distinct territories in the nucleus[2] and that the relative positioning of genes in the chromosome depends on the differentiation state of the cell. In order to understand the mechanisms of gene regulation at the molecular level, one must therefore characterize the structure and dynamics of the genome in its native environment. Because genomic DNA is such a large molecule, this can only be achieved by approximate models which describe the DNA or the chromatin fiber as a simple elastic filament. The molecular details of the DNA or of bound proteins are disregarded in such an approach. Many of the known structural and dynamic aspects of interphase chromosomes, chromatin fibers and DNA can be explained by such models, and predictions can be made about the mechanisms and time scales of intramolecular rearrangements.

We will give here two examples form our recent work: the first one shows measurements of the diameter of superhelical DNA by neutron scattering and comparison with Monte Carlo simulations, the second part of this papers shows an application of polymer models to the problem of chromosome condensation and decondensation.

2. THE STRUCTURE OF SUPERHELICAL DNA

Long-range interactions in DNA can be promoted by folding the DNA into a superhelical structure.[1,3] Here the DNA double strand is wound around itself, such that segments that are far from each other on the primary sequence can come into close neighborhood. Gene regulation by transcription factors binding to enhancer sequences could be mediated through such a folding mechanism. For a quantitative description of enhancer action it is therefore important to understand the structure of the superhelix and its changes in different environments.

To quantify structural properties of DNA free in solution under non-invasive conditions, solution scattering methods or hydrodynamic techniques such as analytical ultracentrifugation or dynamic light scattering have been used.[4–8] The quantities obtained with these methods, such as diffusion or sedimentation coefficients or scattering curves, have been interpreted by using numerical models such as Monte Carlo and Brownian Dynamics simulations.[3,8–15]

The models predict a superhelix diameter of about 10 nm. For verifying the model predictions, structural details of the superhelix in solution on a nanometer length scale have to be measured. One possible technique for doing this is the scattering of thermal neutron beams. This method offers the advantage that unlike X-rays it does not damage the DNA by ionization, and can therefore be used to study superhelical DNA where one single strand break would remove the superhelicity.

Monte-Carlo simulations of superhelical DNA predict characteristic changes in the form of the neutron scattering curve with increasing ionic strength. In the following we summarize some recent results which show that this salt dependence can be verified experimentally and be used to determine the superhelix diameter of DNA in aqueous solution over a wide range of Na^+ concentration.[16] These results are of particular relevance since the distance between opposing double strands in an interwound superhelix directly determines the interaction probability between, e.g., enhancer and promoter. The neutron scat-

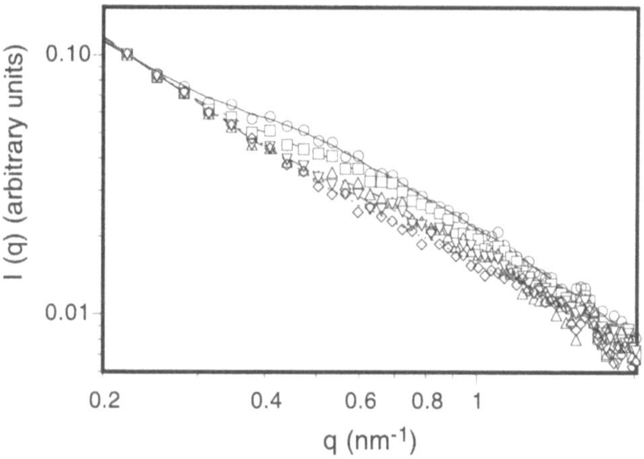

Figure 1. Measured scattering intensity I(q) of pUC18, supercoiled, in H_2O, 10 mM Tris, at different Na^+ concentrations: 0 mM (○), 10 mM (□), 40 mM (△), 100 mM (▽), 500 mM (◇).

tering results shown here are in agreement with the simulations and with recent scanning force microscopy images supporting a superhelix diameter of not less than 10 nm even in the presence of 10 mM Mg^{2+}.[17]

Small angle neutron scattering measurements. As an example, Figure 1 shows the neutron small angle scattering curve of the plasmid DNA pUC18 at different salt concentrations. At low salt (10 mM Tris, 0 mM Na^+) we clearly see an undulation at $q = 0.5\,nm^{+1}$ which decreases and shifts to higher q values with increasing Na^+ concentration ($q = 4\pi/\lambda \sin\theta/2$ is the modulus of the scattering vector, with λ the wavelength of the neutrons and θ the scattering angle). Above 0.1 M Na^+ all scattering curve superimpose; there is no indication for any further structural change.

For quantifying the effect, we computed the ratio of the scattering intensities of the superhelical DNA at different salt concentrations and the relaxed form at 100 mM (Figure 2). The shape of this curve is similar to the form factor of a pair of point scatterers at a

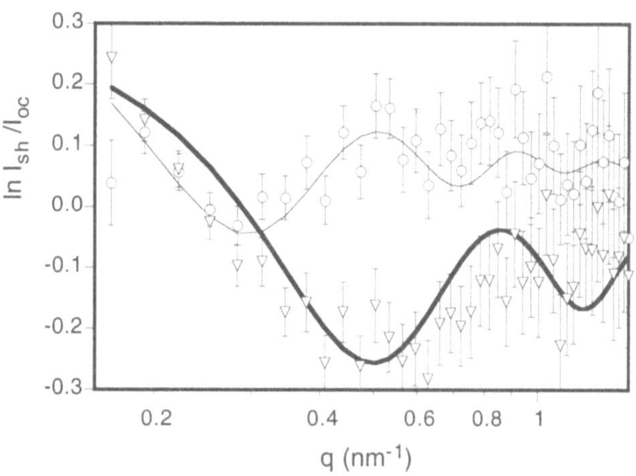

Figure 2. Ratio of the measured scattering intensities I(q) of pUC18, supercoiled, in H_2O, 10 mM Tris, at 0 mM (○) and 100 mM (▽) Na^+ concentration and relaxed pUC18 DNA at 100 mM Na^+ concentration. The solid lines are the scattering form factors of a pair of point scatterers at a distance r = 16.0 nm (thin line) and r = 9.0 nm (thick line).

constant distance d (solid lines in Figure 2). This behavior is to be expected if a certain intramolecular distance occurs with high probability in the superhelical, but not in the relaxed DNA.

We therefore interpret the distance d computed from the data in Figure 2 as a measure of the diameter of the superhelical regions of the DNA. Its value decreases from 16.0 ± 0.9 nm at 0 mM to 13.8 ± 1.1 nm at 10 mM, 11.5 ± 0.7 nm at 40 mM, 9.0 ± 0.8 nm at 100 mM. No further change at 500, 1000 or 1500 mM NaCl can be detected. We regard this as a direct measure of the interstrand separation in interwound regions of the DNA superhelix, since the curve undulation only appears in supercoiled DNA and intermolecular interference effects can be excluded.

DNA simulations. In parallel to the neutron scattering measurements, the salt-dependent change in the static form factor was predicted in simulated scattering curves of pUC18. The relative decrease in the scattering intensity from 10 mM to 100 mM salt concentration is of similar size as in the measured curves. The superimposed measured and simulated scattering curves of pUC18 are shown in Figure 3.

Since the simulated scattering functions agree very closely with the measured ones for both plasmids, we may conclude that the Monte Carlo simulation procedure is a valid representation of the structure of superhelical DNA in solution. Thus, conclusions on structural features of the DNA in free solution are possible from the simulated conformations. We calculated the average distribution function $p_n(r)$ of the distance r between each segment on the DNA chain and its nearest neighbor, with the constraint that the two segments are separated by at least 10 segments along the chain. This function (Figure 4) shows a pronounced peak which can be interpreted as the average superhelix diameter, its position at (18.0 ± 1.5) nm at 10 mM moves to (14.1 ± 1.5) nm at 20 mM and to (9.4 ± 1.5) nm at 100 mM salt concentration. Again, there was no further significant change above 100 mM salt concentration.

The measured and simulated superhelix diameters agree very well above 10 mM salt concentration (Figure 5). The superhelix diameters simulated for a 7 kb plasmid in ref. 8, which also show good agreement with the simulations, are shown for comparison. At 100 mM, our value agrees with the diameter of superhelical regions (9.2 ± 3.3) nm of p1868 as determined by scanning force microscopy in aqueous solution in the presence of 10 mM $MgCl_2$ and 30 mM NaCl.[17] A lateral collapse of the DNA superhelix as

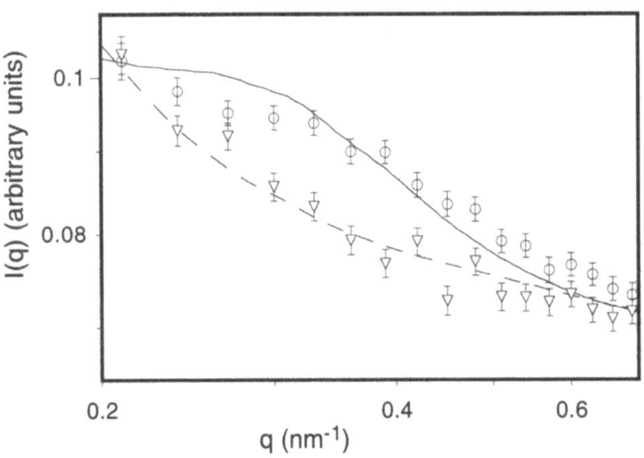

Figure 3. Measured and simulated scattering intensity I(q) of pUC18, in D_2O, at 10 mM (Exp.: O, Sim.: solid line) and at 100 mM (Exp.: ∇, Sim.: dashed line) salt concentration (NaCl + Tris).

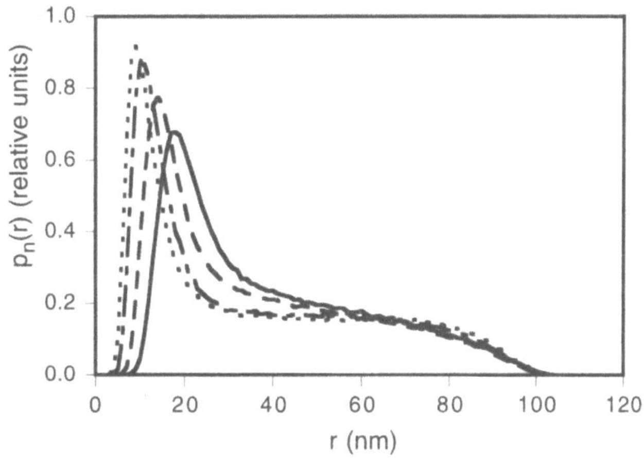

Figure 4. Distance distribution function $p_n(r)$ of the nearest neighbor of each segment, calculated from and averaged over the simulated configurations of pUC18, at 10 mM (——), 20 mM (— —), 50 mM (- · -) and 100 mM (. . . . .) salt concentration.

postulated by Bednar et al. from cryo-electron microscopy studies[11] could not be observed here.

3. MODELLING OF CHROMOSOME STRUCTURE

Decondensation and condensation of chromosomes comprise an important part of the cell cycle whose details are still unknown, and studying the physical principles of these processes can help a lot to understand the role of chromosome structure and dynamics in the function of the cell. Here we show first studies of chromosome dynamics based on a polymer chain model.

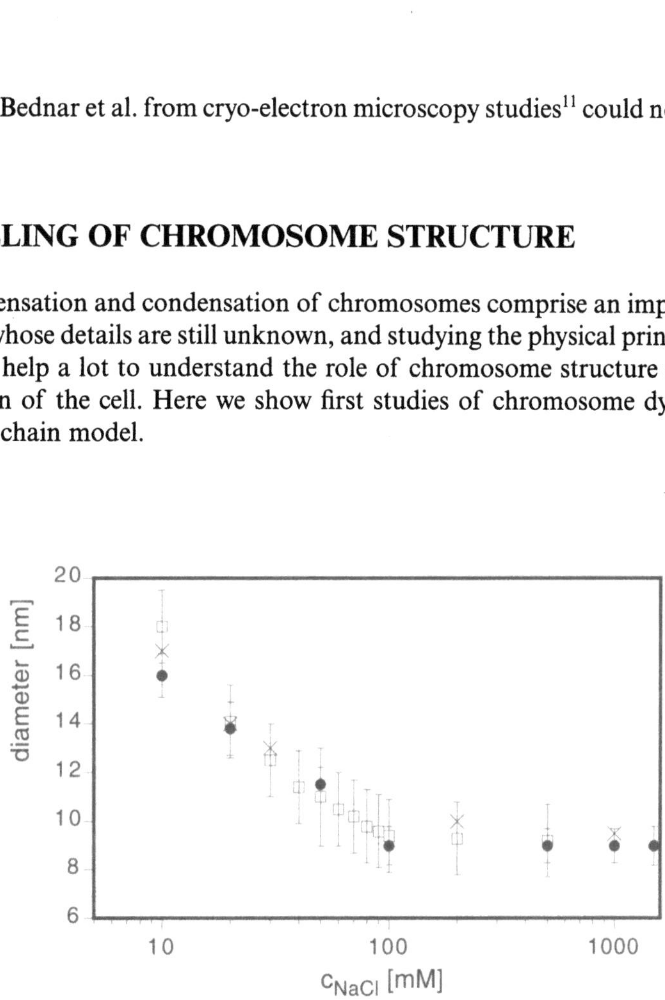

Figure 5. Measured (●) and simulated (□) superhelix diameter of pUC18 vs. salt concentration. The values calculated by (Rybenkov et al., 1997) are shown for comparison (×).

The basic structural element of the model is the chromatin fiber. Its average geometrical properties on length scales above some thousand base pairs can be described by a polymer consisting of rigid segments of a certain length (Kuhn length) and an excluded volume interaction representing the diameter of the chromatin fiber. The chromatin fiber is then folded into a locally compacted structure[18] which is motivated by the observed compartmentalization of metaphase chromosomes into bands and interphase chromosomes into "chromosomal foci"[19] (Figure 6). Following the ideas of the Lämmli group for the metaphase chromosome structure[20] each chromosome band is modeled by a "rosette" consisting of several chromatin loops of about 100 kbp each. The amount of chromatin in each band is taken proportional to the size of the band as determined in high resolution metaphase studies.[21] Adjacent rosettes are connected by a piece of "linker chromatin"; in these studies, its length was taken to be 1200 nm. In the condensed metaphase state, the bases of the neighboring rosettes are held at a distance of 60 nm by a stiff spring and the linker chromatin forms a loop connecting the rosette bases. For simulating the decondensed interphase state, the connecting spring is released and the

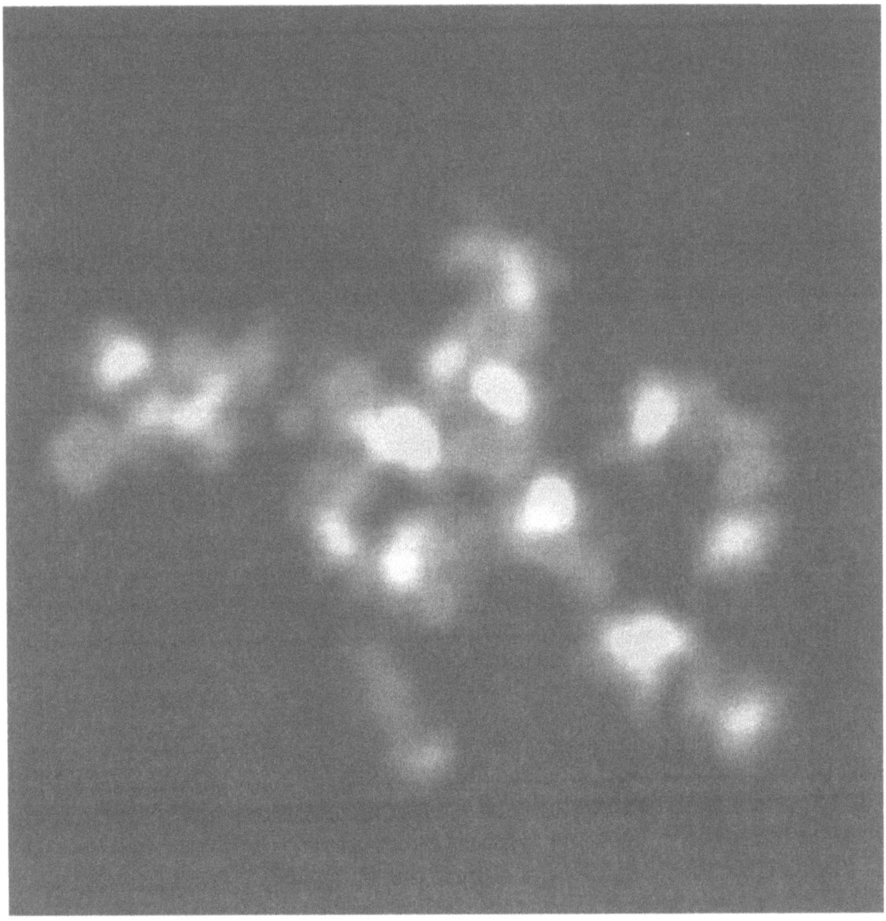

Figure 6. Simulated equilibrium configuration of chromosome 15 in metaphase (decondensed). Successive bands form "chromosomal foci" and are indicated in different shades of gray.

structure relaxed by a Monte-Carlo procedure or, if information about time-dependent structural changes is desired, by Brownian dynamics. In the latter case, the simulated polymer chain is subjected to random thermal motion and the solvent is approximated as a homogeneous viscous fluid.

The simulations can be expanded to whole nuclei by starting with a randomly positioned set of 46 chromosomes in a spherical volume representing the nuclear envelope, each chromosome holding a length of chromatin chain that is proportional to its DNA content.

Chromosome organization. The interphase structure of the MLS model has already been confirmed by a variety of experimental results, notably by a comparison of experimentally observed and simulated distances between pairs of genetic markers. These distances can be measured by fluorescence in situ hybridization (FISH) and compared with their linear separation on the genome. Yokota et al.[22] and Sachs et al.[23] interpreted such data to suggest a random walk configuration of the interphase chromosome. In their model the chromatin chain could freely intermingle to form a random coil which is structured into giant loops of MBp size (random-walk/giant-loop or RWGL model). In a reinterpretation of this data which is discussed in detail in[18] and,[19] we could show equally good agreement with the experimental data using the MLS model with a loop length of 120 kB and a bending rigidity of the chromatin chain given by a persistence length of 200 nm.

While the mean interphase distances are explained equally well by the MLS and RWGL models, the predicted internal organization of chromosomes into subdomains is quite different. For comparison with experiments where chromosomal subdomains are analyzed, we generated "virtual" multi-color images of FISH experiments by simulating the scanning of fluorescent markers attached to the chromatin in a confocal microscope, which allows us to quantify the structure of chromosome territory boundaries, subdomains etc. In accordance to experimental findings, "virtual" confocal sections of our simulated structures show very little overlap of chromosome arms. In contrast, most chromosomes simulated with the RWGL model overlap extensively. In other simulations, R- and G-bands of chromosome 15 were labeled in red and green and computed confocal sections compared with recent work in which early and late replicating bands of chromosomes were marked with modified bases during replication. Here again, the overlap between bands is negligible both in simulation and experiment, supporting the internal organization of interphase chromosomes into "foci".[24]

Chromosome dynamics. Since the parameters of the MLS model have been calibrated by comparison with experimental data, the model can now be used to simulate structural transitions within chromosomes as they are occurring during condensation, decondensation or by mechanical stress.

Starting from the equilibrium structures of an MLS-modeled interphase chromosome (see above) we can compute the diffusion coefficient of the entire structure or its parts by Brownian dynamics simulation from the slope of a plot of the mean square displacement of the center of mass vs. time. On a 200 ms simulated trajectory of chromosome 15 we obtained $D_t = (4.37 \pm 0.03) \cdot 10^{-15}\,m^2s^{-1}$ for the entire chromosome and $D_i = (3.65 \pm 0.03) \cdot 10^{-13}\,m^2s^{-1}$ for the subcompartment. The center-of-mass diffusion coefficient for the entire chromosome is of the same order as the diffusion coefficient of fluorescent markers on the genome of Drosophila observed by video microscopy.[25]

Decondensation. Starting from a rodlike metaphase conformation, the decondensation was simulated by opening one chromatin loop at its base in each subcompartment, such that the length of the linker chromatin between two subcompartments is increased

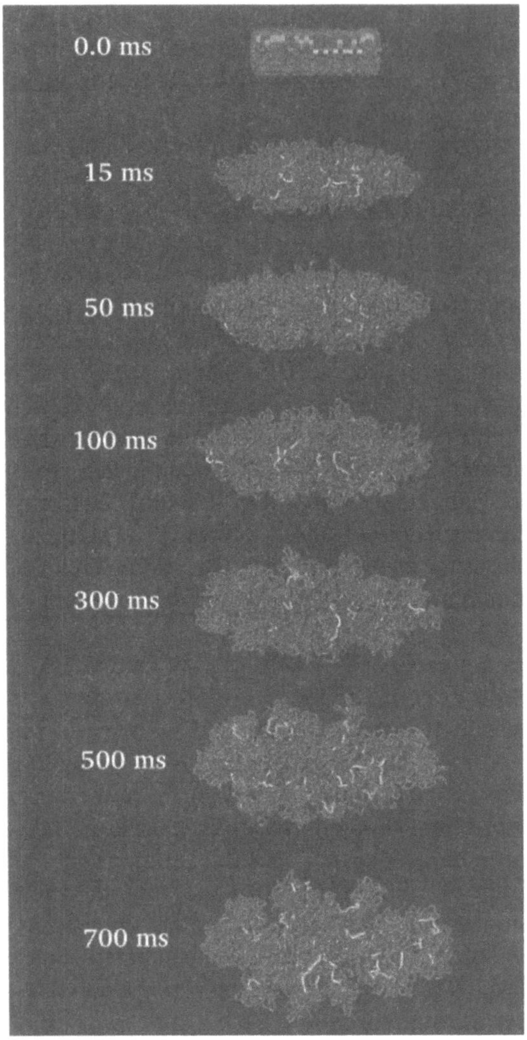

Figure 7. Brownian dynamics simulation of chromosome 15 decondensation.

from 60 to 1200 nm. The resulting structure is then relaxed by Brownian dynamics. Figure 7 shows the kinetics of decondensation of chromosome 15 as computed by our model. A detailed analysis of the decondensation process will be given elsewhere (Mehring, Wedemann and Langowski, in preparation). Here we summarize the main features of the decondensation process as follows:

For the first 100 ms the chromosome expands about twofold longitudinally. During this process, the subcompartments relax to their equilibrium conformation in about 50 ms. Following the fast initial relaxation, the structure very slowly expands in the lateral direction and shrinks again longitudinally until the globular interphase structure is achieved. This process takes about 200 s. Evidently, this time is only a crude estimate of the time scale of chromosome decondensation since it does not take into account active processes or interactions with other nuclear components. Nevertheless, the estimated time is in the range of actual chromosome decondensation processes and the observed structures such as chromosome bands and foci correspond to features that are actually observed in experiment. We can therefore use this model to develop concepts of the mechanism of chromosome decondensation and make quantitative predictions of structural changes during this process.

Condensation. A number of ideas have been forwarded to explain chromosome condensation,[26,27] but so far no experimentally provable model exists. Here we tried to simulate the condensation mechanism based on the MLS model, allowing for quantitative predictions. The condensation was simulated by pulling with a constant force between the attachment points of the linker chromatin connecting two subcompartments. This force would correspond to the action of motor proteins as have already been postulated in the cell.[27] The force was assumed to be 10 pN, comparable to that of known motor proteins.[28]

Figure 8 shows the time course of the simulated condensation of a chain of eight chromosome subcompartments, using an excluded volume barrier of 0.1 kT for the cross-

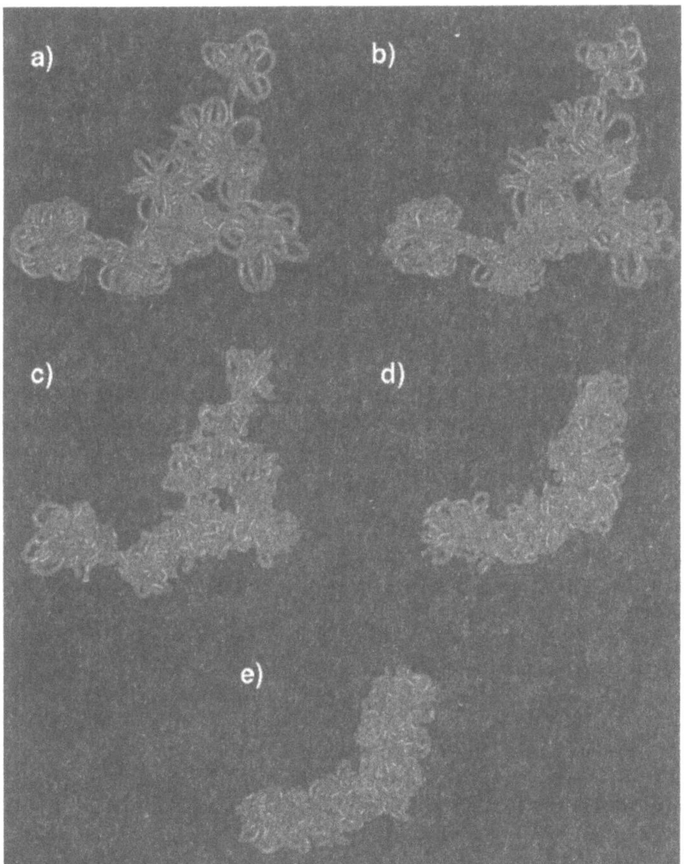

Figure 8. Brownian dynamics simulation of the condensation of a subchain of a chromosome (8 subcompartments) to a metaphase-like structure. Time scale from a) to e): 0, 0.02, 1.4, 11.4, 25.0 ms.

ing of two DNA double strands. After 25 ms a state is reached whose compaction is similar to the metaphase. This constitutes a lower time limit for the condensation of a chromosome since only a small part of an actual chromosome could be simulated with our available computing resources, and because the viscosity of the nuclear environment as well as entanglement with other parts of the genome[26] will probably slow down these processes.

ACKNOWLEDGEMENTS

This work was supported by grant 01KW9620 by the German ministry of education, science and technology (BMBF) to J.L.

REFERENCES

1. Rippe, K., von Hippel, P.H., and Langowski, J. (1995) *Trends Biochem Sci*, **20**(12), 500–506.
2. Cremer, T., Kurz, A., Zirbel, R., Dietzel, S., Rinke, B., Schrock, E., Speicher, M.R., Mathieu, U., Jauch, A., Emmerich, P., and et al. (1993) *Cold Spring Harb Symp Quant Biol*, **58**, 777–792.
3. Klenin, K.V., Frank-Kamenetskii, M.D., and Langowski, J. (1995) *Biophys J*, **68**(1), 81–88.
4. Schurr, J.M. (1977) *Chemical Physics*, **30**, 243–247.
5. Langowski, J. and Giesen, U. (1989) *Biophys Chem*, **34**(1), 9–18.
6. Langowski, J., Giesen, U., and Lehmann, C. (1986) *Biophys Chem*, **25**(2), 191–200.
7. Langowski, J. (1987) *Biophys Chem*, **27**(3), 263–271.
8. Rybenkov, V.V., Vologodskii, A.V., and Cozzarelli, N.R. (1997) *J Mol Biol*, **267**(2), 299–311.
9. Klenin, K.V., Vologodskii, A.V., Anshelevich, V.V., Klishko, V.Y., Dykhne, A.M., and Frank-Kamenetskii, M.D. (1991) *Journal of Molecular Biology*, **217**, 413–419.
10. Vologodskii, A.V., Levene, S.D., Klenin, K.V., Frank-Kamenetskii, M.D., and Cozzarelli, N.R. (1992) *Journal of Molecular Biology*, **227**, 1224–1243.
11. Bednar, J., Furrer, P., Stasiak, A., Dubochet, J., Egelman, E.H., and Bates, A.D. (1994) *Journal of Molecular Biology*, **235**, 825–847.
12. Langowski, J., Chirico, G., and Kapp, U. (1994) In Sarma, R.H. and Sarma, M.H. (eds.), Structural Biology: The State of the Art. Adenine Press, Schenectady, NY, Vol. 1, pp. 175–189.
13. Gebe, J.A., Allison, S.A., Clendenning, J.B., and Schurr, J.M. (1995) *Biophysical Journal*, **68**(2), 619–633.
14. Gebe, J.A., Delrow, J.J., Heath, P.J., Fujimoto, B.S., Stewart, D.W., and Schurr, J.M. (1996) *J Mol Biol*, **262**(2), 105–128.
15. Hammermann, M., Steinmaier, C., Merlitz, H., Kapp, U., Waldeck, W., Chirico, G., and Langowski, J. (1997) *Biophys J*, **73**(5), 2674–2687.
16. Hammermann, M., Brun, N., Klenin, K.V., May, R., Toth, K., and Langowski, J. (1998) *Biophys J*, **75**(6), 3057–3063.
17. Rippe, K., Mücke, N., and Langowski, J. (1997) *Nucleic Acids Res*, **25**(9), 1736–1744.
18. Münkel, C. and Langowski, J. (1998) *Physical Review E*, **57**(5-B), 5888–5896.
19. Münkel, C., Eils, R., Zink, D., Dietzel, S., Cremer, T., and Langowski, J. (1999) *Journal of Molecular Biology*, **285**(3), 1053–1065.
20. Saitoh, Y. and Laemmli, U.K. (1994) *Cell*, **76**, 609–622.
21. Francke, U. (1994) *Cytogenet Cell Genet*, **65**, 206–219.
22. Yokota, H., van den Engh, G., Hearst, J., Sachs, R.K., and Trask, B.J. (1995) *The Journal of Cell Biology*, **130**(6), 1239–1249.
23. Sachs, R.K., van den Engh, G., Trask, B., Yokota, H., and Hearst, J.E. (1995) *Proceedings of the National Academy of Sciences of the USA*, **92**, 2710–2714.
24. Zink, D., Cremer, T., Saffrich, R., Fischer, R., Trendelenburg, M.F., Ansorge, W., and Stelzer, E.H. (1998) *Hum Genet*, **102**(2), 241–251.
25. Marshall, W.F., Straight, A., Marko, J.F., Swedlow, J., Dernburg, A., Belmont, A., Murray, A.W., Agard, D.A., and Sedat, J.W. (1997) *Curr Biol*, **7**(12), 930–939.

26. Sikorav, J.-L. and Jannink, G. (1994) *Biophysical Journal*, **66**, 827–837.
27. Koshland, D. and Strunnikov, A. (1996) *Annu Rev Cell Dev Biol*, **12**, 305–333.
28. Nishizaka, T., Miyata, H., Yoshikawa, H., Ishiwata, S., and Kinosita, K., Jr. (1995) *Nature*, **377**(6546), 251–254.

11

ANALYSIS OF CHROMOSOME TERRITORY ARCHITECTURE IN THE HUMAN CELL NUCLEUS

Overview of Data from a Collaborative Study

H. Bornfleth,[1,2] C. Cremer,[1,2] T. Cremer,[3,2] S. Dietzel,[4] P. Edelmann,[1,2] R. Eils,[2] W. Jäger,[2] D. Kienle,[4] G. Kreth,[1,2] P. Lichter,[5] G. Little,[2] C. Münkel,[5] J. Langowski,[5] I. Solovei,[3] E. H. K. Stelzer,[6] and D. Zink[3]

[1]Institut für Angewandte Physik
Universität Heidelberg
[2]Interdisziplinäres Zentrum für Wissenschaftliches Rechnen (IWR)
Universität Heidelberg
[3]Institut für Anthropologie und Humangenetik
LMU, München
[4]Institut für Humangenetik
Universität Heidelberg
[5]Deutsches Krebsforschungszentrum (DKFZ)
Heidelberg
[6]European Molecular Laboratories
EMBL, Heidelberg

INTRODUCTION

The genome in human cell nuclei is partitioned into mutually exclusive chromosome territories (for review see Cremer et al., 1993; Lamond and Earnshaw, 1998). Detailed knowledge concerning the three-dimensional (3D) structure of these territories, their dynamic changes with time, their intranuclear arrangements and interactions with each other at territory surfaces is indispensable to understand the functional, cell cycle and cell type specific 4D (space-time) architecture of the human genome in situ. With this goal in mind several groups with interdisciplinary expertise founded the Heidelberg 3D Human Genome Study Group, including laboratories experienced in molecular

Genomics and Proteomics, edited by Sándor Suhai.
Kluwer Academic / Plenum Publishers, New York, 2000.

cytogenetics (T. Cremer, P. Lichter), new developments in high resolution fluorescence microscopy (C. Cremer, E.H.K. Stelzer), quantitative modeling of chromatin and chromosome territories (C. Münkel, J. Langowski) and 3D and 4D image analyses (R. Eils, W. Jäger). In close collaboration these groups have established approaches to perform a quantitative analysis of chromosome territory and nuclear architecture. In this report we present an overview of our findings. For an extensive review of the literature the reader is refered to the references cited in our original papers.

VARIABLE STRUCTURE AND COMPARTMENTALIZATION OF CHROMOSOME TERRITORIES INTO MUTUALLY EXCLUSIVE ARM AND BAND DOMAINS

Fluorescence in situ hybridization (FISH) with microdissection probes from human chromosomes 3 and 6 was applied to visualize arm and subregional band domains in 3D-conserved, human amniotic fluid cell nuclei (Dietzel et al., 1998a). Confocal laser scanning microscopy and quantitative 3D-image analysis revealed a pronounced variability of p- and q-arm domain arrangements and shapes. Apparent intermingling of neighboring arm and band domains was limited to the domain surface demonstrating the compartmentalization of chromosome territories into mutually exclusive arm and band domains. 3D-distance measurements with pter- and qter probes supported a high flexibility of chromosome territory folding. Multicolor fluorescence in situ hybridization (FISH) with paint probes for the entire X-chromosome, Xp and Xq arms, as well as Xp terminal, Xq terminal and X centromer specific probes demonstrated distinctly separated Xp arm and Xq arm domains. In accordance with our findings on autosome territories, 3D distance measurements revealed a high variability of intrachromosomal distances between Xpter, Xcen and Xqter specific probes within both X-territories. These data argue against the hypothesis of Walker et al. (1991) that a looped structure of the inactive X territory is formed by tight telomere-telomere associations (Dietzel et al., 1998b). A three-dimensional reconstruction of painted active and inactive X-chromosome territories revealed that both X-territories have similar volumes but differ in shape and surface structure: Xa-territories had a flatter shape and a larger, apparently more complexly folded surface than Xi-territories (Eils et al., 1996).

THE 3D-POSITIONING OF SPECIFIC DNA AND RNA SEQUENCES WITHIN CHROMOSOME TERRITORIES

The intranuclear position of three coding (DMD, MYH7, and HBB) and two non-coding sequences (D1Z2 and an anonymous sequence) was analyzed in the respective chromosome territories of three-dimensionally preserved nuclei of four different human cell types, including cells where DMD and MYH7 were actively transcribed (Kurz et al., 1996). In this analysis the genes were preferentially located in the periphery of chromosome territories independent from the activity of the genes. In contrast, the non-expressed anonymous fragment was found randomly distributed or localized preferentially in the interior of the corresponding chromosome territory.

In other experiments, the three-dimensional positions of the X-located adenine nucleotide translocase genes, ANT2 and ANT3, were compared in the active and inactive X-chromosome territory (Xa and Xi) of female human amniotic fluid cell nuclei (Dietzel et al., 1999). ANT2 is located in Xq24-q25 and is transcriptionally active on Xa, but inactive on Xi. ANT3 is located in the pseudoautosomal region Xp22.3 and escapes X-inactivation. Multicolor fluorescence in situ hybridization (FISH) was combined with spectrally discriminating high resolution confocal laser scanning microscopy and image analysis to measure distances between different subchromosomal regions and to determine the relative 3D positions of these specific genes within the X-chromosome territories with highly improved accuracy (Bornfleth et al., 1998; Dietzel et al., 1998b). Our analysis revealed that transcriptionally active ANT2 and ANT3 genes were positioned more exterior in both chromosome territories in contrast to the inactive ANT2 gene whose position showed a significant shift towards the Xi-territory interior. Although the volumes of both X-territories were similar, 3D-distances between ANT2 and ANT3 were significantly smaller in Xi- as compared to Xa- territories reflecting different territory shapes (see above). These data suggest the possibility that gene position in chromosome territories might be correlated with their genetic activity.

Specific species of RNA were found in accumulations either spherical or track-like in morphology. The transcripts contained within some of these RNA structures appear to have been released from a discrete genomic site (Lampel et al., 1997). The morphology of such RNA signals suggests that the released transcripts are channeled within the nucleoplasm. In order to analyze the status of the RNA within these accumulations, RNA transcripts derived from EBV genes were localized along the genomically integrated EBV genome in human Namalva cells (Lampel et al., 1997). It was shown that sequences from both ends of the EBV genome were all distributed along the entire length of the RNA signals. Furthermore, removal of labeled RNA sequences and subsequent visualization of DNA confirmed the confinement of the genomic sequences to one end of the RNA signal. Therefore, the data support the view of RNA accumulations as a stream of molecules, delineating a path from a dot-like gene locus towards the nuclear envelope for export into the cytoplasm.

The genome of mammals is a mosaic of isochores, defined as long segments of DNA that are homogeneous in base composition. Among them the H3 isochores (located in T bands) represent the GC-richest fraction of genomic DNA with the highest gene density, whereas the AT-richest fractions L1 + L2 (mainly in G bands) have been proven to possess a low gene density. Investigation of the three-dimensional organization of these isochores within the interphase chromosome territories by means of FISH, confocal mocroscopy and 3D image analysis revealed a significant difference in localization (Tajbakhsh et al., 1999). While GC-richest/gene-richest sequences did not show any preferential intraterritorial localization, simultaneously detected AT-richest/gene-poorest DNA fragments were found more to the interior of the same chromosome territories. While these data show that the results obtained with individual genes cannot be generalized, there is clear evidence for differences in the intraterritorial distribution of GC- and AT-rich sequences.

TUMOR SPECIFIC CHROMATIN STRUCTURES ANALYZED IN THE CELL NUCLEUS OF NEUROBLASTOMAS

The detailed analysis of the cell nuclear architecture can provide new clues to the understanding of the altered functions of tumor cells. As an example we investigated the nuclear architecture of neuroblastomas, which belong to the most highly destructive tumours in childhood (Solovei et al., manuscript in preparation). A strong amplification of the cancer gene N-myc in neuroblastoma cells leads to pathological chromatin structure and suggests an unfavourable diagnosis. Amplified N-myc genes appear in two types of aberrant chromatin structures "double minute" chromosomes (DMCs) and "homogenously stained regions" (HSRs). While DMC were mainly found in direct preparations of neuroblastomas, HSR is typically found in established cell lines. We noted that DMCs were preferentially localised at the periphery of chromosome territories and sometimes also in deep invaginations. HSRs in tumor cell nuclei were organised either as chromosome territory like structures or as extended structures which penetrated throughout the major part of the cell nucleus. Three dimensional reconstructions of HSR territories showed a complex structure with many invaginations, although a clear separation from neighbouring chromosome territories was noted. The N-myc gene was distributed over the whole HSR territory volume.

ORGANIZATION OF EARLY AND LATE REPLICATING DNA IN HUMAN CHROMOSOME TERRITORIES

Incorporation of the halogenated thymidine analogs IdU and CldU during distinct time segments of S-phase was used to differentiate between early and late replicating DNA of human diploid fibroblasts (Visser et al., 1998). On mitotic chromosomes the amount and spatial distribution of early and late replicating DNA corresponded to R/G banding patterns (Zink et al., 1999). At the second and further mitotic events both replication labeled and unlabeled chromatids were distributed into the daugher nuclei resulting in a fraction of nuclei exhibiting a few individually distinguishable, labeled chromosome territories. Replication labeled territories of chromosomes 13 and 15 were identified by additional chromosome painting. The territories displayed a structural rearrangement in G_1 cells compared to quiescent (G_0) cells resulting in the compaction of the territories (Zink et al., 1999; Bornfleth et al., 1999a). The distribution of early and late replicating DNA was analysed for both chromosome territories in both G_1 and G_0. Early and late replicating DNA formed distinctly separated chromatin foci within these chromosome territories. These foci displayed diameters of some 400 to 800 nm with a median overlap of only 5%–10%. Chromatin foci corresponding to early replicating R-bands and late replicating G-/C-bands appeared as a persistent feature of chromosome territory organization during all stages of the cell cycle and corresponded with the R- and G-band structure of mitotic chromosomes. These foci apparently were also maintained during subsequent cell cycles as distinctly separated units of chromosome organization. In combination with findings from other groups it appears that DNA organized into replication foci during S-phase remains stably aggregated in non S-phase cells and that these stable aggregates provide fundamental units of nuclear or chromosome architecture, called subchromosomal foci (see Zink et al., 1998a, 1999 and refs. therein).

QUANTITATIVE ANALYSIS OF IN VIVO DYNAMICS OF CHROMOSOME TERRITORIES AND SUBCHROMOSOMAL FOCI IN LIVING CELL NUCLEI

A new approach for the *in vivo* visualization of indivdual chromosome territories in the nuclei of living human cells was recently developed (Zink et al., 1998a). The fluorescent thymidine analog Cy3-AP3-dUTP was microinjected into the nuclei of cultured human cells, such as human diploid fibroblasts, HeLa cells and neuroblastoma cells and incorporated during S-phase into the replicating genomic DNA. Labeled cells were further cultivated for several cell cycles in normal medium. This scheme yielded sister chromatid labeling *in vivo* at the second mitosis after microinjection. Random segregation of labeled and unlabeled chromatids into daughter nuclei during this and subsequent mitotic events resulted in nuclei exhibiting individual *in vivo* detectable chromatid territories. Each *in vivo* labeled territory contained a number of subchromosomal foci with a diameter of 400–800 nm as noted above for fixed cells labeled with halogenated thymidine analogs. This approach has made possible a 4D *in vivo* analysis of the shape and position of chromosome territories and subchromosomal foci (Zink and Cremer, 1998b; Bornfleth et al., 1999b). The evolution of these foci over time was analyzed quantitatively in time-lapse series of three-dimensional (3D) confocal images. To enable a quantitative analysis of the dynamics of the foci, the 3D image stacks had to be aligned for each individual cell nucleus, since nuclei moved for several μm during the observation period of 4 h. The best results were obtained by a correlation function analysis (CFA) with 4 degrees of freedom (3 translational and 1 rotational). Its accuracy was tested using a model data set with 5 simulated territories in a nuclear volume. Even for the low light levels estimated from the noise in the experimental data (about 14 photons were collected in a voxel with maximum intensity), the motion could be reconstructed by the CFA. After image alignment, several parameters describing the morphology and dynamics of territories were investigated. These included their overall morphology, the number of foci found, the distances between bary centers of intensity of individual territories in a nucleus, the diffusion of foci and occasional directed motion of foci inside the territories. The overall morphology of territories did not show considerable changes over the observation period. Small-scale motions of foci (<1 μm) caused the merging or division of clusters of foci in the microscopic images, leading to reversible changes in the number of detected foci. Large-scale motions of foci or whole territories (>1 μm) were observed only for a subset of the chromosomes investigated. We assume that subchromosomal foci provide an important higher order structure of human genome architecture *in vivo* (see above for the demonstration of chromatin foci in nuclei of fixed cells) and expect that the further analysis of their dynamics will shed light on the functional organization of the cell nucleus.

MODELING OF THE STRUCTURAL AND FUNCTIONAL ARCHITECTURE OF CHROMOSOME TERRITORIES

The finding that chromosomes are organized in the cell nucleus in mutually exclusive territories has important implications for the understanding of the functional nuclear architecture and, in particular, for the process of chromosome aberration formation (Cremer et al., 1996; Kreth et al., 1998). An attempt was undertaken to integrate the above findings into a model of the functional architecture of the cell nucleus, termed the interchromosomal domain (ICD) compartment model (Zirbel et al., 1993; Cremer et al., 1993, 1995, 1996): According to this model we predict that macromolecular complexes for transcription, splicing, DNA replication, and DNA repair are located in the ICD compartment. This hypothetical compartment starts at nuclear pores, extends between the chromosome territories (Zirbel et al., 1993) and further expands as a branching network of ICD channels from the chromosome territory periphery into the territory interior (Cremer et al., 1995, 1996). The finest branches may extend between chromatin domains which may reflect the experimentally observed subchromosomal foci described above and may consist of clusters of individual chromatin loops. The recently proposed multiloop subcompartment (MLS) model (Münkel and Langowski, 1998; Münkel et al., 1999) assumes a folding of the chromatin fiber into 120 kb loops and an arrangement of these loops into rosette-like subcompartments. Chromosome territories consist of such subcompartments connected by short pieces of chromatin. Number and size of subcompartments correspond to chromosome bands in early prophase. The structural features of chromosome territories predicted by the MLS model fits very well the experimental data described above for the compartmentalization of chromosome territories in arm, band and subband domains. In the light of the ICD model the location of DMCs at the surface of the chromosome territories brings them in close neighbourhood to large accumulations of splicing factors, the so-called speckles, which are also localized in the periphery of chromosome territories (Zirbel et al., 1993; Solovei et al., manuscript in preparation). We also propose that huge HSRs which penetrate throughout the major part of some neuroblastoma cell nuclei (see above) reflect many repetitive amplicons, which expand within the ICD space and thus facilitate the close contact with the protein aggregates for transcription and splicing located in this space. Further experiments are under way to support or reject this hypothesis.

EXPERIMENTAL TESTING OF THE INTERCHROMOSOMAL COMPARTMENT

In order to explore the interchromosomal nucleoplasmic domains experimentally, filament forming proteins were introduced into the nucleus by transfecting cells with a vimentin gene engineered to contain a nuclear localization signal (NLS-vimentin). In stably transfected human cells incubated at 28°C, the Xenopus NLS-vimentin assembled progressively with time to form strictly orientated intranuclear filamentous arrays (Bridger et al., 1998). Quantitative analysis based on 3D imaging microscopy revealed that these arrays were localized almost exclusively outside of chromosome territories. The filaments also colocalized with specific nuclear RNAs, coiled bodies and PML bodies, all situated outside of chromosome territories, thereby interlinking these structures. This strongly implies that these nuclear entities coexist in the same interconnected nuclear compartment.

ACKNOWLEDGMENTS

These studies were supported by BMBF grant 01 KW 9620 (German Human Genome Project) and the Biomed 2 program of the European community (BMH4-CT95-1139).

PUBLICATIONS FROM THE HEIDELBERG 3D HUMAN GENOME STUDY GROUP

(Note: For the purpose of this overview of current data from the Heidelberg 3D Human Genome Study Group we listed only publications from the participants of this study group. The many, important studies published from other groups have been cited in our original papers. Readers of this overview are refered to these citations and to two recently published excellent reviews: Bridger, J.M. and Bickmore, W.A. (1998) Putting the genome on the map, Trends Genet., 14, 403–409; Lamond, AI, and Earnshaw, WC (1998) Structure and function in the nucleus. Science 280, 547–553.)

REFERENCES

Bornfleth H., Sätzler K., Eils R., and Cremer C. (1998) High-precision distance measurements and volume-conserving segmentation of objects near and below the resolution limit in three-dimensional confocal fluorescence microscopy. J Microsc 189, 118–136.

Bornfleth H., Edelmann P., Zink D., and Cremer C. (1999a) Three-dimensional analysis of genome topology, in: Handbook of Computer Vision and Applications, Vol. III. B. Jähne, H. Haußecker, P. Geißler editors, Academic Press, San Diego/New York, 859–878.

Bornfleth H., Edelmann P., Zink D., Cremer T., and Cremer C. (1999b) Quantitative motion analysis of sub-chromosomal foci in living cells using four-dimensional microscopy. Biophysical J 77, 2871–2886.

Bridger J.M., Herrmann H., Münkel C., and Lichter P. (1998) Identification of an interchromosomal compartment by polymerization of nuclear-targeted vimentin. J Cell Sci 111, 1241–1253.

Cremer T., Kurz A., Zirbel R., Dietzel S., Rinke B., Schröck E., Speicher M.R., Mathieu U., Jauch A., Emmerich P., Scherthan H., Ried T., Cremer C., and Lichter P. (1993) The role of chromosome territories in the functional compartmentalization of the cell nucleus. Cold Spring Harbor Symp Quant Biol 58, 777–792.

Cremer T., Dietzel S., Eils R., Lichter P., and Cremer C. (1995) Chromosome territories, nuclear matrix filaments and interchromatin channels: a topological view on nuclear architecture and function. Kew Chromosome Conference IV (Eds P.E. Brandham and M.D. Bennett), pp. 63–81, Royal Botanic Gardens, Kew.

Cremer C., Münkel C., Granzow M., Jauch A., Dietzel S., Eils R., Guan X.Y., Meltzer P.S., Trent J.M., Langowski J., and Cremer T. (1996) Nuclear architecture and the induction of chromosomal aberrations. Mutat Res Rev Genet Toxicol 366/2, 97–116.

Dietzel S., Jauch A., Kienle D., Qu G., Holtgreve-Grez H., Eils R., Münkel C., Bittner M., Meltzer P.S., Trent J.M., and Cremer T. (1998a) Separate and variably shaped chromosome arm domains are disclosed by chromosome arm painting in human cell nuclei. Chromosome Res 6, 25–33.

Dietzel S., Eils R., Sätzler K., Bornfleth H., Jauch A., Cremer C., and Cremer T. (1998b) Evidence against a looped structure of the inactive human X-chromosome territory. Exp Cell Res 240, 187–196.

Dietzel S., Schiebel K., Little G., Edelmann P., Rappold G.A., Eils R., Cremer C., and Cremer T. (1999) The 3D-positioning of ANT2 and ANT3 genes within female X-chromosome territories correlates with gene activity. Exp Cell Res 552, 363–375.

Eils R., Dietzel S., Bertin E., Granzow M., Schröck E., Speicher M.R., Ried T., Robert-Nicoud M., Cremer C., and Cremer T. (1996) Three-dimensional reconstruction of painted human interphase chromosomes: active and inactive X-chromosome territories have similar volumes but differ in surface and shape. J Cell Biol 135, 1427–1440.

Kreth G., Münkel C., Langowski J., Cremer T., and Cremer C. (1999) Chromatin structure and chromosome aberrations: modeling of damage induced by isotropic and localized irradiation. Mutat Res 404, 77–88.

Kurz A., Lampel S., Nickolenko J.E., Bradl J., Benner A., Zirbel R.M., Cremer T., and Lichter P. (1996) Active and inactive genes localize preferentially in the periphery of chromosome territories. J Cell Biol 135, 1195–1205.

Lampel S., Bridger J.M., Zirbel R.M., Mathieu U.R., and Lichter P. (1997) Nuclear RNA accumulations contain released transcripts and exhibit specific distributions with respect to Sm antigen foci. DNA Cell Biol 16, 1133–1142.

Münkel C. and Langowski J. (1998) Chromosome structure predicted by a polymer model. Phys Rev E 57, 5888–5896.

Münkel C., Eils R., Dietzel S., Zink D., Mehring C., Wedemann G., Cremer T., and Langowski J. (1998) Compartmentalization of interphase chromosomes observed in simulation and experiment. J Mol Biol 285, 1053–1065.

Solovei I., Kienle D., Little G., Eils R., Schwab M., Cremer C., and Cremer T. (1999) Topology of tumor specific chromatin structures in Neuroblastoma cell nuclei. Manuscript submitted.

Tajbakhsh J., Luz H., Bornfleth H., Lampel S., Cremer C., and Lichter P. (2000) Spatial distribution of GC- and AT-rich DNA sequences within human chromosome territories Exp Cell Res 255, 299–237.

Visser A.E., Eils R., Jauch A., Little G., Bakker P.J.M., Cremer T., and Aten J.A. (1998) Spatial Distributions of Early and Late Replicating Chromatin in Interphase Chromosome Territories. Exp Cell Res 243, 398–407.

Zink D., Cremer T., Saffrich R., Fischer R., Trendelenburg M.F., Ansorge W., and Stelzer E.H.K. (1998a) Structure and dynamics of human interphase chromosome territories *in vivo*. Hum Genet 102, 241–251.

Zink D. and Cremer T. (1998b) Chromosome dynamics in nuclei of living cells. Current Biology 8, 321–324.

Zink D., Bornfleth H., Visser A., Cremer C., and Cremer T. (1999) Organization of early and late replicating DNA in human chromosome territories. Exp Cell Res 247, 176–188.

Zirbel R.M., Mathieu U.R., Kurz A., Cremer T., and Lichter P. (1993) Evidence for a nuclear compartment of transcription and splicing located at chromosome domain boundaries. Chromosome Res 1, 93–106.

12

FROM SEQUENCE TO STRUCTURE AND FUNCTION

Modelling and Simulation of Light-Activated Membrane Proteins

Jerome Baudry,[1] Serge Crouzy,[2] Benoit Roux,[3] and Jeremy C. Smith[1,4]

[1]SBPM/DBCM, CEA-Saclay
91191 Gif-sur-Yvette cedex
France
[2]DBMS, CEA-Grenoble
Avenue des Martyrs, Grenoble
France
[3]Département de Chimie, Université de Montréal
Succursale centre ville
Montréal, Canada
[4]Lehrstuhl für Biocomputing, IWR der Universität Heidelberg
Im Neuenheimer Feld 368, D-69120 Heidelberg
Germany

INTRODUCTION

Genes code for protein sequences which in turn encode information on three-dimensional protein structures. An important challenge for the future is to be able to understand this sequence-structure pathway, and to develop methods and algorithms for calculating protein structure from sequence. The general protein folding problem is not yet solved, but some progress has been made in calculating structures homologous in sequence to a protein of known structure. We have applied this procedure to examine light-driven membrane proteins.

Genomics and Proteomics, edited by Sándor Suhai.
Kluwer Academic / Plenum Publishers, New York, 2000.

Many of the functions of biological membranes are performed by proteins bound to them. Among the roles of these proteins are the reception/transmission of messages and/or the transport of materials. However, due to difficulty in their crystallization only a small number of atomic-detail three-dimensional structures exist for membrane-spanning proteins. Among the few known structures are those of the light-transducing proteins, the photosynthetic reaction centre and bacteriorhodopsin.[1–4] In the present paper we briefly review some recent progress in the modelling and simulation of light-activated membrane proteins before presenting some new, preliminary results on bacteriorhodopsin.

The paucity of crystallographic structures has led to a bottleneck in structural membrane protein research and adds impetus to the development of computer modelling techniques for determining their structures. One of these techniques is homology modelling: it can be possible to determine an unknown protein structure by using a known X-ray structure as a three-dimensional template, if there is sequence homology between the two. The higher the sequence homology the higher the probability of obtaining a reliable model structure.

An example of homology modelling at high sequence identity is the recent work on the photosynthetic reaction centre protein from the bacterium *Rhodobacter capsulatus*.[5] This protein has been the subject of a considerable amount of molecular biological and spectroscopic work aimed at improving our understanding of the primary steps of photosynthesis. A structural model was derived by combining information from the experimental structure of the highly homologous (54% sequence identity) reaction centre from *Rhodopseudomonas viridis*[1] with molecular mechanics and simulated annealing calculations. In the *Rb. capsulatus* model the orientations of the bacteriochlorophyll monomer and bacteriopheophytin cofactors on the pathway inactive in electron transfer differ significantly from those in the reaction centre of *Rps. viridis*. The orientational difference was found to be in agreement with linear dichroism measurements.[6] Moreover, the pattern of cofactor hydrogen-bonding to the protein was found to be in agreement with optical spectroscopic experiment (Mattioli, T. personal communication). The *Rb. capsulatus* model was used to provide an explanation as to why a partially-symmetrized mutant *Rb. capsulatus*, which has been of particular interest for experiments on primary excited states in photosynthesis, lacks an electron acceptor bacteriopheophytin (BPh_L).[7–9] Conformational energy calculations on the partially symmetrised mutant and several BPh_L-binding revertants also provided an explanation for the relative BPh_L-binding properties of the proteins, in terms of interactions involving two residues in the binding pocket, these being a tryptophan and a methionine.[9]

Modelling at lower sequence homology, although less reliable, can be useful for suggesting experiments as part of an iterative procedure to obtain structural information on a membrane protein of particular interest. An example of this is the recent modelling of the photosystem II reaction centre core in plants for which a model was constructed by exploiting homology existing with the bacterial reaction centre proteins.[10]

In the rare cases where high-resolution experimental structures do exist modelling and simulation can be undertaken so as to refine structural detail and to understand physically how structure leads to function. A good example of such a system is bacteriorhodopsin (bR), a protein that functions as a light-driven proton pump in the purple membrane of the bacterium *Halobacterium salinarium*.[11] The light-absorbing chromophore in bR is a retinal molecule that is covalently bonded *via* its Schiff base to the ε-amino group of Lys 216.[12] The characteristic purple colour of bacteriorhodopsin is due to absorption by the chromophore. The absorption is red-shifted with respect to that of related model compounds in solution, an effect that has been proposed to originate from interactions between the retinal and its polar environment in the protein.[13]

The retinal interactions may include hydrogen bonds with the Schiff base. Structures for bR at high resolution have been obtained.[2,3] These revealed a channel through the protein that includes the Schiff base. Site-directed mutagenesis experiments suggest that the channel contains the pathway for proton transfer through bR.[14–17] A considerable amount of data exist that suggest that the proton transfer channel is at least partially hydrated. Low resolution neutron diffraction using contrast variation has indicated that about four water molecules are present in the neighborhood of the Schiff base although their positions in the direction perpendicular to the membrane plane could not be accurately determined.[18] There is, however, considerable other evidence that water molecules are directly associated with the Schiff base. A resonance Raman study suggests that a negatively charged counterion located near the Schiff base group is stabilized by water molecules.[19] Solid state ^{13}C and ^{15}N NMR experiments led to a model being proposed in which a water molecule is directly hydrogen-bonded to the Schiff base.[20] Other solid state ^{1}H and ^{15}N NMR experiments suggest that there is a direct exchange of the Schiff base NH hydrogen with bulk water.[21] Another resonance Raman study, of the Schiff base hydrogen/deuterium exchange, also led to the conclusion that a water molecule is directly hydrogen bonded to the Schiff base NH proton.[22] Finally, the recent crystallographic structure of Pebay-Peyroula et al. has directly identified some water molecules associated with the Schiff base.[3]

Clearly, a detailed understanding of Schiff base hydrogen bonding in the various stages of the photocycle will be required for a complete description of bR function. Computational chemistry has an important role to play in resolving such questions, by identifying and quantifying hydrogen-bonding geometries and energies of pertinent model systems. For example, quantum chemistry and molecular mechanics techniques have been combined to determine the geometries and energetics of retinal–water interactions.[23,24] *Ab initio* molecular orbital calculations were used to determine potential surfaces for water–Schiff base hydrogen bonding and to characterize the energetics of rotation of the C-C single bond distal and adjacent to the Schiff base NH group. The *ab initio* results were combined with semiempirical quantum chemistry calculations to produce a data set used for the parameterization of a molecular mechanics energy function for retinal. Using the resulting molecular mechanics force field the hydrated retinal and associated bR protein environment were energy minimized and the resulting geometries examined. Two distinct sites were found in which water molecules can make hydrogen-bonding interactions: one near the NH group of the Schiff base in a polar hydrophilic region directed towards the extracellular side, and the other near a retinal CH group in a relatively hydrophobic region directed towards the cytoplasmic side.

To enable further investigations of internal hydration in bR and other systems a statistical mechanical formulation was derived that can be employed using molecular dynamics (MD) simulation to calculate the free energy of transfer of a small molecule from one environment to a specific site in another using molecular dynamics simulation.[25] The method was used to calculate the free energy of transfer of water molecules from the bulk to individual sites in the proton transfer channel of bR.[25] The channel contains a region lined primarily by nonpolar side-chains. The results obtained indicate that the transfer of water molecules from bulk water to this apparently hydrophobic region is thermodynamically favorable. The presence of two water molecules in direct hydrogen-bonding association with the Schiff base was also found to be thermodynamically allowed.

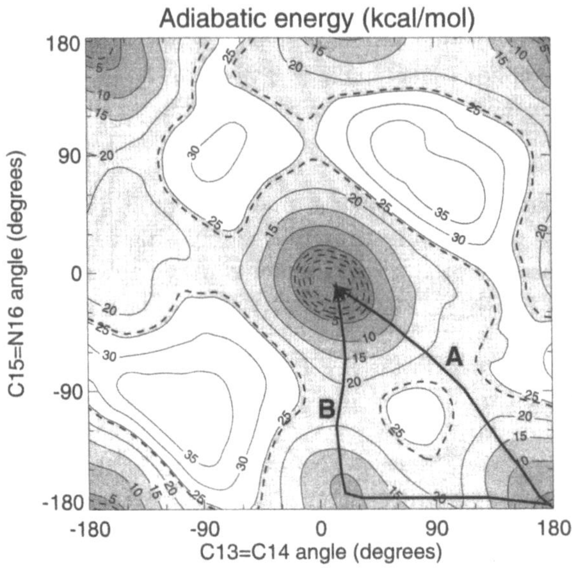

Figure 1. Potential energy map for rotation around two retinal double bonds, C13 = C14 (Φ_1) and C15 = N16 (Φ_2), in bacteriorhodopsin. The map was calculated "adiabatically" *i.e.*, constraining the double bonds while relaxing the other degrees of freedom of the system. Arrows indicate the lowest energy pathways for *trans* to *cis* isomerization of the two double bonds.

Once a complete structural model of bR is obtained theoretical investigations into the photocycles of this protein can be envisaged. One interesting aspect of this in bR is the phenomenon of dark-adaptation. When left in the dark for about one hour, bacteriorhodopsin reaches an equilibrium state. In this dark-adapted state, two species of bR can be found. One of these, which makes up ~1/3 of the total protein population, contains *all-trans* retinal. The other species, which makes up the remaining ~2/3, contains an isomerized *(13,15)cis* retinal, in which the C13 = C14 and C15 = N16 double bonds are *cis*.[26] The relative populations of the two forms in dark-adapted bR suggest that the difference between their free energies is lower than thermal energies at 300K < k_BT = 0.6 kcal/mol, where k_B is Boltzmann's constant and T is the temperature.

Under such circumstances molecular simulation can be used to examine factors influencing the free-energy difference between *all-trans* and *(13,15)cis* and to quantify factors determining the preferred pathway for conformational isomerization between the two forms.

A two-dimensional energy map as a function of rotation around the C13 = C14 and C15 = N16 bonds of retinal in bR is shown in Figure 1.[27–29] The map was calculated with an empirical, molecular mechanics energy function. The function contains a sinusoidal term for "intrinsic" rotation, which contributes a barrier to rotation due mainly to double-bond twist deformation, and other terms, which collectively represent the "environmental" effect on the rotational barrier. This map was calculated with an intrinsic dihedral term in the energy function that leads to a rotational barrier of about 15 kcal/mol around each of the two double bonds.[30] According to Figure 1 two pathways for the transition are possible, with approximately the same potential energy barrier. The first (labelled A) is a "bicycle-pedal" mechanism,[30,31] in which the two dihedral angles are isomerized simultaneously. The second (labelled B) involves sequential isomerization, in

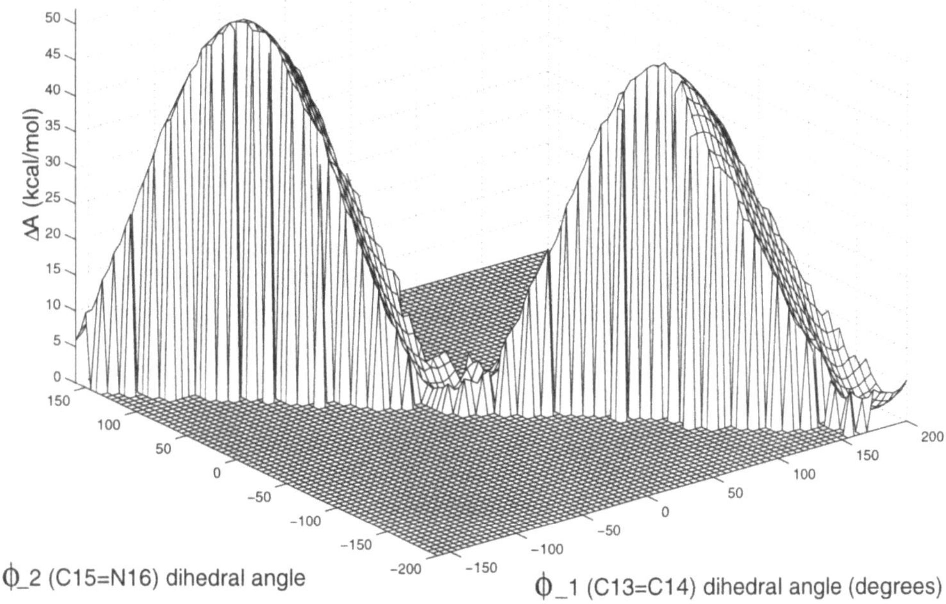

Figure 2. Free energy surface along the bicycle-pedal diagonal.

which one of the bonds is isomerized before the other. Changes of the intrinsic diedral term for rotations around C13 = C14 and C15 = N16 will modify the energy barriers for isomerization of retinal. A higher barrier will favour path B whereas a lower barrier will favour path A. Which of these two paths is actually taken in bR is unknown at present, due to uncertainities in the quantification of the intrinsic rotational term. The present results indicate that the path taken depends critically on the balance between environment and intrinsic rotational propensity.

Calculations of the free energy difference (ΔA) between the two conformers of retinal in dark-adapted bR have also been performed.[28,29] These involved "umbrella-sampling" simulations along the bicycle-pedal pathway *i.e.*, path A in Figure 2, although the end result is, in principle, independent of the pathway taken. The resulting free energy profile is shown in Figure 2. The free energy difference of interest is that between the minima in Figure 2. The calculated *(13,15)cis—all-trans* ΔA is –1.1 kcal/mol, a value within k_BT of experiment. This lower free energy of the *(13,15)cis* species in bR contrasts with the results of calculations *in vacuo*, where the *all-trans* species was found to be more stable by 2.1 kcal/mol.[27]

The influence of the retinal environment was investigated by calculating the free energy profile for modified or mutated bRs. A bR model where the water molecules present in the retinal binding site were deleted gave a ΔA of –4.2 kcal/mol *i.e.*, with the *(13,15)cis* species more stable than in hydrated bR by about 3 kcal/mol. On the other hand, mutation of resdues Asp212 and Asp85 to Ala, led to a ΔA of +1.9 kcal/mol, *i.e.*, with the *all-trans* species more stable than the *(13,15)cis*, also by about 3 kcal/mol compared to the wild-type bR.

CONCLUSIONS

The above calculations show how various groups of atoms can influence the conformational equilibrium in dark-adapted bR. In particular, the calculations suggest that the opposing action of the water and aspartate residues approximately cancels in wild-type, hydrated bR, leading to a similar free energy of these two species. More generally, the modelling and simulation of membrane protein structures and dynamics is still in its infancy. However, this field will be of growing importance as more and more sequences of membrane proteins are determined. And as structural research progresses, the investigation in turn of associated dynamical and functional properties can also be expected to gain importance.

ACKNOWLEDGMENTS

We thank E. Pebay-Peroula for providing the crystallographic structure of bR.

REFERENCES

1. J. Deisenhofer, O. Epp., K. Miki, R. Huber, and H. Michel, *Nature*, 1985, **318**, 618.
2. R. Henderson, J.M. Baldwin, T.A. Ceska, F. Zemlin, E. Beckmann, and K.H. Downing, *J. Mol. Biol.*, 1990, **213**, 899–929.
3. E. Pebay-Peyrouola, G. Rummel, J.P. Rosenbusch, and E.M. Landau, *Science*, 1997, **277**, 1676–1681.
4. Y. Kimura, D.G. Vassylyev, A. Miyazawa, A. Kidera, M. Matsushima, K. Mitsuoka, K. Murata, T. Hirai, and Y. Fujiyoshi, *Nature*, 1997, **389**, 206–211.
5. N. Foloppe, M. Ferrand, J. Breton, and J.C. Smith, *Proteins: Structure, Function and Genetics* 1995, 22(3), 226–244.
6. J. Breton, E.J. Bylina, and D.C. Youvan, *Biochemistry*, 1989, **28**, 6423–6430.
7. M.H. Vos, F. Rappaport, J.C. Lambry, J. Breton, and J.L. Martin, *Nature*, 1993, **363**, 320.
8. S.J. Robles, J. Breton, and D.C. Youvan, *Science*, 1990, **248**, 1402.
9. N. Foloppe, M. Ferrand, and J.C. Smith, *Chem. Phys. Lett.*, 1995, **242**, 238–243.
10. B. Svensson, C. Etchebest, P. Tuffery, P. van Kan, J. C. Smith, and S. Styring *Biochemistry* 1996, **35**, 14486–14502.
11. D. Oesterhelt and W. Stoeckenius, *Nature (London), New Biol.*, 1971, **233**, 149–152.
12. K.J. Rothschild, P.V. Argade, T.N. Earnest, K.-S. Huang, E. London, M.-J. Liao, H. Bayley, H.G. Khorana, and J. Herzfeld, *J. Biol. Chem.*, 1982, **257**, 8592–8595.
13. R.A. Mathies, S.W. Lin, J.B. Ames, and W.T. Pollard, *Annu. Rev. Biophys. Biophys. Chem.*, 1991, **20**, 491–518.
14. T. Mogi, L.J. Stern, N.R. Hackett, and H.G. Khorana, *Proc. Natl. Acad. Sci. USA.*, 1987, **85**, 5595–5599.
15. T. Mogi, L.J. Stern, T. Marti, B.H. Chao, and H.G. Khorana., *Proc. Natl. Acad. Sci. USA.*, 1988, **84**, 5595–5599.
16. L.J. Stern and H.G. Khorana, *J. Biol. Chem.* 1989, **264**, 14202–14208.
17. T. Marti, H. Otto, T. Mogi, S.J. Rösselet, M.P. Heyn, and H.G. Khorana, *J. Biol. Chem.*, 1991, **266**, 6919–6927.
18. G. Papadopoulos, N. Dencher, G. Zaccaï, and G. Büldt, *J. Mol. Biol.* 1990, **214**, 15–19.
19. P. Hildebrandt and M. Stockburger, *Biochemistry*, 1984, **23**, 5539–5548.
20. H.J.M. De Groot, S.O. Smith, J. Courtin, E. van der Berg, C. Winkel, J. Lugtenburg, R.G. Griffin, and J. Herzfeld, *Biochemistry*, 1990, **29**, 6873–6882.
21. G.S. Harbison, J.E. Roberts, J. Herzfeld, and R.G. Griffin, *J. Am. Chem. Soc.*, 1988, **110**, 7221–7227.

22. H. Deng, L. Huang, R. Callender, and T. Ebrey, *Biophys. J.*, 1994, **66**, 1129–1136.
23. M. Nina, B. Roux, and J.C. Smith, in *Structures and Functions of Retinal Proteins*. J.L. Rigaud editor. Colloque INSERM/John Libbey Eurotext Ltd., 1992, **221**, 17–20.
24. M. Nina, J.C. Smith, and B. Roux, *J. Mol. Struc. (THEOCHEM)*, 1993, **286**, 231–245.
25. B. Roux, M. Nina, R. Pomes, and J.C. Smith, *Biophysical Journal*, 1996, 670–681.
26. G.S. Harbison, H.J.M. de GRoot, R. Geghard, J.M.L. Courtin, J. Lugtenburg, J. Herzfeld, R. Mathies, and R.G. Griffin. *Proc. Natl. Acad. Sci. USA*. 81 1706–1709 (1984).
27. J. Baudry, S. Crouzy, B. Roux, and J.C. Smith. *J. Chem. Inf. Comp. Sci.* 37 1018–1024 (1997).
28. J. Baudry, S. Crouzy, B. Roux, and J.C. Smith. Biophys. J. 76 1909 (1999).
29. S. Crouzy, J. Baudry, J.C. Smith, and B. Roux. J. Comp. Chem. 20 1644 (1999).
30. I. Logunov and K. Schulten. *J. Am. Chem. Soc.* 118 9727–9735 (1996).
31. A. Warshel. *Nature*. 260 679–683 (1976).
32. B. Roux, M. Nina, R. Pomes, and J.C. Smith. *Biophys. J*, 69 1554–1563 (1996).

SHOX HOMEOBOX GENE AND TURNER SYNDROME

E. Rao and G. A. Rappold

Institute of Human Genetics
University of Heidelberg
Im Neuenheimer Feld 328, 69120 Heidelberg

Growth is a fundamental aspect in the development of an organism. Height represents a multifactorial trait, influenced by both environmental and genetic factors. Recently, research has focused on the genetic aetiologies of height. With an incidence of 3 in 100, growth failure is fairly frequent and accounts for a large number of cases that require medical attention. Growth is a highly complex process controlled by many gene products. Consequently, many different mutant genes may lead to growth failure. Growth hormone deficiency and growth hormone receptor defects, as well as mutations in genes leading to skeletal disorders have been shown to cause a short stature phenotype. Taken together, however, these disorders account for only a small percentage of cases, leaving the vast majority unexplained to date, termed "idiopathic" short stature.

A role for the human sex chromosomes in growth has been suggested and at least two different loci controlling growth have been described. This was deduced from genotype-phenotype correlations in male and female individuals and in patients with sex chromosome abnormalities. A well-known and frequent (1:2500 females) chromosomal disorder, Turner syndrome or Ullrich-Turner syndrome (45, X), is also consistently associated with short stature. Significant clinical variability exists in the phenotype of females with Turner syndrome; among the more frequent somatic features are congenital lymphedema, webbed neck, aortic coarctation and renal anomalies. Short stature, however, is a consistent finding and together with gonadal dysgenesis is considered to be the leading symptom of this disorder. Thus patients with TS suffer from a spectrum of problems which are thought to be due to monosomy of genes common to the X and Y chromosome. Consequently, haploinsufficiency of such genes would be the most likely mechanism of this syndrome.

In 1959 Ford et al. showed that the loss of one X chromosome leads to the phenotype of Turner syndrome. In 1965 it became obvious that the loss of only the short arm of the X chromosome already showed the full-blown phenotype of Turner syndrome,

Genomics and Proteomics, edited by Sándor Suhai.
Kluwer Academic / Plenum Publishers, New York, 2000.

suggesting that genes for gonadal dysgenesis, short stature and the various somatic Turner abnormalities would reside on Xp. In 1984, it was shown that the growth-controlling locus would resign on Xp22.3 (Figure 1). In the following years cytogenetic studies have provided further evidence that terminal deletions of the short arms of either the X or the Y chromosome consistently lead to short stature. Fifty chromosomal rearrangements involving Xp22 have been reported that localize the gene responsible for short stature to the pseudoautosomal region (PAR1). We have narrowed down the critical interval for a short stature locus to a 170 kb region within PAR1 by deletion mapping of patients with short stature. This interval was deleted in 36 individuals with short stature and different rearrangements on Xp22 or Yp11.3. The deletion was not detected in any of the relatives with normal stature and in a further 30 individuals with rearrangements on Xp22 or Yp11.3 with normal height (Rao et al., 1997). We have established a cosmid contig encompassing the 170 kb interval for short stature. By exon trapping a novel homeobox-containing gene, SHOX (short stature homeobox-containing gene) was identified (Figure 2).

SHOX generates at least two different transcripts designated SHOXa and b by alternative splicing (Rao et al., 1997). The structure of the isoform SHOXa was confirmed by another group (Ellison et al., 1997). The alternatively spliced products differ in the C-terminal end and probably modify the phosphorilation and binding of the protein. The two different isoforms exhibit a distinct tissue distribution. SHOXa is widely expressed whereas SHOXb expression is more restricted and predominantly found in bone marrow fibroblasts. A complete alignment of the two cDNAs with the genomic DNA established the exon-intron boundaries. The gene is composed of 7 exons ranging in size between 58 bp and 1.1 kb (Figure 3). Exon 2 contains a putative CpG island, the start codon and part of the 5′ UTR. A stop codon as well as the 3′ UTR is present in each of the

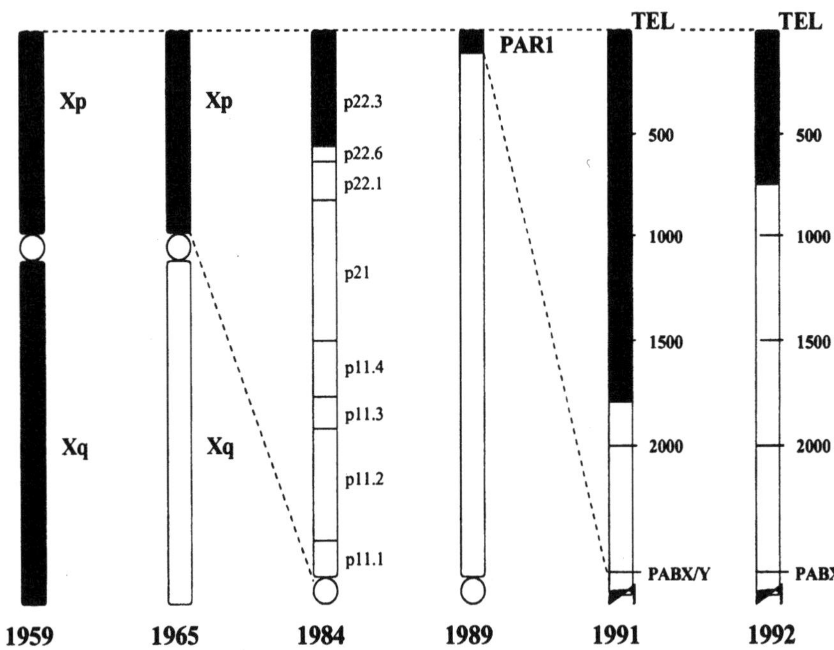

Figure 1. Search for the small stature locus on the human X chromosome. Deletion mapping and genotype—phenotype correlations on patients with sex chromosome anomalies have assigned a growth controlling loci into the pseudoautosomal region (PAR1).

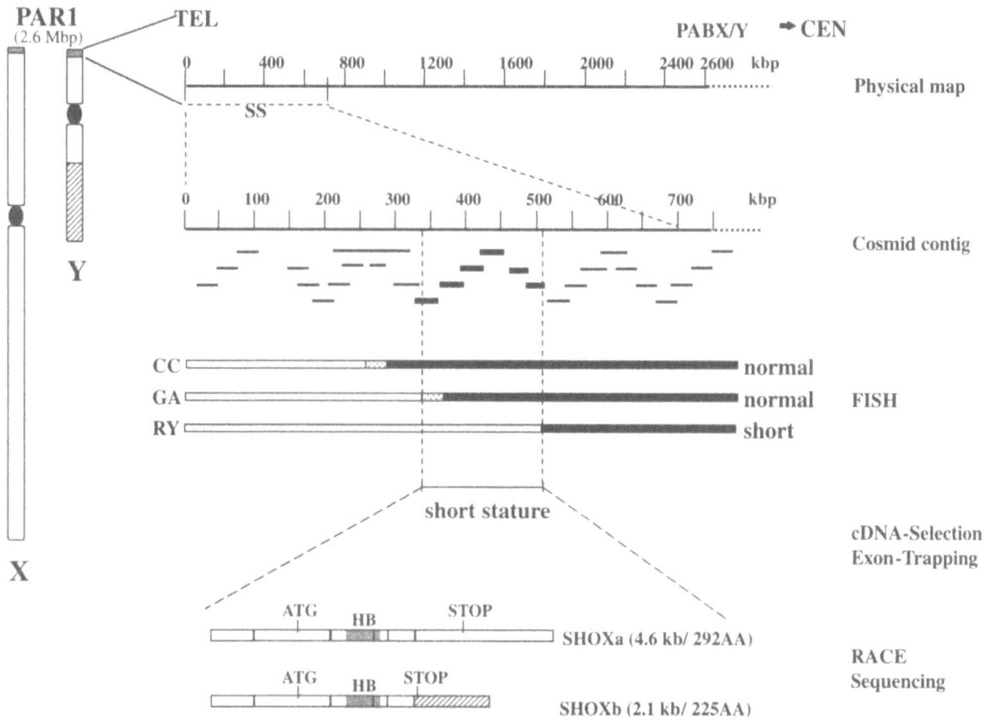

Figure 2. Scheme outlining the positional cloning project of the short stature gene SHOX. On the left side, the sex chromosomes X and Y are depicted. They share a 2.6 Mbp large identical region, termed pseudoautosomal region. The obligate recombination between the X and Y is restricted to PAR1. On the top, the physical extension of PAR1 is shown. The distal 700 kb adjacent to the telomere were considered as critical interval for the growth controlling locus. Cosmids from the cosmid contig covering this region were used as probes for FISH mapping on metaphase chromosome of patients with a partial monosomy of the PAR1 region. The 3 most crucial patients CC, GA and RY are shown; black bars denoting the regions where the respective cosmids are present in double dose; white bars indicating single dose (haploidy). Patient GA with a terminal deletion (white bar) and normal height defined the distal boundary of the critical region (340 kb from the telomere). The short stature gene can not reside in the terminal region because monosomy of this locus has no phenotypical consequence. Patient RY with a larger terminal deletion and short stature defined the proximal boundary at 510 kb from the telomere.

To search for transcription units within the smallest critical interval cDNA selection and exon trapping was carried out. Three exons were isolated which all belonged to the same gene, termed SHOX. 5′ and 3′ RACE and sequencing was carried out to isolate the full-length gene.

alternatively spliced exons 6a and 6b. As the name SHOX indicates, the gene contains a homeobox encoded by exons 3 and 4. The homeobox is 180 bp in size and encodes for 60 aminoacids of a homeodomain. The predicted homeodomain of SHOX shows a typical helix-turn-helix structure, a DNA-binding motif found in a number of developmental and tissue-specific transcription factors. Homeodomain proteins have been characterised extensively as transcriptional regulators involved in pattern formation in both invertebrate and vertebrate species. Because of its predicted structure SHOX probably plays a role as transcriptional regu-lator as well.

The criteria imposed on a Turner gene are escape of X-inactivation and X-Y homology. This indicates that a Turner gene has to be expressed in double dosis, from the active

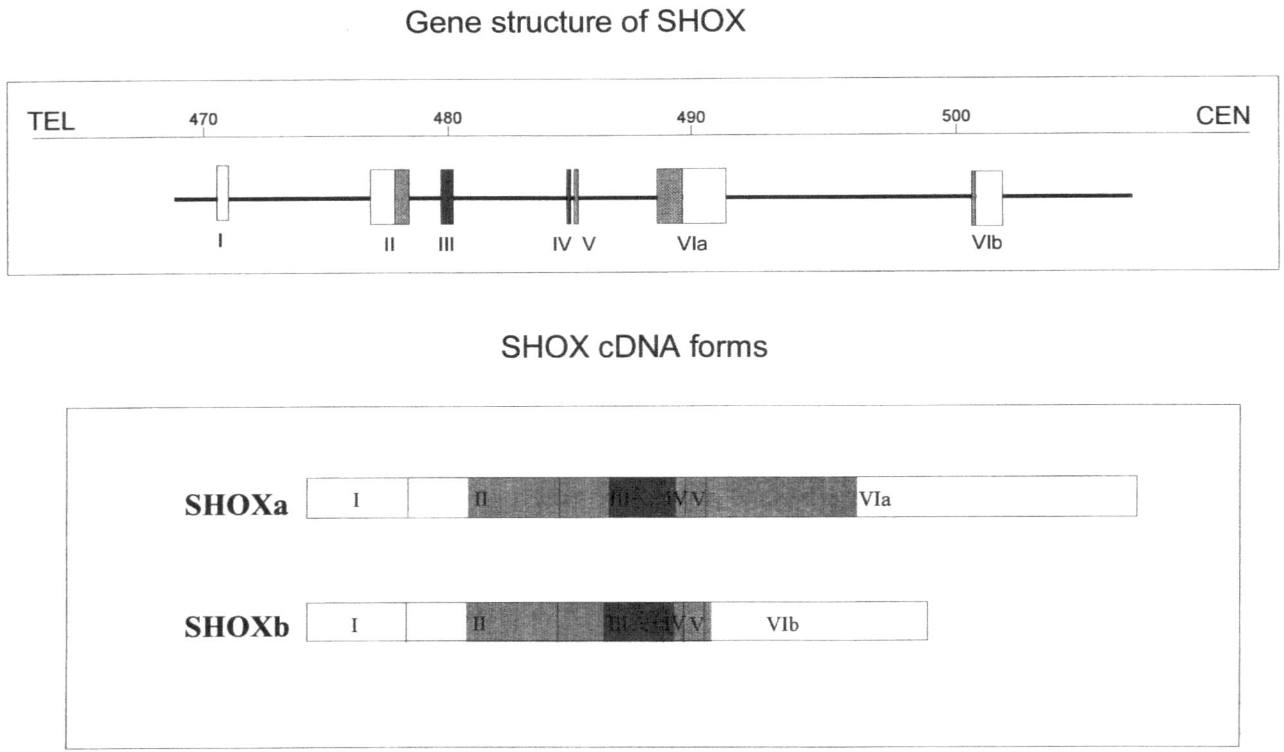

Figure 3. Gene structure of SHOX. The SHOX gene is composed of 7 exons and spans a genomic region of 40 kbp (top). SHOX generates two alternatively spliced products called SHOXa and SHOXb, that differ only at their 3′ end (bottom).

and the inactive X chromosome, and from the Y chromosome. To assess the transcriptional activity of SHOXa and SHOXb on the X and Y-chromosome we used RT-PCR. Various hybridoma cell lines containing as the only human chromosome the active X-chromosome (4X, 2X, 1X), or only the Y-chromosome, or only the inactive X-chromosome were used. All cell lines revealed the expected amplification product providing clear evidence, that both SHOXa and SHOXb escape X-inactivation and are expressed from both the X- and the Y-chromosome.

Deletion analysis, gene expression and escape of X-inactivation gave us convincing evidence, but no definite proof for SHOX as short stature gene. In order to demonstrate the causal relationship between SHOX and short stature, we needed to test for SHOX mutations in individuals with idiopathic short stature, normal karyotype and normal hormonal parameters. The assumption was that in this patient group a small percentage would present SHOX point mutations or deletions.

To define the frequency and mutation spectrum of SHOX we have started a large study looking for SHOX mutations in patients with idiopathic short stature. Based on our present results we estimate that 1% of all patients with idiopathic short stature may carry a SHOX mutation (unpublished results). The SHOX mutations found so far represent mutations leading to protein truncation (Figure 4).

How frequent are SHOX mutations in the general population and how frequent within the population of idiopathic short stature patients? Point mutations were detected in patients with idiopathic short stature with a frequency of 2 in 240. Considering an incidence of 2–3% for idiopathic short stature in the population, one may postulate an incidence of 1 SHOX point mutation in 4000 individuals in the population. Gene deletions due to large deletions in the PAR1 or terminal deletions of Xp or Yp or the complete loss of an X-chromosome in Turner patients occur with a frequency of 1 in 2500 females. Consequently, it is possible that one out of 2000 living individuals may carry a SHOX mutation.

Recently, haploinsufficiency of SHOX has been shown to also lead to short stature in Leri-Weill syndrome (Belin et al., 1998; Shears et al., 1998). Leri-Weill syndrome or dyschondrosteosis represents a mesomelic short stature syndrome with a characteristic

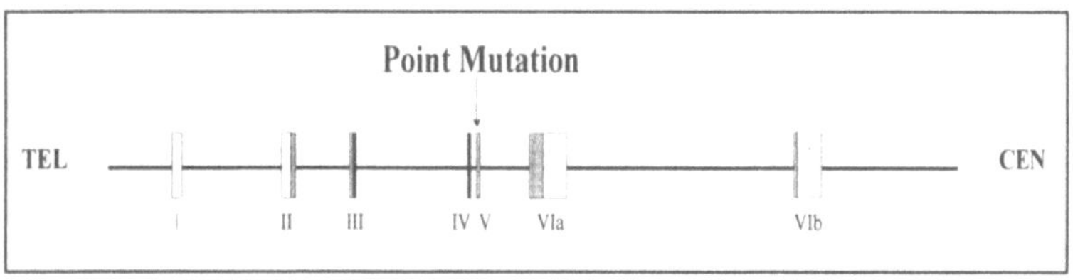

Exon V:

C → T (674) : Arg → Stop Idiopathic short stature
C → T (674) : Arg → Stop Idiopathic short stature
C → T (674) : Arg → Stop Leri-Weill Syndrome
C → G (688) : Tyr → Stop Leri-Weill Syndrome

Figure 4. Identical point mutations in the SHOX gene leading to either idiopathic short stature or Leri-Weill syndrome.

deformity of the forearm (Madelung deformity). Homozygous SHOX deletions lead to Langer dwarfism, an extreme form of short stature (Belin et al., 1998; Shears et al., 1998). It was shown that the identical mutation at bp 674 can lead to idiopathic short stature or to Leri-Weill syndrome (Figure 4).

SHOX is widely conserved between species. The most striking similarity of SHOXa (72%) was found to a mouse gene, OG12X, encoding a protein with unknown function (Rovescalli et al., 1996). Both proteins have identical homeodomains, suggesting that both mouse and human genes represent true homologs. The homology of SHOX is, however, not restricted to the homeodomain, but extends at a lower level into the C- and N-terminal regions. At the N-terminus the homology is rather low (50%), whereas the C-terminus shares a homology of 86%. Putative sites for phosphorilation, a recognition sequence for an SH3 domain and amino acids stretches similar to other transactivating transcription factors are highly conserved. At the C-terminus a so-called "Aristaless-domain" can be found, which was previously seen in a number of homeodomain proteins which was suggested to play a role as transactivating domain. SHOX therefore meets all requirements of transcription factors: it contains a homeodomain to bind DNA and a putative C-terminal transactivating domain for interaction with other proteins for example involved in the transcription machinery.

We have mapped OG12X and it does not reside on the mouse X chromosome, similar to two other genes from the human PAR1 region, CSF2RA and IL3RA. The autosomal location of this gene in mice and the lack of short stature in X0 mice is consistent with SHOX having a role in human short stature. Mutations within OG12X would be predicted to produce mice with growth failure in the heterozygous state. By gene knock-out experiments, this hypothesis can now be readily tested. More insights into the functioning of this gene will also be gained by generating mice with homozygous mutations.

In man, we have also isolated a SHOX-related gene termed SHOX2 (formerly called SHOT). The homeodomains of SHOX, SHOT and OG12X (mouse) were shown to be identical, suggesting that the three proteins bind to equivalent DNA elements and therefore trigger similar physiological pathways (Blaschke et al., 1998). In situ hybridization of the mouse gene OG12X on sections from staged mouse embryos detected highly restricted transcripts in the craniofacial, brain and heart tissues, with the strongest expression in the developing limb (Blaschke et al., 1998). The expression analysis of OG12X using two different probes has not revealed any differences in the spatial or temporal expression pattern of the two isoforms, OG12Xa and OG12Xb. Expression of OG12X was detected during embryonic development in mesoderm derivates that contribute to bone and cartilage formation and in ectodermal tissues including brain, spinal cord, and ganglia. The highest levels of expression were found in mesodermal tissues of the face involved in nose and palate formation, the developing eyelid and tissue surrounding the optic nerve, as well as in the developing heart mesoderm and in the mesoderm condensing around the chondrification centers of the limb.

REFERENCES

Belin, V., Cusin, V., Viot, G., Girlich, D., Toutain, A., Moncla, A., Vekemans, M., LeMerrer, M., Munnich, A., and Cormier-Daire, V. (1998) SHOX mutations in dyschondrosteosis (Leri-Weill syndrome). Nature Genet. 19, 67–69.

Blaschke, R.J., Monaghan, A.P., Schiller, S., Schechinger, B., Rao, E., Padilla-Nash, H., Ried, T., and Rappold, G.A. (1998) SHOT, a SHOX-related homeobox gene is implicated in craniofacial, brain, heart and limb development. PNAS 95, 2406–2411.

Ellison, J.W., Wardak, Z., Young, M.F., Robey, P.G., Laig-Webster, M., and Chiong, W. (1997) PHOG, a candidate gene for involvement in the short stature of Turner syndrome. Hum. Mol. Genet. 6, 1341–1347.

Rao, E., Weiss, B., Fukami, M., Rump, A., Niesler, B., Mertz, A., Moroya, K., Binder, G., Kirsch, S., Winkelmann, M., Heinrich, U., Breuning, M.H., Ranke, M., Rosenthal, A., Ogata, R., and Rappold, G.A. (1997) Pseudoautosomal deletions encompassing a novel homeobox gene cause growth failure in idiopathic short stature and Turner syndrome. Nature Genet. 16, 54–63.

Rovescalli, A.C., Asoh, S., and Nirenberg, M. (1996) Cloning and characterization of four murine homeobox genes. Proc. Natl. Acad. Sci. USA 93, 10691–10696.

Shears, D.J., Vassal, H.J., Goodman, F.R., Palmer, R.W., Reardon, W., Superti-Furga, A., Scambler, P.J., and Winter, R.M. (1998) Mutation and deletion of the pseudoautosomal gene SHOX cause Leri-Weill dyschondrosteosis. Nature Genet. 19, 70.

14

A FEATURE-BASED APPROACH TO DISCRIMINATION AND PREDICTION OF PROTEIN FOLDING

Boris Mirkin[1,2] and Otto Ritter[1,3]

[1]Department of Molecular Biophysics, DKFZ, Heidelberg, Germany
[2]DIMACS, Rutgers University, Piscataway, New Jersey
[3]Biology Department, Brookhaven National Laboratory
Upton, New York

1. INTRODUCTION

Proteins are the fundamental molecules of life, they form both the structural and the functional building blocks of cells in all living organisms. Proteins are also the most complex molecules. Their three-dimensional structures are extremely diverse, and so are their biological functions. Elucidation of the relationship between protein sequence, structure, and function is one of the most significant open problems in science and technology. Despite the growing body of observed data and partial theories, there are still no satisfactory methods which would enable us to analyze, characterize, and predict a protein's structure and/or function with regard to its known primary sequence.

There are two fundamentally different approaches to the problem, and a continuum of their combinations. The first approach is based on theory and models from physics and chemistry, the other is based on computational and statistical analysis of observed data. This paper belongs to the second direction.

From the data analysis perspective, we can observe, first, that there has not been much achieved in terms of producing reasonable feature spaces except for those directly related to amino acid sequence similarity data, and, second, the specifics of emerging problems yet have not been addressed in full. One of the specific issues is related to difference between prediction and description. Typically, prediction is based on a description of the phenomenon in question in the framework of a prediction model. However,

Genomics and Proteomics, edited by Sándor Suhai.
Kluwer Academic / Plenum Publishers, New York, 2000.

such a description not always can be exploited for substantive analyses. For instance, the neural network based predictions of protein folding classes involve neural net models that are useless from the point of view of the biologist: they may give a good prediction, but the net structure and corresponding weights can provide no insights into the biological nature of the genomic processes, at least currently. That means that the problem of description in proteonomics becomes a problem on its own, which must be addressed accordingly.

In this paper, we restrict ourselves to considering the protein spatial structure learning problem as a problem in the framework of an existing classification of proteins, SCOP by Hubbard, Murzin, Brenner and Chothia (1997), that is suitable for our purposes because of both its substantive contents and availability in the Internet. The problem of learning SCOP classes is discussed in section 2 along with a review of most popular machine learning approaches. Then we suggest a strategy for solving the problem to involve the following four dimensions: (a) proteins are considered in terms of a feature space rather than in terms of their sequence/structure similarities; (b) the subgroups are supposed to be logically described by the features; (c) resampling is used as a learning tool rather than a testing device; and (d) multiscale representation of a sequence for searching through the feature space is used for learning. Although at least some of these dimensions already have been exploited in literature, their combination and, moreover, application to proteins have never been undertaken, to our knowledge. Discussion of the issues related to the four dimensions is done in four sections 3 through 6. In section 7, the suggested method, APPCOD, is formulated as a computational algorithm, and, in section 8, examples of computations performed with it are presented. Conclusion, section 9, contains a brief discussion of future work.

2. THE PROBLEM OF LEARNING SCOP CLASSES

The problem of learning of a folding group can be put in a more or less statistical way by exploiting a system that has incorporated some knowledge of structural and evolutionary similarities among proteins. Such a system is Structural Classification Of Proteins (SCOP) currently maintained as a website, http://scop.mrc-lmb.cam.ac.uc/scop/. SCOP is a multi-level classification structure involving currently (in its version 1.37, we have been dealing with) over a dozen thousand protein domains from proteins in Protein Data Bank (see Figure 1).

There have been a number of works reported recently with regard to learning the upper classes of SCOP, the structural α/β classes (first level of the hierarchy) and folding classes (second level of the hierarchy), see Chou et al. (1998), Dubchak et al. (1995, 1999) and references therein.

We consider the problem of learning SCOP classes in the following setting. Let M be an interior node in the SCOP hierarchy (the root, a class, fold, superfamily, or a family) and S another node descending from M (a class, fold, superfamily, family, or a protein domain, respectively). The problem is to find a rule separating protein domains in S from other proteins in M. If, for instance, M is the root (all proteins) and S the Globins family, the problem is to learn what separates Globins from the other proteins. Finding such a rule can be trivial: for instance, the definition of S in SCOP separates it in M, but it is based on the spatial structure information which should not be involved in the learning rule. In this paper, we concentrate on those separating rules that involve only terms related to the primary structure of proteins.

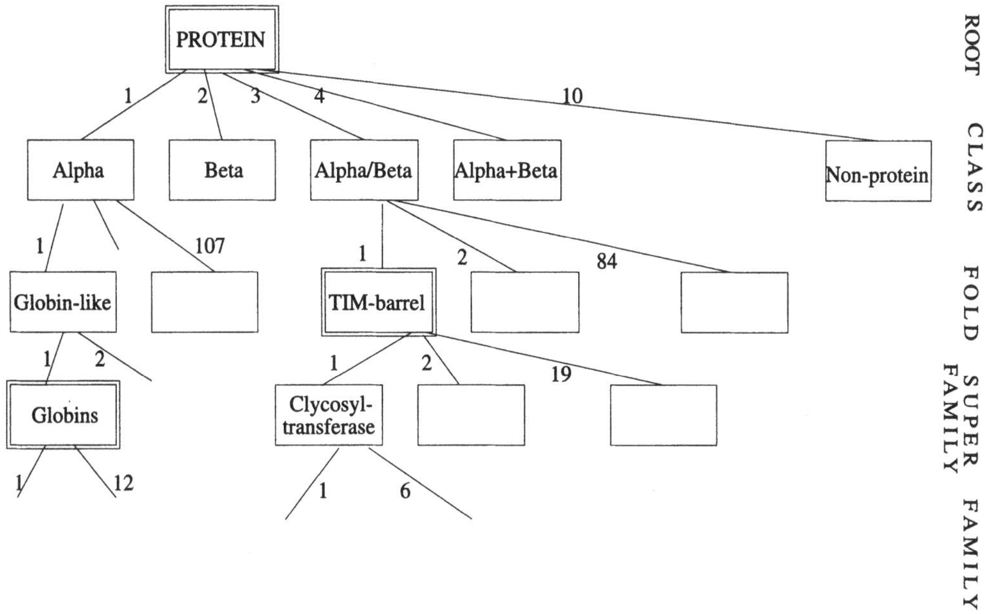

Figure 1. Upper levels of SCOP with some of the classes labelled by their names, some by their numbers (at the links).

To formalize the problem, we have selected an approach that is based on features of the amino acids constituting the sequence. This way the entire issue of sequence/structure alignment, that attracts much of attention in the literature, becomes irrelevant. Thus all the gapping and weighting problems in comparing sequences are dropped out with the amino-acid-feature-based approach. Moreover, within this approach we apply a procedure that allows for formulation of an explicit and easy-to-interprete separation rule in a logical format that differs from most work in the field. Yet one more distinctive feature of our approach is that reliability of the rule is achieved by using the standard cross-validation techniques in the process of learning, not just in testing as usually. These four aspects along with related references will be reviewed in the following four sections.

3. FEATURE-BASED DESCRIPTIONS

The problem of prediction of 3D structure of a protein based on its sequence has been met by researchers with a number of developments (for reviews, see Fisher and Eisenberg (1996), Jones (1997), Westhead and Thornton (1998) and paper by W. Taylor in this volume). These developments are mostly based on what is called threading: aligning a target sequence along a known 3D structure (fold) with a scoring function reflecting the energy needed for the sequence to fold into the 3D structure. The best match from a library of known folds then is considered to be the "parent structure" which is then updated according to specifics of the target sequence, to get a predicted fold (comparative modelling). Knowledge of (predicted) secondary structure elements is helpful in threading as well as use of the multialignment approach.

With the development of extensive classifications of the protein 3D structures, such as SCOP by Hubbard et al. (1997) or CATH by Orengo et al. (1997), a possibility emerged of exploring a less challenging task of predicting just a folding class, a node in the classification tree, not a fold. The difference is that a folding class can be represented with just the list of its members, without any explicit reference to their spatial structures. Most efforts so far have been devoted to prediction of the four major structural folding classes (all-α, all-β, $\alpha + \beta$, and α/β) by the amino acid composition. The amino acid composition of a sequence is the vector of relative contents (per cent) of each of the 20 amino acids in the sequence. Most recent results (75% of correct predictions overall in a jackknife experiment) and a review of the subject can be found in Chou et al. (1998). The state of the art in machine learning of the structural folding classes in the space of percentages of the secondary structure elements is described in Zhang and Zhang (1998).

More recently, an effort has been made in predicting other folding classes as, say, of the level of Fold in SCOP (see Figure 1) by the team led by S.-H. Kim (see Dubchak et al., 1995, 1999). For prediction, they use a neural network based procedure for learning a class in a number of feature spaces with consequent averaging predictions based on a majority rule. Each of the feature spaces exploited in this approach consists of 21 features derived from an amino acid property categorized priorly in three categories. For instance, the amino acids can be categorized in categories of hydrophobic, hydrophilic, or neutral residues (according to their hydrophobicity), or in helix, extended (stranded), or coiled residues (according to the element of the secondary structure they belong to), etc. Having such a categorisation of a feature done, three variables are defined as percentage of each of the categories along the sequence; three more variables are defined as relative frequencies of change from one to another category along the sequence. Each of the remaining fifteen features is defined as the percentage of the length of the sequence when one of the three categories reaches one of the following five points: the beginning, the 25%, 50%, 75% of the number of residues in the category (quantile), and the end position of the category's occurrences. With thus produced feature spaces the method fared relatively well in the CASP2 experiment (Levitt, 1997).

Some other feature spaces have been exploited in Bachinsky et al. (1997) and Hobohm and Sander (1995), though for less unambigous problems in maintaining of protein family structures. A most comprehensive set of features have been considered so far by Wei, Chang and Altman (1998): five classes of protein features have been specified as those based on atom, residue, chemical group, secondary structure, and other (such as B-factor or solvent accessibility). However, these spaces have been employed in Wei, Chang and Altman (1998) for within protein structure learning problems, such as recognition of protein sites, rather than for fold recognition.

In our study, we selected a residue-based feature space involving the order of residues in the sequence. Of four hundred features of amino acids available in the data base AAindex (see Kawashima et al., 1998), we selected six features related to size or charge of an amino acid in ProtScale library of the ExPASy website (see Appel et al., 1994). More specifically, the variables are molecular weight, MW, bulkiness, BU, (consensus) hydrophobicity, HY by Eisenberg-Schwarz-Komarony-Wall (1984), polarity [two scales], PG by Grantham (1974) and PZ by Zimmerman–Eliezer–Simha (1968), and membrane buried helix parameter, MB by Rao-Argos (1986), all from ProtScale at http://www.expasy.ch/tools/#primary. These features, averaged over arbitrary intervals of the protein sequences, constitute our feature space. The sequence intervals are measured per cent as, for instance, the intervals [0,20] and [50,75] related to the initial 20% of the sequence length and positions between its midpoint and the three-quarter point, respectively. An advantage of this space, as well as of other spaces mentioned, is that no alignment of proteins is needed to measure the features. On the other hand, the features are relevant to some biochemical and biophysical properties of the sequences and their sites. Of course, these properties cannot express such intuitively defined features as "hydrophilic residues are allowed in the barrel interior but not between the barrel and helices" in Murzin and Bateman, A. (1997), p. 108–109, that involve primary, secondary and tertiary structure elements. However, a somewhat simpler feature like "hydrophilic residues are allowed in the middle of a sequence but not in its end" can be easily formulated in terms of our feature space: just the average hydrophobicity of an interval at the end must be larger than that of an interval in the middle. The comparison can be done by arithmetically combining the features involved. It means, actually, that the ratio of the features must be larger than 1. This way employing more complex logic expressions such as in PROGOL by Finn et al. (1998) can be avoided.

4. DESCRIPTION AS A TOOL IN MACHINE LEARNING

4.1. The Problem

Having a protein feature space specified, every protein can be represented as a multidimensional point in this space. Thus our learning problem can be reformulated in terms of Figure 2: given a set of points M (all circles), find a way to separate those representing a subset $S \subset M$ (black colour).

The separation rule should allow to test any sequence to appear and assign it with black or white colour depending on its predicted belongingness to S. The quality of the separation rule can be characterized by the misclassification rate expressed with two quantities: the numbers/proportions of false positives and false negatives. The false negative is an instance of entity from S erroneously diagnosed by the separation rule as being from outside, that is, a black circle recognized as a white one. In contrast, the false

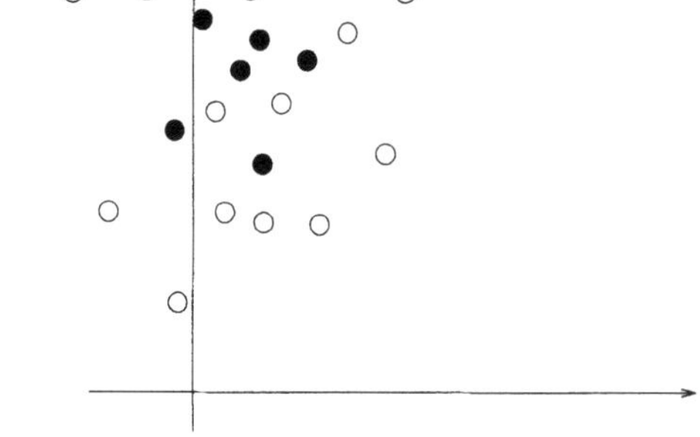

Figure 2. Black circles to be discriminated from the set of all/other circles.

positives are entities from outside of S that have been recognized as belonging to S. In Kubar, Bratko, and Michalski (1998) these errors are referred, respectively, as the error of omission and error of commission. The less the misclassification rates, the better the rule.

The rule should be as good as possible in terms of the misclassification rates. Yet one more requirement is that the rule should be interpretable in such terms that the biologists could use it for further advancements in understanding the nature of proteins.

Of a number of machine learning approaches developed so far, let us discuss in the subsequent three subsections those most popular: (1) discriminant functions, (2) neural networks, and (3) conceptual descriptions.

4.2. Discriminant Function

This is an intensional construction in the feature space R^n: a function $G(x)$, $x \in R^n$, is referred to as a discriminant function (separating surface) for a subset $S \subset M$ if $G(y_i) \geq \pi > 0$ for all $i \in S$ while $G(y_i) < \pi$ for all $i \in M - S$, where π is a threshold. Here, y_i are the feature space representations of the proteins $i \in M$.

Usually, the discriminant function is a hyperplane $G(x) = \Sigma_k c_k x_k$ where x_k are components of x. Linear functions can separate only convex sets, which relates the theory of discriminant hyperplanes to the theory of convex sets and functions.

Such a hyperplane is shown on Figure 3. Obviously, the solution is too simplistic to be used in practical calculations for general fold recognition.

The theory of discriminant functions, developed by R. Fisher before the World War II, was initially a part of the mathematical multivariate statistics heavily loaded with probabilistic estimates, but currently it is moving into somewhat less rigid area of machine

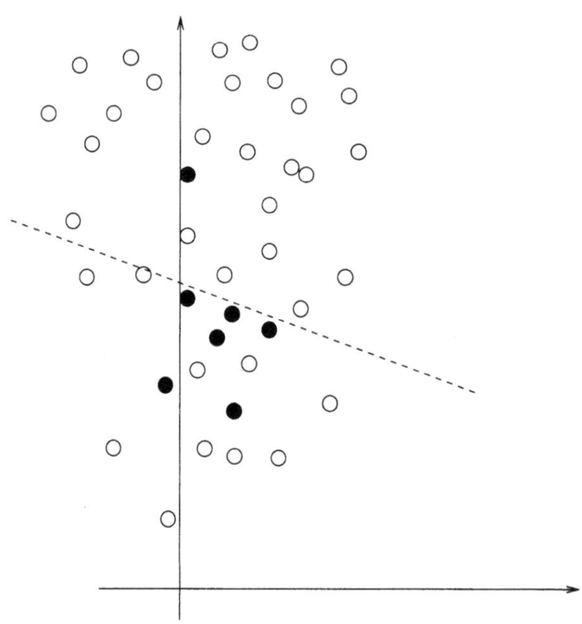

Figure 3. A linear discriminant function admitting one false negative and many false positives.

learning to involve more heuristical approaches for transformation of the variables (for a review and references, see Hand, 1997).

There exists an approach related to finding convenient transformation rules, formerly called the *potential function* method, which is quite general and, at the same time, reducible to linearity. The potential function $\psi(x, y)$ (currently, *kernel*) reflects similarity between x and y and, usually, is considered a function of the squared Euclidean distance between x and y such as $\psi(x, y) = 1/(1 + ad^2(x, y))$ or $\psi(x, y) = exp(-ad^2(x, y))$ where a is a positive constant.

The potential discriminant function for a class $S \subseteq M$ is defined then as the average potential with respect to the points of S as the prototype points: $G_S(x) = \Sigma_{i\in S}\psi(x, y_i)/|S|$. The class of such potential functions is quite rich. It appears, for instance, that using $\psi(x, y) = exp(-ad^2(x, y))$ with sufficiently large a as the potential function in $G_S(x)$, the function $G_S(x)$ can separate any S from $M - S$, see Andrews (1972). This shows that the approach is theoretically justified as applicable to any set learning problem.

This approach is related with appropriate transformations of the feature space, as follows. Potential functions depending on x and y through the Euclidean distance between them, can be represented as $\psi(x, y) = \Sigma_p\lambda^2{}_p g_p(x)g_p(y)$ where $\{g_p(x)\}$ is a set of the so-called eigen-functions. This allows transforming the classification problem into a so-called "straightening space" based on the transformed variables $z_p = \lambda_p g_p(x)$. In this straightening space, the potential function becomes the scalar product, $\psi(x, y) = (z(x), z(y))$, which makes all the constructions linear. This approach is being currently modified to the format of machine learning in terms of the so-called support vectors that serve as entities "modelling" the discriminant plane, see Schlkopf et al. (1998), Vapnik (1998). It seems that eventually, when more is known of the relevant feature transformations and computationally effective methods, the approach should be explored in the framework of the general fold learning.

4.3. Neural Networks As a Learning Tool

Artificial neural networks provide for an extremely effective learning framework.

A formal neuron is a model for the neuron, a nerve cell working like an information-transforming unit. The neuron provides a transformation of an input vector $x = (x_k)$ into the output signal, $y = \theta(\Sigma_k c_k x_k - \pi)$, where $\theta(v) = 1$ if $v > 0$ and $\theta(v) = 0$ if $v \leq 0$. Actually, the neuron discriminates between two half-spaces separated by the hyperplane $\Sigma_k c_k x_k = \pi$. Frequently, the threshold output function θ is substituted by the so-called *sigmoid* function $\theta(v) = 1/(1 + e^{-v})$ which is analogous to the threshold function but is smooth and more suitable for mathematical derivations. An interpretation of the formal neuron: the components k represent synapses excited on the level x_k; the weight c_k shows relative importance of the synapse to the neuron; the neuron fires output if the total charge $\Sigma_k c_k x_k$ is higher than the neuron threshold π.

Sometimes the neuron is considered to have the identity output function $\theta(v) = v$ thus performing just linear transformation $\Sigma_k c_k x_k$; this is called *linear neuron*.

A single hidden layer neural net (see Figure 4) defines a double transformation, $\theta[\Sigma_{t=1}^{T} w_t\theta_t]$ where $\theta_t = \theta(\Sigma_k c_k^t x_k^t - \pi_t)$, of the input vectors x^t, t = 1, . . . T, into an output through T "hidden" neurons. Such a net has two important properties: 1) it can be used to approximate any continuous function; 2) it can be used to separate any subset S of a given set of normed vectors y_i, $i \in M$. To resolve the latter problem, let us take $T = |M|$ to consider any $t = 1, \ldots, T$ as a corresponding element of M; then, for any neuron $t \in$

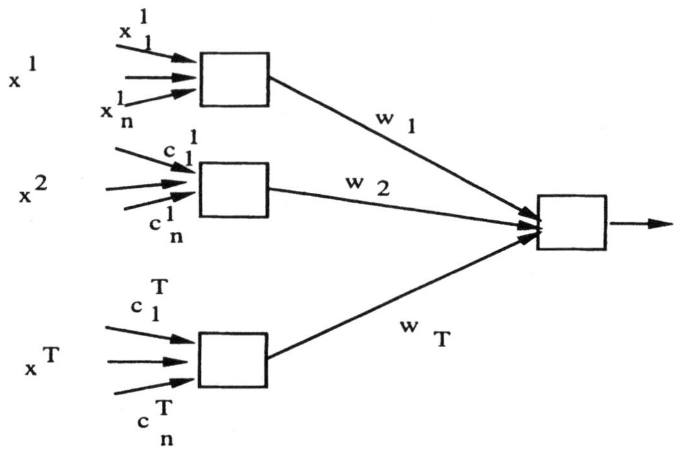

Figure 4. A neural network with a hidden layer.

M, let its weight vector $c^t = y_t$. Obviously, the maximum of the scalar products (c_t, y_i) with regard to $i \in M$, in this case, is reached for the only $i = t$. Thus, fixing π_t between this maximum value and the second maximum, we have $\phi(t, y_i) = \theta(\Sigma_k c_k^t y_k^i - \pi_t) = 1$ if and only if $i = t$; $\phi(t, y_i) = 0$ when $i \neq t$. Then, taking $w_t = 1$ for $t \in S$ and $w_t = -1$ for $t \in M - S$, we get the desired output.

The first neuron-like learning algorithm, the *perceptron*, was proposed by F. Rosenblatt (see Nilsson, 1965) to learn a linear separating surface when S can be linearly separated. The perceptron perceives the entity points coming in sequence, starting from an arbitrary coefficient vector c and changing it after every try with the following rule: add to c (respectively, subtract from c) all erroneously classified points from S (respectively, from $M - S$), thus turning c toward a direction between the summary points of S and $M - S$. This guarantees that the method converges.

In a multilayer perceptron, a similar learning idea requires sequential weight changes layer-by-layer starting from the output layer (back-propagating). The *back-propagation* learning process proposed by Rumelhart, Hilton and Wilson (1986) is, actually, a version of the method of steepest descent (a.k.a. hill-climbing) in the theory of minimization as applied to the square-error criterion $E(c) = \Sigma_i(x_i - d_i)^2$ where x_i and d_i are actual and ideal (shown by the "teacher") outputs of the neuron network, respectively. Let the output of the p-th neuron in a layer equal $x_{pi}(c) = \theta_p((c_p, y_i))$, where y_i is the input to the neuron from the preceding layer when the input to the network is i-th given point. Change of the weight vector c_p in the neuron is controlled by the equation $\Delta_i c_p = \alpha\delta_{pi}y_i$ where α is the step size factor (usually, constant) and $\delta_{pi} = -\theta_p'((c_p, y_i))(x_i - d_i)$ if this is the output layer, or $\delta_{pi} = -\theta_p'((c_p, y_i))\Sigma_q\delta_{qi}c_{qi}$ for a hidden layer where q represents the next (more close to the output) layer's suffix.

Similar ideas, involving though different objective functions and propagation formulas, are put in the so-called Kohonen networks approach (see Kohonen, 1995).

As any local optimization method, in a particular computational environment, the back-propagation method can converge to any local optimum or even not converge at all, which does not hamper its great popularity.

However, in the protein fold recognition problem, this method suffers of a major draw-back: the solution is quite difficult to interpret because the network coefficients are not that easy to express in operational terms related to proteins.

4.4. Conceptual Description

The most popular conceptual description tool, the regression or classification or decision tree, divides the entity set M in a tree-like manner, each division done by dividing a feature's range into two or more categories (for a review and references, see Salzberg, 1998). For instance, Figure 5 presents a decision tree found with initially partitioning the vertical axe (feature y) in two intervals, A and B, and then by partitioning A into $a1$ and $a2$ along the same axe while dividing B into $b1$ and $b2$ along the horisontal axe (feature x). The four regions of the space have simple conceptual descriptions such as "$a2$ of y" or "B of y and $b1$ of x". The separating rule defined by this tree is obvious: black circles are in "$a2$ of y", and white circles are in other regions, as shown in Figure 5. Each of the regions provides for one false positive.

Deciding on which partition and by which of the variables to do is based on a scoring function that measures the degree of uniformity of the resulting regions with regard to S and $M - S$ as shown in Figure 5. Such a measure is provided in most popular programs, C4.5 by Quinlan (1993) and CART (see Breiman et al., 1984), by the entropy or index Gini, respectively; the program OC1 in Salzberg (1998) allows for any of seven goodness measues. These measures take into account both, the group to separate, S, and the rest, $M - S$, as equally important so that the tree is designed to maximize overall accuracy of the tree. This may lead to somewhat skewed results since the trees "tend to optimize accuracy on the larger class" (Salzberg, 1998, p. 190), which is especially important in the problem under consideration where S typically is smaller than $M - S$.

The conceptual description techniques developed so far do not pay much attention to the cases like ours. We are interested in getting a description for the S only: $M - S$ can be very much nonhomogeneous and its conceptual description may have no meaning at all! For instance, a somewhat better decision tree in Figure 5 could have been found if the second split of the axe y ($a1$ and the rest) had been the first one to collect as many as possible of the black circles in the same region. Some of the recent data mining techniques do cover our case as those developed in Klösgen (1996), Srikant, Vu and Agraval

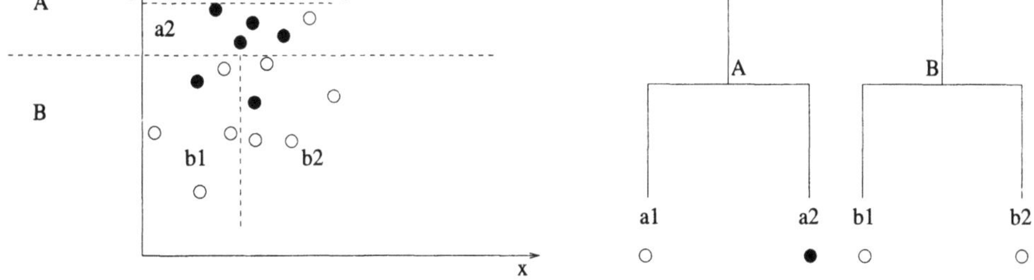

Figure 5. Regression tree shown both ways, in the feature space (on the left) and as a decision tree (on the right).

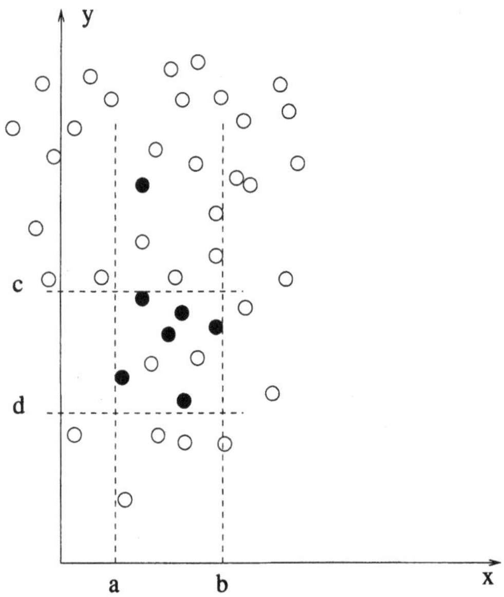

Figure 6. Conjunctive description: (1) false negative, (2) false positives.

(1997), Kubar, Bratko and Michalski (1998); however, the algorithms proposed in these publications such as Explora in the latter reference do not come out with a description of S: they are oriented just on finding conceptually described potentially overlapping subsets of S, however many of them and how complex the overlaps are.

In our view, most relevant to our problem would be getting a conceptual description such as that presented in Figure 6. The black circle set, according to this description, is in the rectangle $a \leq x \leq b$ & $c \leq y \leq d$. The errors, according to this separating rule, are one false negative and two false positives.

4.5. Approximate Conjunctive Description

An algorithm for finding a conceptual description of S in M without getting a description of $M - S$, as just a conjunction of feature intervals, has been described in Mirkin (1999). This algorithm follows a traditional idea in data mining: the larger the difference between within-S average of a feature and its grand mean, the more important the feature is for separating S Klösgen (1996). Moreover, in Mirkin (1999), it is shown that a proper measure of difference is the difference squared, because it is an additive item of the Pythagorean decomposition of the data scatter into the part explained by the group S and the unexplained part, thus bearing a proportion of the overall data variance.

However, the difference squared may be not compatible with the criterion of accuracy of the description (minimum of errors), which implies that the simple ordering of the features according to decrease of their within-S and grand mean differences is not enough to get a good description. An algorithm suggested in Mirkin (1999) for finding a conceptual description of S within M uses both of the criteria, the differences and the errors, in the following way.

According to this procedure, a number of within-cluster-range-based terms is to be initially collected into a logical conjunction (first phase), after which redundant terms are excluded one-by-one (second phase). The first phase goes along the ordering of features according to their difference-between-averages squared, starting with the empty set of conjunctive terms. Any particular feature is considered along the ordering to decide whether or not it should be included in the conjunction. It is included only if this decreases the number of false positives. The process stops when there are no features left in the ordering or when the number of false positives becomes zero (or any other prespecified threshold value). The second phase goes in the opposite direction along the terms collected at the first phase to decide whether a single term under consideration can be removed from the collection or not. It is removed if its removal does not change the number of false positives. This method of consecutively applying of what is called forward search and backward search selection strategies in machine learning, builds an APProximate COnjunctive Description and thus will be called APPCOD-1 in the remainder.

APPCOD-1 performs rather well when the classes are located in different zones of the original feature space. The method works poorly in the domains where group S is spread over the feature space so that it cannot be separated into that box-like cylinder volume which corresponds to an output conjunction.

However, the method's performance can be improved by transforming and combining the variables. To do this, denote by A the set of original features and B a set of features from A participating in the APPCOD-1 generated conjunctive description. Obviously, B is subject to stopping thresholds in APPCOD-1: the number of items in the resulting conjunction and the minimum error level admitted. Denote by S(A, B) the set of all pair-wise products, xy, ratios, x/y and y/x (the latter being denoted as $x\backslash y$), sums, $x + y$, and differences, $x - y$, for all $x \in$ A and $y \in$ B. Then iteratively perform APPCOD-1 on A, S(A, B), S(A, B(S(A, B)), etc. This way of iteratively combining the variables has two properties: first, it exploits the best APPCOD derived features for combining; second, it does not lose the original wealth of features since at each step all original features are involved in the combining process, too.

Such a process, called APPCOD-2, usually leads to drastic reduction of the number of false positives in the APPCOD-1 results.

To the stopping thresholds of APPCOD-1, APPCOD-2 adds one more threshold: a bound on the number of combining operations involved in a compound variable. The more operations, the more complex a variable. Thus, the same question emerges as in all other separation techniques: how complex a separation rule can be accepted by the learner? Potential remedies related to such "simplicity" principles as the minimum description length principle by Rissanen (1989) should be tried.

Still, both APPCOD-1 and APPCOD-2 may produce unsatisfactory descriptions when group S, on average, does not differ from the entire set M; that is, when no feature has its within-S average different from its grand mean. In such cases, other features should be tried or S can be divided into its "extreme parts" to be described separately.

5. RESAMPLING AS A TOOL FOR MODEL SELECTION

Resampling is a procedure oriented at testing and improving reliability of data based estimators and rules. Generically, it is very simple: take a sample of the data set under consideration, apply your algorithm to the sample (training phase) and test its result on the rest (testing phase). Perform this many times and average the results. Quality of the average test results shows performance of the algorithm in question in changing environments. The better the average test results, the more reliable is the algorithm. The estimates and rules derived from random samples (with replacement) are called sometimes bootstrap estimates/rules, Efron and Tibshirani (1993). Two particular schemes of resampling have become most popular: the k-fold cross-validation and leave-v-out cross-validation. In the k-fold cross-validation, the data set is divided into k subsets of (approximately) equal size. The algorithm is trained k times, each time leaving one of the subsets out of training, and using only the omitted subset to test. If k equals the sample size, this is called "leave-one-out" cross-validation. "Leave-v-out" is a more elaborate and expensive version of cross-validation that involves leaving out all possible subsets of v cases.

When the cross-validation is performed over just randomly generated subsets, it is frequently called the "split-sample" or "hold-out" method. Leave-one-out sampling scheme is frequently used for what is called jackknife estimating, when an estimate is calculated not just from the data set as usually, but from each of the subsets involved and then is averaged over the subsets. The difference between the jack-knife estimate and that standard one shows the bias of the latter.

Leave-one-out cross-validation often works well for continuous error functions such as the mean squared error, but it may perform poorly for discontinuous error functions such as the number of misclassified entities. In the latter case, k-fold cross-validation is usually preferred (with a value of 10 for k, typically).

To apply APPCOD to a sample, we need to solve an estimation problem: find left and right bounds of within-S feature intervals. When S is small, its samples are even smaller and they may and do give much biased estimates of the within-S feature ranges: in a sample from S, the left bound increases while the right bound decreases, which leads to a smaller within-S feature interval. As shows Figure 7, where all circles represent set M, black circles its subgroup S, and the double black circles, a sample from S, the sample-based within-S interval (in this case, (c, d)) is always smaller than the entire-set based within-S interval (in this case, (a, b)).

Changes will also occur in the intervals of white, $M - S$, circles located to the right and to the left of the within-S interval.

This effect can be utilised for correcting the reduced sizes of the feature intervals by exploiting the points located between them or their quantiles. This is a problem that has not been yet properly addressed. In our algorithm, we use a somewhat rough estimate suggested by I. Muchnik (1998): the sample-based within-S interval is scaled with a constant factor (depending on the set and sample sizes) to make it up to the entire set estimate.

Since the number of misclassified entities may be different over different samples, resampling can be used not only for testing, but for learning, too: when a model or algorithm or parameters of an algorithm are selected to minimize the error over a number

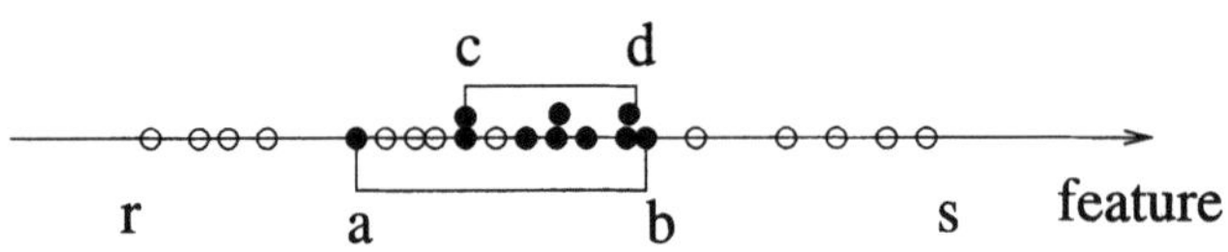

Figure 7. Decreasing the within-group interval under sampling: from (a, b) to (c, d).

of computations. This is called model selection. Examples of model selection can be seen in Efron and Tibshirani (1993), pp. 243–247 (selection of decision tree sizes), and Salzberg (1998), pp. 193–194 (averaging probabilites obtained with split-sample-based decision trees).

Model selection is especially applicable to APPCOD found decision rules because of their simplicity: any rule is just a set of feature intervals. Yet implementation of this idea is not straightforward because of the stage of combining the features, APPCOD-2: different samples may and do lead to diferent compound features, which may effectively prevent comparing and, thus, averaging different APPCOD decision rules.

To overcome this hurdle, the process of learning via resampling can be divided into two phases. The first phase is to collect most effective combined variables over different samples by applying APPCOD-2 to them. These most effective combined variables are included then in the feature space, so that the initial set of features is supplemented with the set of most effective combined variables.

The updated feature set serves as an input to the second phase, which starts with multiple application of APPCOD-1 (no combined variables this time!) to samples from the updated data set. Then all APPCOD-1 found decision rules are comparable subsets of features from the same feature set.

To average these subsets, a majority rule is applied: a feature is put in the averaged decision set if and only if it occurs in more than 50% of the decision subsets. As seems to be well known, this majority rule, actually, follows the optimality principle for a related reconciling problem: for a given set of subsets $A_k \subseteq B$ ($k = 1, \ldots, n$), find such an $A \subseteq B$ that minimizes $\Sigma_{k=1}^{n} d(A, A_k)$. The distance $d(A, A_k)$ here is just the number of features that belong to either A or A_k but not both. As can be easily proven, the majority rule generates a global solution to the reconciling problem.

Still one more step is to be performed: averaging intervals of the majority-selected features by averaging their left and right bounds over all occurrences of the features in the last series of APPCOD-1 found solutions.

This two-phase process involving both APPCOD-2 and APPCOD-1 will be referred to as APPCOD-3.

The two latest steps in APPCOD-3 (averaging feature subsets and averaging interval bounds) can be done over not all of APPCOD-1 generated solutions, but only with regard to those of them that have shown a better performance. For instance, in one of APPCOD-3 versions, the program does twenty APPCOD-1 computations with the updated feature set, and then excludes the worst ten of the results, thus leaving only best ten solutions for further averaging, which greatly improves the error rate in the final conjunction.

One more comment about the presented two-phase implementation of the resampling based model selection is that at each of the phases, the best solutions are selected according to an error rate criterion. These are selecting the best combined features, at the first phase, and selecting the best descriptions, at the second phase. To define what is the best, we need to specify a scoring function as a weighted sum of the number of false positives and the number of false negatives. The weights are compromise factors of the two items. The larger the factor at, say, the number of false negatives, the more important this part of the criterion is in the selected solution. Also, when sizes of S and $M - S$ are much different, the relative numbers, proportions of false positives and false negatives, can be taken to make these two comparable.

6. FEATURE REFINING APPROACH

With the scale of protein sequence length measured per cent, the number of distinguishable intervals of the scale is about 5050: 1 interval of length 100, 2 intervals of length 99, . . . , and 100 intervals of length 1. Thus, each feature of amino acids may be enveloped at the sequence level into 5050 interval features (the averages along the intervals), which leads to about 30,000 sequence features in total (for we have six amino acid features to use as described in section 3).

Although this size of the feature space is rather challenging at the present time, it is not only the computation time that prevents us from brute force search through the space as is: the feature weighting, the pillar of APPCOD computations, is easy to do as it involves comparing only two averages, the grand mean and within-S mean, per feature. More important an issue, in this case, is that abundance of variables may be misleading when the data in hand are as ambiguous as the data base in SCOP which lacks clear criteria of classification. In such cases, typically, redundant features may get not well-grounded importance and lead to false conclusions. This is why we prefer here to search through the feature space by modelling the scientific approach of refining interesting features by exploring them in greater detail. This way, a considerable reduction of computation is achieved, too.

In the beginning, each feature is measured rather roughly along only three intervals covering 0–40, 60–100, and 30–70 percents of the sequence. The denotation a–b here refers to an interval starting at a% of the sequence and ending at b% of it. The sequences are considered long enough to use the percentage scale without much distortion. Given the six amino acid features, this makes an eighteen-dimensional feature space. If a distinctive description of S in this space is reached, the computation stops. Otherwise, if the APPCOD-3 based feature intervals give a too high misclassification rate, the features occurred in the description should be refined to get a look ot the situation in more detail. Let A be the set of features in the space to which APPCOD is applied and $B(A)$ denotes the set of features that occurred either in APPCOD-3 generated description or in formulas for combined features added at the first stage of APPCOD-3 to the original feature set A. Then each feature from $B(A)$ is to be supplemented in A by three features defined over detailed intervals. If an $f \in B(A)$ is measured over interval a–b, which can be denoted by (f, a, b), the three new features are defined as $(f, a, a + c)$, $(f, b - c, b)$ and $(f, a + c/2, b - c/2)$ where $c = (b - a)/2$, the half length of a–b. This way, new features added to A cover the starting, middle and ending halves of the interval a–b. The process of adding the three refined features to the feature space will be referred to as the *refining*. If, with A thus updated, APPCOD-3 produces nothing better in terms of the misclasssification rate, the process stops and the previous result is considered as the final one. Otherwise, with the APPCOD description of S improved, next iteration of the same feature refining process is executed.

The misclassification rate traditionally is defined as the total number of errors divided by the size of set M of all entities under consideration. This can be misleading when the size of group S is significantly smaller than that of $M - S$. If, for instance, there are 30 false positives and 20 false negatives when $|M| = 1000$, then the misclassification rate is $(30 + 20)/1000 = 5\%$ which correctly reflects the individual rates of errors when $|S| = 500$: the relative commission error, the number of false positives related to $|M - S|$, is $30/500 = 6\%$, and the relative omission error, the number of false negatives related to $|S|$, is $20/500 = 4\%$, which is in concordance to the total missclassification rate. However, when $|S| = 40$ so that $|M - S| = 960$, the individual error rates are

30/960 = 3.1%, keep with previors for false positives, and 20/40 = 50%, for false negatives. In this situation, the total error rate, 5%, just hides the fact of unsatisfactory level of recognition of the entities from S. This is why we use the averaged individual rate,

$$a\frac{fn}{|S|}+(1-a)\frac{fp}{|M-S|},$$

as the scoring (quality) function of the conjunctive descriptions. Here, fp and fn are the numbers of false positives and false negatives, respectively, and a is the coefficient of compromise between the individual rates. In all further examples, $a = 0.5$.

To cope with the increase of the size of the current feature space in the feature refining process, the features that have not been involved in updating the space for two or more consecutive iterations, can be removed from the space at all.

This process of iteratively applying APPCOD-3 to the refined feature spaces will be called APPCOD-4.

The APPCOD-4 resembles methods of processing images on the multiscale basis, when some parts are looked at in greater detail than others are. Recently, this area of research has received a great impetus in mathematical developments to make the image and signal processing algorithms more effective (see Starck, Bijaoui and Murtagh, 1998). Although genomic data bases yet have not as many entities as visual data, some analogues of the techniques described in Starck, Bijaoui and Murtagh (1998) may be applicable to multiscale processing genomic data, too.

7. FOUR STAGES FOR APPCOD

Let us collect and put the four APPCOD stages described in previous sections, each on top of the preceding one, more formally. The input consists of a set of protein sequences, M, its subset S to be separated with a conjunctive description, and a feature set A represented by amino acid features and intervals of the sequence, over which the features are averaged.

APPCOD-1: Finding a conjunctive description

Step 0. (Initial Setting) Set prior constraints on the maximum number of terms and the number of false positives admitted.

Phase 1. (Forward Feature Space Search) Collecting COD, within-S feature ranges, in a conjunctive description according to the feature weight ordering with skipping those not-decreasing the number of false positives.

Phase 2. (Backward Feature Space Search) Removing from COD those feature ranges that do not or least affect the number of false positives, to get the number of terms (feature ranges) less than or equal to the prespecified in Step 0 quantity.

APPCOD-2: Combining features for conjunctive description

Step 0. (Initial Setting) Set a prior constraint on the maximum number of operations involved in a combined variable.

Phase 1. (Feature Selection) Apply APPCOD-1.

Phase 2. (Combining Features) Combine the APPCOD-1 selected features with the original ones by using arithmetic operations and goto Phase 1.

APPCOD-3: Getting a reliable conjunctive description

Step 0. (Initial Setting) Set a prior proportion of the entities for random sampling, number $n1$ of the samples, compromise coefficients for numbers of false positives and false negatives, and number $n2$ of voting descriptions.

Phase 1. (Producing Descriptions) Apply APPCOD-2 to each of $n1$ samples.
Phase 2. (Updating the Feature Space) Add to A combined features (if any) from the best of $n1$ descriptions at Phase 1.
Phase 3. (Finding Averaged Description) Apply APPCOD-1 to the best $n2$ samples from the updated data set and take the averaged description (majority features with averaged intervals).

APPCOD-4: Multiscale refining search through protein features.
Step 0. (Initial Setting) Set the original feature space A: a number of amino acid features averaged over three rough intervals, 0–40, 30–70, and 60–100, covering all the sequence length. Put counter of the number of iterations, count-iter = 1, and the error score, 100%.
Phase 1. (Producing Descriptions) Apply APPCOD-3 with A the space. Compare results with those kept from the preceding iteration. If the current error score is smaller, go to Phase 2. If not, end with the results of iteration number count-iter – 1.
Phase 2. (Updating the Feature Space) Add to A features obtained by refining the features occurred in the conjunctive description or in combined features added to A in Phase 1. Remove features that have not been involved neither in resulting conjunctive descriptions nor in added combined variables through t consecutive iterations of the alforithm. Add 1 to count-iter and go to Phase 1.

In further experiments we, typically, set the description complexity (maximum number of the items in a conjunction) equal to 3, the feature complexity (the maximum number of arithmetic operations involved in a compound variable) equal to 1, while maintaining the sample size (in APPCOD-3) on the level of 40% of the entire set and the interval extension factor (also in APPCOD-3) on the level of 55%.

8. EXAMPLES OF PROTEIN GROUPS DESCRIBED

In this section, results of the following three experiments will be presented:

1. Comparison of the results on separating TIM-barrel from the root of SCOP according to a secondary-structure-based feature set;
2. Separating TIM-barrel and α/β-chains in our feature space;
3. Separating subgroups on the lower levels of SCOP.

Though far from a complete assessment of the approach, this may give some preliminary insights into its comparative advantages and shortcomings.

Although currently available protein data bases may contain a significant part of all proteins in living organisms, their contents reflect rather interests of substantive researchers in genomics than the distribution of proteins in life and, in this sense, are biased: some proteins are overrepresented with their much similar multiple versions while the others are underrepresented or not presented at all. This is why most research in bioinformatics is done via so-called non-redundant data bases. A non-redundant data base contains only mutually different proteins: each two have no more than, for instance, 30% (or, 40%) of the sequence identity for the aligned subsequences longer than, for instance, 80 residues. A non-redundant data base, thus, has only one copy of each (sub)family of similar protein sequences, which yeilds one more feature: the cardinality of the non-redundant data base is relatively small amounting to just several hundred, not many thousand as in an original data base, entities.

Such a non-redundant data base was built by Dubchak et al. (1999) to test their approach in recognition of 128 SCOP folding classes. One of the feature spaces designed in Dubchak et al. (1999) to reflect the distribution of the secondary structure elements (helix, strand and coil) along the amino acid primary sequences (as reviewed in section

3), has been used in our study. We restricted ourselves to only one (from 128) folding class considered in Dubchak et al. (1999), TIM-barrel (see Figure 1), for it is the most numerous in the non-redundant data base of SCOP folding classes (24 entities against the entire set of 516 proteins; in the original data base studied in 1996, the number of proteins was 607 (29, TIM-barrel), but 91 was then removed as corresponding protein structures disappeared from SCOP in 1998). This nonredundant data base is used in most of our experiments.

8.1. TIM-Barrel in a Secondary-Structure-Based Data Set

The neural network based prediction results for TIM-barrel reported in Dubchak et al. (1999): 31% false negatives and 19.8% false positives. These results also involve several feature spaces and a voting rule, in prediction, as well as the bootstrap testing.

In our computations, only the space defined by the categorisation of positions according to the type of the secondary structure element they belong to (helix, strand, or coil) as defined in section 3 is considered.

The results found with the feature complexity equal to 0 (only original features are present) are as follows. The majority averaged description found in a run of APPCOD-3 is this: $8.83 \leq comS \leq 21.21$ & $59.85 \leq quaH \leq 87.10$, where comS is the percentage of strand positions in the sequence and quaH is the length of the sequence, per cent, before 25% (quantile) occurrence of an alpha-helix position. The misclassification rates are: 21.2% of false positives and 8.3% of false negatives. Another run of APPCOD-3 yeilded the same majority features with slightly tightened bounds: $8.96 \leq comS \leq 20.82$ & $60.99 \leq quaH \leq 86.43$. This reduced the rate of false positives to 19.9% without increasing the rate of false negatives, 8.3%. One more run of APPCOD-3 has led to increasing the number of items in the majority conjunction to four (though only three items have been permitted in the individual descriptions) by adding features thrH (the length of the sequence, per cent, before 75% (quantile) occurrence of an alpha-helix position) and firC (the length of the sequence, per cent, before the first occurrence of a coiled position) to the two already appeared. The description this time is: $73.84 \leq thrH \leq 100$ & $36.56 \leq firC \leq 71.28$ & $7.44 \leq comS \leq 23.38$ & $59.36 \leq quaH \leq 89.43$. The accuracy of this description is: 18.2% of false positives and 4.2% of false negatives. This outperforms the results for TIM-barrel reported in Dubchak et al. (1999). The accuracy can be further enhanced by making the bounds (at least for the latter two features) tighten. By reducing the bounds for comS and quaH to those found at the second run, we could have achieved zero false negatives and less than fifteen per cent of the false positives.

The majority conjunction found when one arithmetic operation in combining features has been admitted is this: $54.57 \leq quaS—firS \leq 82.12$ & $0.11 \leq traH/comC \leq 0.36$ & $0 \leq traC/firS \leq 0.012$. Here, firS and quaS are, respectively, the lengths of sequences, per cent, before the first and 25% occurrences of strand positions; traH and traC are percentages of transitions between helix and coil and between strand and coil, respectively; and comC is the percentage of coiled positions in the sequence. The first combined variable has an obvious meaning of the relative length of the interval between first and 25% occurrences of strand positions. The meaning of other combined features should be examined by the specialists. The accuracy of the description is 15% of false positives and 8.3% of false negatives.

What is nice about these descriptions is that they do not require huge numbers of items or operations that are typically needed in neural networks to get good results.

8.2. TIM-Barrel and α/β Proteins in the Feature Space

The APPCOD-3 program applied to the nonredundant set of 516 proteins as M and 24 TIM-barrel proteins within as S, with the eighteen interval features (6 amino acid variables averaged over the three rough intervals), produced the following majority conjunction: $0.02 \le PG(60, 100) * HY(60, 100) \le 1.28$ & $14.97 \le MW(0, 40)/PG(60, 100) \le 23.38$ whose misclassification rates are: 169 false positives (error of commission 34.3%) and 1 false negative (error of omission 4.2%).

By refining the four interval features involved, twelve new variables were added, leading to an improved description: $0.04 \le PG(70, 90)/MW(50, 70) \le 0.07$ & $0.92 \le MW(0, 20)/MW(30, 50) \le 1.06$ & $127.22 \le MW(40, 60) + BU(60, 100) \le 143.34$, with the number of false positives equal to 117 (error of commission 23.8%) and the number of false negatives, 2 (error of omission 8.3%).

By refining the variables involved with simultaneously removing the features that have not been involved in the formulas in neither case, the space dimension has been upgraded to 34 to yield an upgraded majority description: $822.27 \le MW(30, 40) * PG(75, 85) \le 1097.58$ & $0.94 \le MW(0, 20)/MW(30, 50) \le 1.06$ & $116.43 \le MW(40, 50) \le 130.16$. This description involves mostly molecular weight on 10% and 20% intervals and admits 73 false positives (error of commission 14.8%) and 4 false negatives (error of omission 16.7%). No further refining improves this description whose quality is comparable with that reached in the space of the secondary-structure-related variables (in the previous subsection).

However, things become more murky when we move on to treating the set of all α/β protein domains. This set consists of 143 specimens in the nonredundant data set and it has been considered in either capacity, as M (with regard to TIM-barrel as S) or as S (with regard to the entire protein domain set).

The best description of α/β class in the set of all 516 protein domains via APPCOD-4 has been produced on just second step (with intervals of 20% length) and accounts for 216 false positives (error of commission 57.9%) and 3 false negatives (error of omission 2.1%). The enormous misclassification rate suggests that the subset of α/β protein domains is not homogeneous in the feature space. Thus, either the space should be changed or the APPCOD approach upgraded to deal with nonhomogeneous subsets by priorly partitioning them into homogeneous chunks.

Somewhat better results have been found at describing 24 TIM-barrel sequences within 143 α/β sequences. APPCOD-4 has produced the following majority description: $15.48 \le PZ(30, 70)/MB(60, 100) \le 43.83$ & $-6.21 \le MW(60, 80) - MW(50, 70) \le 5.72$ & $116.37 \le MW(30, 50) \le 129.25$ & $6.63 \le PG(70, 90) \le 8.37$ with 32 false positives (error of commission 26.9%) and 3 false negatives (error of omission 12.5%). This description involves features that are similar to those in the description of TIM-barrel within the entire protein domain set, above, which indicates again on the idea that the α/β sequences are dispersed over the entire data set in the space.

8.3. Treating Lower Layers of SCOP

We consider here two problems: description of α-Amylase (36 entities) as a subgroup of the TIM-barrel proteins in SCOP (512 entities) and description of all α- and β-haemoglobin chains (178 entities) within the class of all globin-like proteins (359 entities). The haemoglobin chain subgroup combines all subfamilies with codes 1.1.1.1.16 through 1.1.1.1.32 in SCOP. This experiment also can be considered as testing APPCOD against the bias in protein contents of the contemporary databases. We hope that the bias does not much affect the results since APPCOD is based on rather robust functions of samples such as the averages and proportions.

Alpha-Amylase within TIM-barrel. Starting with the rough three intervals, APPCOD-4 has produced the following averaged description: $1.60 \le \mathrm{HY}(0, 40) * \mathrm{PZ}(0, 40) \le 3.25$ & $0.09 \le \mathrm{HY}(0, 40) * \mathrm{MW}(60, 100) \le 0.12$. The accuracy of the conjunction is: 23 (4.8%) false positives and 0 false negatives.

After upgrading the features involved in the description by dividing the intervals in three parts (for instance, MW(60, 100) has been upgraded into MW(60, 80), MW(60, 80), and MW(70, 90)), the resulting description is: $118.02 \le MW(60, 100) - PZ(0, 20) \le 125.88$ & $1.36 \le PZ(0, 20) * HY(0, 40) \le 2.85$ & $0.18 \le HY(0, 40) \le 0.25$, which is more balanced and exact. There are 8 false positives (1.7%) and 1 false negative (2.8%).

In the nonredundant data set, there are 24 TIM-barrel proteins of which 4 are α-amylase. Applied to this data set, APPCOD-4 leads to description $0.18 \le HY(0, 40) \le 0.21$ that admits no false positives and just 1 (25% though) false negative. This latter description does involve a feature from the above, all-of-SCOP based, description, but it is much more rough than that above.

Haemoglobins in Globin-Likes. The group of 178 haemoglobin α- and β-chains within the globin-like folding class is characterized by APPCOD-4 with the following description: $7.56 \le \mathrm{PG}(60, 100) \le 8.05$ & $0.116 \le \mathrm{BU}(60, 100)/\mathrm{MW}(30, 70) \le 0.127$. This seems rather accurate: no false positives and 2 (1.1%) of false negatives. The meaning of the variables is not clear, especially with regard to the fact that they relate to different parts of the sequence.

9. DIRECTIONS FOR WORK

Traditionally, learning of protein folds is performed with regard to their physical properties expressed in terms of the energy and alignment with homologues. In this paper, an original approach to learning of protein classes has been described as based on different learning strategy, approximate conjunctive description, and different feature spaces, amino acid properties averaged along intervals. This approach appeals to the attractive idea of describing protein groups in a biologically meaningful way. The experimental results show that this strategy is promising and should be further explored.

There are two immediate directions for further developments: (1) enhancement of the APPCOD method and (2) creation and maintaining of a set of descriptions of SCOP classes in such a way that it may become a tool for further spatial and functional analyses.

In the first direction, more restricted analytical tools should be implemented to allow not all potential combinations of the features but only those regarded as appropriate by the user. This can be important when descriptions of other groups are to be taken into account. In particular, the hierarchical structure of SCOP should be incorporated in descriptions so that the APPCOD description of a group S within M could exploit the results of description of a larger group, S' within the same $M(S \subset S')$.

On the more technical side, the approach should involve more attention to compatibility of such criteria as the misclassification rate, simplicity of the description, the depth of scaling, etc. Care should be taken to overcome potential nonhomogeneity of an S versus M by dividing S into separate pieces getting potentially different descriptions.

As to the second direction, the major issue seems to be in creating and maintaining a library of descriptions of all groups in SCOP. Such a library may serve not only as a classification device, but also in explaining the features involved with other feature spaces, for instance, with those related to secondary structure or contact maps. Since the APPCOD descriptions are logical and simple, this may lead to deepening of automatisation in learning structural and functional regularities in proteonomics.

REFERENCES

Andrews, H.C. (1972) Introduction to Mathematical Techniques in Pattern Recognition, New York, Wiley-Interscience.

Appel R.D., Bairoch A., and Hochstrasser D.F. (1994) A new generation of information retrieval tools for biologists: the example of the ExPASy WWW server. Trends Biochem. Sci. **19**, 258–260 (http://www.expasy.ch/cgi-bin/protscale.pl).

Bachinsky, A., Yargin A., Guseva, E., Kulichkov, V., and Nozolenko, L. (1997) A bank of protein family patterns for rapid identification of possible functions of amino acid sequences, Comput. Appl. Biosciences, **13**, 115–122.

Bairoch, A., Bucher, P., and Hoffmann K. (1996) The PROSITE database, its status in 1995, Nucleic Acids Research, **24**, 189–196.

Breiman, L., Friedman, J., Olshen, R., and Stone, C. (1984) Classification and Regression Trees. Wadsworth Intl. Group.

Chou, K.-C., Liu, W.-M., Maggiora, G.M., and Zhang, C.-T. (1998) Prediction and classification of domain structural groups, Proteins, **31**, 97–103.

Dubchak, I., Muchnik, I., Holbrook, S.R., and Kim, S.-H. (1995) Prediction of protein folding class using global description of amino acid sequence, Proc. Natl. Acad. Sci. USA, **92**, 8700–8704.

Dubchak, I., Muchnik, I., Mayor, C., and Kim, S.-H. (1999) Recognition of a protein fold in the context of SCOP classification, Proteins, **35**, 401–407.

Efron, B. and Tibshirani, R.J. (1993) An Introduction to the Bootstrap, Chapman and Hall.

Fayyad, U.M., Piatetsky-Shapiro, G., Smyth, P., and Uthurusamy, R. (Eds.) (1996) Advances in Knowledge Discovery and Data Mining. Menlo Park, Ca, AAAI Press/The MIT Press.

Finn, P., Muggleton, S., Page, D., and Srinivasan, A. (1998) Pharmacophore discovery using the Inductive Logic Programming system PROGOL, Machine Learning, **30**, 241–270.

Fisher, D. and Eisenberg, D. (1996) Protein fold recognition using sequence derived predictions, Protein Science, **5**, 947–955.

Gibrat, J.-F., Madej, T., and Bryant, S.H. (1996) Surprising similarities in structure comparison, Current Opinion in Structural Biology, **6**, 377–385.

Gracy, J. and Argos, P. (1996) Automated protein sequence database classification, Bioinformatics, **14**, 164–187.

Hand, D.J. (1997) Construction and Assessment of Classification Rules. London, Wiley.

Hobohm, U. and Sander, C. (1995) A sequence property approach to searching protein databases, Journal of Molecular Biology, **251**, 390–399.

Hubbard, T., Murzin, A., Brenner, S., and Chothia, C. (1995) SCOP: a structural classification of proteins database, Nucleic Acids Research, **25**, 236–239.

Jones, D.T. (1997) Progress in protein structure prediction, Current Opinion in Structural Biology, **7**, 377–387.

Kawashima, S., Ogata, H., and Kanehisa, M. (1998) AAindex: amino acid index database, http://www.genome.ad.jp/dbget/aaindex.html.

Klösgen, W. (1996) Explora: A multipattern and multistrategy discovery assistant, in "Advances in Knowledge Discovery and Data Mining" (Eds. U.M. Fayyad, G. Piatetsky-Shapiro, P. Smyth, and R. Uthurusamy), Menlo Park, Ca, AAAI Press/The MIT Press, pp. 249–271.

Kohonen, T. (1995) Self-Organizing Maps. Berlin, Springer-Verlag.

Kubar, M., Bratko, I., and Michalski, R.S. (1998) A review of machine learning methods, in "Machine Learning and Data Mining: Methods and Applications" (Eds. R.S. Michalski, I. Bratko, and M. Kubat), Wiley, pp. 3–70.

Levitt, M. (1997) Competitive assessment of protein fold recognition and alignment accuracy, Proteins (Suppl.), **1**, 92–104.

Mirkin, B. (1999) Concept learning and feature selection based on square-error clustering, Machine Learning, **35**, 25–40.

Muchnik, I. (1998) Sample-based intervals can be corrected by constant factor scaling, Personal communication.

Muggleton, S., King, R., and Sternberg, M. (1992) Protein secondary structure prediction using logic. Protein Engineering, **5**, 647–657.

Murzin, A.G. and Bateman, A. (1997) Distant homology recognition using structural classification of proteins, Proteins (Suppl.) **1** (1997) 105–112.

Nilsson, N.J. (1965) Learning Machines, New York, McGraw-Hill.

Orengo, C.A., Michie, A.D., Jones, S., Jones, D.T., Swindells, M.B., and Thornton, J.M. (1997) CATH—A Hierarchic Classification of Protein Domain Structures, Structure, **5**, 1093–1108.

ProtScale: Amino acid scale representation. In: http://www.expasy.ch/tools/#primary.

Quinlan, J.R. (1993) C4.5: Programs for Machine Learning. Morgan Kaufmann Publishers, San Mateo, CA.

Rissanen, J. (1989) Stochastic Complexity in Statistical Inquiry. Singapore, World Scientific Publishing Co.

Rumelhart, D.E., Hilton, G.E., and Wilson, R.J. (1986) Learning internal representation by error propagation. In: D.E. Rumelhart and J.L. McClelland (Eds.) Parallel Distributed Processing: Explorations in the Microstructures of Cognition, Cambridge, MIT Press, pp. 318–362.

Salzberg, S.L. (1998) Decision trees and Markov chains for gene finding, in "Computational Methods in Molecular Biology" (Eds. S.L. Salzberg, D.B. Searls, and S. Kasif), Amsterdam, Elsevier Science B.V., pp. 187–203.

Srikant, R., Vu, Q., and Agraval, R. (1997) Mining association rules with the item constraints, in "Proceedings of Third International Conference on Knowledge Discovery and Data Mining" (Eds. D. Heckerman, H. Manilla, D. Pregibon, and R. Uthrusamy), Menlo Park, CA, AAAI Press, pp. 67–73.

Schlkopf, B., Burges, C., and Smola, A.J. (Eds.) (1998) Advances in Kernel Methods—Support Vector Learning, MIT Press.

Starck, J.L., Bijaoui, A., and Fionn Murtagh (1998) Image Processing and Data Analysis: The Multiscale Approach, Cambridge University Press.

Vapnik, V.N. (1998) Statistical Learning Theory, Wiley and Son.

Wei, L., Chang, J.T., and Altman, R.B. (1998) Statistical analysis of protein structures, in "Computational Methods in Molecular Biology" (Eds. S.L. Salzberg, D.B. Searls, and S. Kasif), Amsterdam, Elsevier Science B.V., pp. pp. 207–225.

Westhead, D.R. and Thornton, J.R. (1998) Protein structure prediction, Current Opinion in Biotechnology, **9**, 383–389.

Zhang, B., Rychlewski, L., Pawlowski, K., Fetrow, J.S., Skolnick, J., and Godzik, A. (1999) From fold predictions to function predictions: Automation of functional site conservation analysis for functional genome predictions, Protein Science, **8**, 1104–1115.

Zhang, C.-T. and Zhang, R. (1998) A new criterion to classify globular proteins based on their secondary structure contents, Bioinformatics, **14**, 857–865.

15

LINKING STRUCTURAL BIOLOGY WITH GENOME RESEARCH

The Berlin "Protein Structure Factory" Initiative

Udo Heinemann,[1,2] Juergen Frevert,[3] Klaus-Peter Hofmann,[4]
Gerd Illing,[2,5] Hartmut Oschkinat,[2,6] Wolfram Saenger,[2]
and Rolf Zettl[7]

[1]Forschungsgruppe Kristallographie
Max-Delbrück-Centrum für Molekulare Medizin, Berlin
[2]Institut für Kristallographie
Freie Universität Berlin
[3]BioteCon, Gesellschaft für Biotechnologie und Consulting mbH,
Potsdam
[4]Institut für Medizinische Physik und Biophysik
Klinikum Charité der Humboldt-Universität zu Berlin
[5]Interdisziplinärer Forschungsverbund Strukturbiologie, Berlin
[6]Forschungsinstitut für Molekulare Pharmakologie, Berlin
[7]Max-Planck-Gesellschaft zur Förderung der Wissenschaften
Ressourcenzentrum im DHGP am MPI für molekulare Genetik, Berlin

1. CHALLENGES TO STRUCTURAL BIOLOGY IN THE GENOME AERA

In the genome aera, the challenge to structural biologists is defined as follows: To determine the three-dimensional structures of a representative set of proteins such that all further studies of protein function, e.g. in a medical-pharmacological context, may be carried out on a firm structural basis. This challenge cannot be met in the conventional way whereby a protein crystallographer or an NMR spectroscopist applies her or his sophisticated methods to the study of that single protein structure that seems the most interesting at the time. For sure, this approach has been tremendously successful over the last decade, filling the Protein Data Bank at an ever increasing speed with structures of ever increasing beauty, complexity and biological relevance. However, in the light of the

Genomics and Proteomics, edited by Sándor Suhai.
Kluwer Academic / Plenum Publishers, New York, 2000.

above challenge, an all-out approach to structure determination is needed in much the same way as it was and is very successfully applied to genome research. This approach has become known as "structural genomics".

1.1. Structural Genomics

The term "structural genomics" has been in use for quite some time, but has acquired a completely new meaning very recently. Traditionally, it represented an effort to characterize the (physical) structure of a complete genome by gene mapping and sequencing.[1] Now, it stands for initiatives inspired by the genome sequencing projects that aim at the determination of three-dimensional protein structures in a systematic way.[2–6] The approaches taken towards this goal fall into two broad categories:

(1) In the first, the emphasis is on determining the structures of a set of proteins or protein domains which would yield a complete representation of all protein (domain) folds present in the biosphere. This approach is based on the notion that the number of folding types (folds) for globular protein domains is not unlimited.[7–9] Very probably, it does not exceed the number of structure entries now present in the Protein Data Bank. One may therefore hope to cover the complete universe of three-dimensional protein structures within a few years, provided that it is possible to identify new folds from protein sequence. Computer-based methods for fold recognition are currently being developed in a number of laboratories.[10–12] In a small bacterial genome, fold assignment with high confidence is possible only for a small subset of coding sequences.[13] However, advances in biocomputing methodology are likely to improve the success rate in the near future.[14] A convenient route towards fast structure determination targets proteins from hyperthermophilic bacteria or archaea, because they can be easily purified from recombinant *Escherichia coli* cells and lend themselves especially to crystallization or NMR structure determination. A number of crystal structures of these proteins has already been determined.[15–17] The knowledge of a representative set of protein domain structures is hoped to enable the complete fold prediction for newly sequenced genomes by homology modelling. The availability of the predicted tertiary folds for most proteins in a genome would in itself be of enormous value for many fields of biological research. In addition, it may considerably facilitate the detailed structure determination by protein crystallography and NMR spectroscopy of those proteins for which this is deemed necessary.

(2) A second approach to structural genomics focusses on structure analysis methodology. Here, the main idea is to closely cooperate with and learn from the genome sequencing projects. The use of the wide variety of available coding sequences and efforts towards parallelisation and automation of structure analysis are unifying features of this approach. As before, bioinformatics will play an important role in this brand of structural genomics for the identification of relevant proteins or protein domains that are amenable to structure analysis. The RIKEN NMR structure determination project[18] exemplifies the technology-oriented structural genomics efforts by attempting to establish a facility for the broad-scale analysis of three-dimensional protein structures in solution. The Berlin "Protein Structure Factory" initiative belongs into the same category of structural genomics. However, by employing both

X-ray diffraction and NMR methods it does not rely on one structure analysis technique exclusively. A main ingredient of the Protein Structure Factory is the close collaboration with the German Human Genome Project (DHGP).

Common to all structural genomics initiatives are efforts to identify and eliminate bottlenecks in the structure determination process. For example, it is generally agreed that the availability of bright synchrotron beamlines is a prerequisite for the successful use of diffraction methods.[19] Membrane proteins, constituting up to 30% of the protein inventory of an organism and against which more than 50% of the currently used and tested drugs are targetted, represent the most persistent bottleneck for all analytical methods, because they are only water-soluble in the presence of detergents and difficult to overproduce in quantities that are required for biophysical studies.

2. THE "PROTEIN STRUCTURE FACTORY": AN INTEGRATIVE APPROACH

The term "Protein Structure Factory" was chosen to represent a common initiative of the DHGP and structural biologists from the Berlin area aimed at the broad-scale analysis of proteins. The Protein Structure Factory will be established to characterize proteins encoded by the genes or cDNAs available at the Berlin Resource Center of DHGP. At a later stage, it may analyze various sets of input proteins selected by criteria of potential structural novelty or medical or biotechnological usefulness. It represents an integrative approach to structure analysis combining the computer-based analysis of genes by bioinformatics techniques, automated gene expression and purification of gene products, generation of a biophysical fingerprint of the proteins and the determination of their three-dimensional structures either in solution by NMR spectroscopy or in the crystalline state by X-ray diffraction. Here we briefly describe the main features of the planned Protein Structure Factory.

2.1. Bioinformatics

Bioinformatics has two main tasks in the Protein Structure Factory: To predict what can be done and to propose what should be done. Predicting what can be done is equivalent to identifying proteins that will permit their three-dimensional structures to be determined by X-ray crystallography or NMR spectroscopy. These proteins will have some properties in common. They will be soluble in aqueous buffers up to a critical concentration, they will have a defined globular structure, and this structure will be stable for at least as long as it takes to grow and expose crystals or to measure the NMR spectra. Proteins that contain long stretches of hydrophobic or charged amino-acid residues, have extended sequence repeats or use a limited repertoire of amino acids over long polypeptide segments often do not display these properties. However, they may still contain single or multiple domains that permit structure analysis. In addition, bioinformatics will provide valuable information aiding the structure determination by predicting sites of post-translational modification and identifying proteins of known, similar tertiary structure. Structural prediction will be used to decide whether a given protein will be studied by NMR spectroscopy or by X-ray diffraction or, for the latter case, whether its structure analysis will require experimental phase determination or can be based on a homologous model.

To propose what should be done is the more challenging task. It is equivalent to finding proteins with interesting properties such as novel folds or a function in biochemical pathways that may be associated with disease. The more interesting a protein appears, the more effort will have to be invested in its structure analysis. Computational tools for functional sequence assignments are currently being developed.[21] This work addresses questions concerning the subcellular localization of proteins, their membership in families defined by function[22–24] and their involvement in pathological states.[25,26]

2.2. Automated Gene Expression

The method of choice to produce recombinant proteins for structural and biophysical studies is the heterologous expression of their genes in *E. coli*. Proteins that cannot be synthesized in *E. coli* may alternatively be made in *Saccharomyces cerevisiae* or *Pichia pastoris*. For structure analysis by X-ray diffraction methods, the methionine residues of many proteins will have to be replaced by selenomethionine. Likewise, NMR structure determination will often require that the proteins be labelled with ^{13}C and/or ^{15}N which can be introduced through cell growth on media containing these isotopes in the form of ^{13}C glucose or $^{15}NH_4Cl$.

Within the Protein Structure Factory, gene expression systems will be obtained either by the cloning of PCR products or by the direct construction of cDNA libraries in expression vectors (expression libraries).[27] Both techniques will rely on the automated manipulation of clones in multi-well microtiter plates or on high-density membrane filters. Methods for the detection of protein coding or novel clones with antibodies directed against protein tags or by oligonucleotide fingerprinting are available.[27,28]

2.3. Purification of Tagged Proteins

The concept of the Protein Structure Factory requires the high-throughput production of highly pure proteins in about 50 mg quantities for structure analysis by NMR spectroscopy and X-ray crystallography. This is accomplished in two production units for the parallel fermentation and online purification of recombinant organisms (one for *E. coli* and one for *S. cerevisiae* or *P. pastoris*). *E. coli* is the organism of first choice, since it can be cultivated easily and offers a large number of readily available expression systems. Genes exhibiting low expression in *E. coli* or yielding proteins which are produced as inclusion bodies are expressed in yeast.

The recombinant organisms will be cultivated synchronously in a battery of fermenters (Figure 1). The cells from the different fermenters are homogenised successively with a high pressure homogeniser. The solubilised proteins are separated from the biomass by microfiltration and the processed filtrate is then concentrated by ultrafiltration. The following purification of the recombinant proteins takes advantage of two tags of these proteins: a His_6-tag and a strep-tag[29,30] whose corresponding DNA sequences are fused to the 3′- and 5′-terminus of the protein-coding gene. This allows a highly efficient separation of the recombinant protein from host cell protein. In the first step, the recombinant proteins are successively bound to a Ni-NTA column and eluted with imidazole. A second affinity chromatography on a streptavidin matrix is applied for the final purification. This semi-automated production and online purification will require two days for proteins synthesized in *E. coli* or three days for proteins from yeast. This production unit is designed to provide several homogeneous proteins for structure analysis per day.

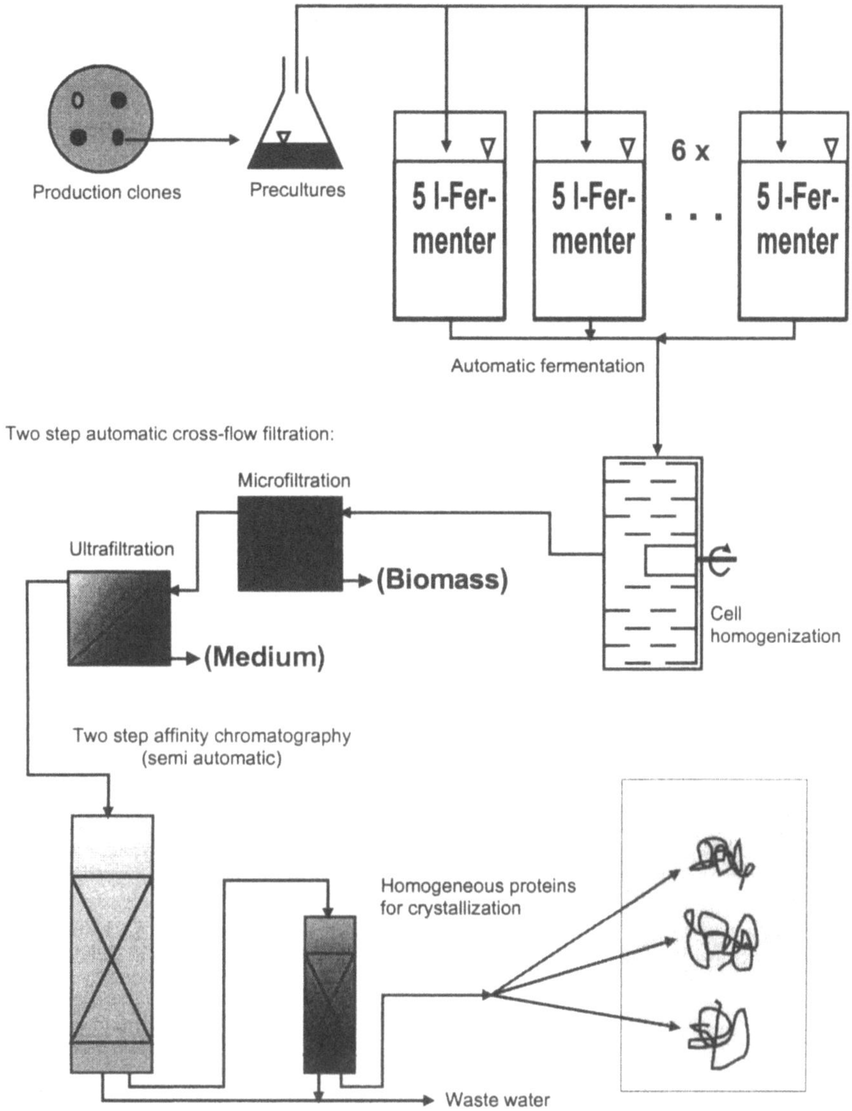

Figure 1. Fermenter setup for the high-throughput production of proteins from *E. coli* and yeast.

2.4. A Biophysical Protein Fingerprint

The goal of this unit is to characterize the proteins, as they become available from expression and purification, by conventional spectroscopic and calorimetric techniques. It will mainly serve to confirm and to complement the information obtained from bio-computing for further structure determination. The proteins will be analysed with respect to their secondary structure and stability, in dependence on temperature and pH.

The following techniques will be employed:

- Fourier-transform infrared spectroscopy (FTIR), to obtain secondary structure information by analysing the amide bands,

- circular dichroism spectroscopy (CD), to confirm the data obtained by FTIR,
- fluorescence spectroscopy, to investigate stability as a function of pH,
- differential scanning calorimetry (DSC), to measure thermal stability.

Automated routines for the data acquisition and evaluation procedures will be necessary to keep pace with the expected throughput of proteins. In part, these routines are already available, some have to be developed.

In summary, this unit will furnish biophysical parameters concerning secondary structure and conformational stability of proteins, independent of and preliminary to the determination of high-resolution structures. It will help to establish experimental conditions for protein crystallization and NMR studies. The biophysical data may also be useful in those cases where high-resolution protein structures cannot be obtained.

2.5. NMR Spectroscopy

The role of NMR will be in the structure determination of protein domains and of their functional complexes, and in the investigation of ligand binding to help in the design of bioactive small molecules. For this purpose, it is necessary to automate the key steps in the NMR structure determination procedure. These include data acquisition, sequence-specific resonance assignments and structure calculation. Currently, it takes weeks to months for the spectral assignments, especially those of the NOESY spectrum, to be accomplished. In order to be able to determine the structures for all three steps, some concepts and algorithms for automating the procedures exist, and more need to be developed.

Automated data acquisition is probably the easiest task in this project. It includes the definition of a data set which is suited for automated interpretation. Most modern NMR spectrometers already provide features which allow one to automate the data acquisition itself. The critical step for being able to determine the structures of a large number of proteins is in the necessary automation of the assignment procedure. To date, a number of computer programs for this purpose are available,[31] but, in any case, manual interference is required. Most of these software packages will require peak lists obtained from the multi-dimensional spectra, which usually contain false peaks generated from noise or artifacts. The logics of the program are not then capable of handling this problem. In the context of the Protein Structure Factory, it is required to generate a new piece of software which works directly on the spectra and is already able to recognize peaks, noise and artifacts as such. On the basis of a data set comprising CBCA N NH, CBCA (CO) NNH, HCCH-COSY, HCCH-TOCSY, and amino-acid-sensitive experiments, it is expected that the program will generate a list of chemical shifts comprising those of all protons, carbons and nitrogens present in the protein that can be used to evaluate the three-dimensional NOESY spectra.

This peak list is then subjected to an automated structure calculation protocol proposed by M. Nilges,[32] which essentially allows one to assign the NOESY spectra during the structure calculation. In this manner, it is expected that approximately three months of manual work can be saved per structure. It is expected that the NMR structures of proteins with up to 120 amino acids can be solved routinely, if their solubility is high enough, and that sufficient signal-to-noise can be obtained in the 2- and 3-dimensional spectra. The Protein Structure Factory also provides means to exploit the structural information generated. In this context, NMR spectroscopy will be used to study ligand-protein interactions in screening campaigns to detect binding in a site-specific manner. This information will be used to optimize ligands.

2.6. Protein Crystallization

At present, the crystallization of proteins is still the bottleneck in the structure determination by means of X-ray diffraction. There is no simple correlation between properties of proteins and the large number of parameters that have to be considered during crystallization. Consequently, the crystallization of proteins is mostly an empirical process that requires a broad screening of different crystallization conditions. In the Protein Structure Factory, it is planned to have available a large number of purified proteins or protein domains per year that are considered for crystallization. Since a manual optimization of crystallization conditions on the projected scale is not feasible, the development and the utilization of a crystallization robot is a key issue of the crystal structure determination within the Protein Structure Factory. The necessary innovations will rely on two well established groups with ample experience in protein crystallography and in the construction of robots.

It is planned to build a crystallization robot that is pipetting protein solutions and a buffer screen consisting of about 100 different conditions (pH, buffer, salt, polyethylene glycol, alcohols, salts) for "hanging drop" vapor diffusion experiments:[33] a drop consisting of protein and buffer is equilibrated against the buffer at about twice the concentration, so that the protein solution in the drop is brought to supersaturation and eventually to crystallization. This is set up in trays with 24 wells, and the trays are automatically stored at two temperatures, preferably 4°C and 18°C. The robot examines the trays by light scattering to monitor aggregation of protein and, if possible, nucleation, and in later stages the trays will be observed by microscopes with suitable software to automatically recognize crystalline material.

2.7. Acquisition of X-Ray Diffraction Data using Synchrotron Radiation

The use of synchrotron radiation will be crucial to the Protein Structure Factory: high brilliance and tuneable wavelengths are prerequisites for fast data collection, the use of small crystals and multiwavelength anomalous diffraction (MAD) phasing.[19] An example for a diffraction image obtained from a small crystal at a synchrotron is shown in Figure 2. With the opening of BESSY II, direct access to a third-generation XUV storage ring source with excellent conditions is available nearby. However, to shift the maximum of the emitted spectrum towards the X-ray range, a high-field multipole wiggler has to be installed as has been done at other medium energy storage rings (ALS,[34] MAX II,[35] ELETTRA).

Two beamlines are planned within the Protein Structure Factory: the central beamline is optimized for rapidly measuring high resolution MAD data sets. This MAD beamline will be equipped with a focussing premirror, a double crystal monochromator and a refocussing mirror to serve in the wavelength range from 0.7 Å to 2.75 Å which covers the absorption edges of all commonly used heavy atoms.[36] To make use of the expected short exposure times a state-of-the-art CCD detector with fast bus and high capacity storage system will be installed at the MAD station. This will be especially useful in cases when fine slicing down to 0.1° is employed.

The other beamline is designed as a constant-energy station with a selectable wavelength around 0.9 Å and will be used for the fast checking of crystal quality and further

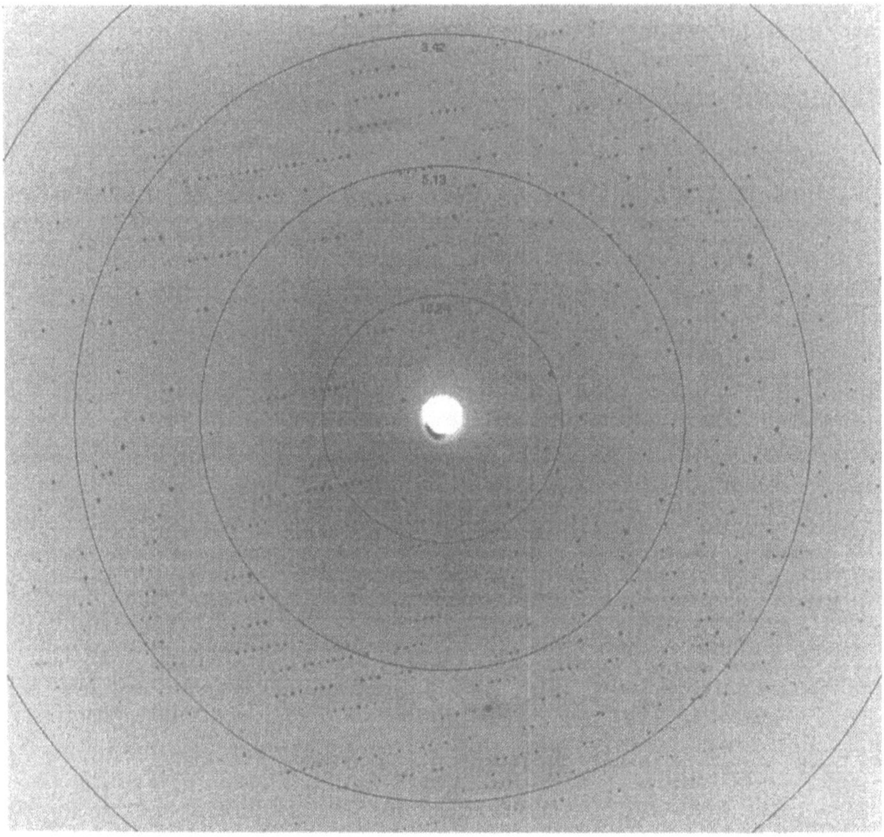

Figure 2. Part of the diffraction pattern from a small crystal of the *Sarcocystis muris* lectin SMLx. 1° oscillation photograph taken at beamline BW7A at the EMBL Outstation at DESY, Hamburg. The unit cell axis running horizontally has a length of 158.2 Å. Note the sharp and well resolved reflections arising from a synchrotron beam with small angular divergence.

preliminary examinations. It will accept radiation from the the side portion of the wiggler fan and will be equipped with a premirror and a bent crystal monochromator to select the appropriate wavelength and to focus and deflect the X-ray beam. Both stations will be equipped with gaseous nitrogen cooling and both need highly automated beamline control, efficient software protocols and organization schemes to act as high-throughput system.

2.8. Crystal Structure Determination

The high-throughput determination of three-dimensional protein structures based on the X-ray diffraction data collected at the synchrotron beamlines (see above) will have to employ robust and efficient methods at four essential steps: Phasing, model building, refinement and quality control. In some cases it will be possible to use homologous protein or domain structures for molecular replacement phasing. As the Protein Data Bank grows and the techniques for detecting homology at the level of three-dimensional structure improve, the frequency with which such search models are available will increase

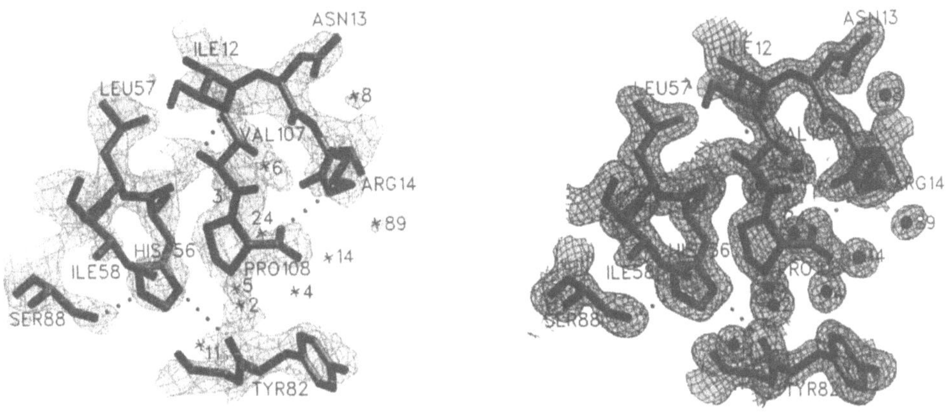

Figure 3. Electron density around the C-terminal residue Pro-108 of the bovine adrenodoxin Adx(4-108).[38] The experimental (MAD) electron density map based on the anomalous scattering of the iron atoms from the protein's [2Fe-2S] cluster (left) clearly shows the protein backbone and the orientation of side chains and even indicates the positions of some bound water molecules shown as numbered asterisks (left). For comparison, the final difference density map calculated at 1.85 Å resolution using model phases is shown on the right.

substantially. Crystal structures can be solved easily if the structural similarity of a search model is high enough.

The analysis of protein structures with unpredicted fold requires experimental phase determination. Once dreaded because of the tedious trial-and-error searching for isomorphous derivatives, phasing has become a routine process with the advent of MAD methods.[37] All proteins produced in recombinant *E. coli* can be labelled with heavy-atom markers in the form of selenomethionine and thus subjected to MAD phasing. The power of MAD phasing may be appreciated from Figure 3 comparing the experimental electron density (from MAD) with the final, refined density in a portion of the structure of a bovine adrenoxin, Adx(4-108).[38] Here, the two iron atoms of the protein were sufficient for MAD phasing to produce density that not only clearly reveals the protein atoms around the C-terminus of Adx(4-108) but even some of the water molecules bound in this region.

Currently, methods for semi-automated model building into electron-density maps[39] and structure refinement[40] are being developed in a number of laboratories. These methods will be incorporated into the crystal structure determination process of the Protein Structure Factory. Finally, it will be necessary to stringently assess the quality of the determined structures[41] before they are allowed to enter a database.

3. CONCLUSIONS

Genomics does not end when all base pairs of DNA have been sequenced. In contrast, it may be argued that the interesting part of the work—aimed at understanding whole organisms by starting from the molecules of life—is the one involving studies of structure and function of the gene products. Structural genomics approaches as the one described above and and large-scale, high-throughput functional studies, functional genomics,[42] are starting to provide the tools to performing these analyses.

ACKNOWLEDGMENT

We are grateful to Jürgen J. Müller (Max-Delbrück-Centrum) for providing figures 2 and 3. Supported by the Bundesministerium für Bildung und Forschung through the Leitprojekt Proteinstrukturfabrik.

REFERENCES

1. McKusick, V.A. (1997) Genomics: Structural and functional studies of genomes. Genomics **45**, 244–249.
2. Šali, A. (1998) 100,000 protein structures for the biologist. Nature Struct. Biol. **5**, 1029–1032.
3. Terwilliger, T.C., Waldo, G., Peat, T.S., Newman, J.M., Chu, K., and Berendzen, J. (1998) Class-directed structure determination: Foundation for a protein structure initiative. Protein Sci. **7**, 1851–1856.
4. Shapiro, L. and Lima, C.D. (1998) The Argonne Structural Genomics Workshop: Lamaze class for the birth of a new science. Structure **6**, 265–267.
5. Gaasterland, T. (1998) Structural genomics taking shape. Trends Genet. **14**, 135.
6. Koonin, E.V., Tatusov, R.L., and Galperin, M.Y. (1998) Beyond complete genomes: from sequence to structure and function. Current Opinion Struct. Biol. **8**, 355–363.
7. Finkelstein, A.V. and Ptitsyn, O.B. (1987) Why do all globular proteins fit the limited set of folding patterns? Prog. Biophys. Mol. Biol. **50**, 171–190.
8. Chothia, C. (1992) One thousand protein families for the molecular biologist. Nature **357**, 543–544.
9. Orengo, C.A., Jones, D.T., and Thornton, J.M. (1994) Protein superfamilies and domain superfolds. Nature **372**, 631–634.
10. Bork, P. and Eisenberg, D. (1998) Sequences and topology. Deriving biological knowledge from genomic sequences. Current Opinion Struct. Biol. **8**, 331–332.
11. Fischer, D. and Eisenberg, D. (1996) Protein fold recognition using sequence-derived predictions. Protein Sci. **5**, 947–955.
12. Rice, D.W. and Eisenberg, D. (1997) A 3D-1D substitution matrix for protein fold recognition that includes predicted secondary structure of the sequence. J. Mol. Biol. **267**, 1026–1038.
13. Fischer, D. and Eisenberg, D. (1997) Assigning folds to the proteins encoded by the genome of *Mycoplasma genitalium*. Proc. Natl. Acad. Sci. USA **94**, 11929–11934.
14. Huynen, M., Doerks, T., Eisenhaber, F., Orengo, C., Sunyaev, S., Yuan, Y., and Bork, P. (1998) Homology-based fold predictions for *Mycoplasma genitalium* proteins. J. Mol. Biol. **280**, 323–326.
15. Kim, K.K., Hung, L.-W., Yokota, H., Kim, R., and Kim, S.-H. (1998) Crystal structures of eukaryotic translation initiation factor 5A from *Methanococcus jannaschii* at 1.8 Å resolution. Proc. Natl. Acad. Sci. USA **95**, 10419–10424.
16. Lim, J.-H., Yu, Y.G., Han, Y.S., Cho, S.-j., Ahn, B.-Y., Kim, S.-H., and Cho, Y. (1997) The crystal structure of an Fe-superoxide dismutase from the hyperthermophile *Aquifex pyrophilus* at 1.9 Å resolution: Structural basis for thermostability. J. Mol. Biol. **270**, 259–274.
17. Kim, K.K., Kim, R., and Kim, S.-H. (1998) Crystal structure of a small heat-shock protein. Nature **394**, 595–599.
18. Saegusa, A. (1998) Japan's genome programme goes ahead, with protein analysis. Nature **392**, 219.
19. Kim, S.-H. (1998) Shining light on structural genomics. Nature Struct. Biol. **5**, 643–645.
21. Bork, P. and Koonin, E.V. (1998) Predicting functions from protein sequences—where are the bottlenecks? Nature Genetics **18**, 313–318.
22. Schultz, J., Milpetz, F., Bork, P., and Ponting, C.P. (1998) SMART, a simple modular architecture research tool: identification of signaling domains. Proc. Natl. Acad. Sci. USA **95**, 5857–5864.
23. Bork, P., Dandekar, T., Eisenhaber, F., and Huynen, M. (1998) Characterization of targeting domains by sequence analysis: glycogen-binding domains in protein phosphatases. J. Mol. Med. **76**, 77–79.
24. Yuan, Y., Schultz, J., Mlodzik, M., and Bork, P. (1997) Secreted Fringe-like signaling molecules may be glycosyltransferases. Cell **88**, 9–11.
25. Museghian, A.R., Bassett, D.E., Jr., Boguski, M., Bork, P., and Koonin, E.V. (1997) Positionally cloned human disease genes: New motifs and evolutionary conservation. Proc. Natl. Acad. Sci. USA **94**, 5831–5836.
26. Bork, P., Hofmann, K., Bucher, P., Neuwald, A., Altschul, S.F., and Koonin, E.V. (1997) A superfamily of conserved domains in DNA damage-reponsive cell cycle checkpoint proteins. FASEB J. **11**, 68–76.
27. Maier, E., Maier-Ewert, S., Bancroft, D., and Lehrach, H. (1997) Automated array technologies for gene expression profiling. Drug Discovery Today **2**, 315–324.

28. Maier, E., Maier-Ewert, S., Ahmadi, R., Curtis, J., and Lehrach, H. (1994) Application of robotic technology to automated sequence fingerprint analysis by oligonucleotide hybridisation. J. Biotech. **35**, 191–203.
29. Hochuli, E., Bannwarth, W., Dobeli, H., Gentz, R., and Stüber, D. (1988) Genetic approach to facilitate purification of recombinant proteins with a novel metal chelate adsorbent. Bio/Technology **6**, 1321–1325.
30. Schmidt, T.G.M. and Skerra, A. (1994). One-step affinity purification of bacterially produced proteins by means of the "Strep-tag" and immobilized recombinant core streptavidin. J. Chromatogr. A **676**, 337–345.
31. Oschkinat, H. and Croft, D. (1994). Automated assignment of multidimensional nuclear magnetic resonance spectra. H. Meth. Enzymol. **239**, 308–318.
32. Nilges, M., Macias, M.C., O'Donoghue, S.I., and Oschkinat, H. (1997). Automated NOESY interpretation with ambiguous distance restraints: the refined NMR solution structure of the pleckstrin homology domain from β-spectrin. J. Mol. Biol. **269**, 408–422.
33. Weber, P.C. (1997) Overview of protein crystallization methods. Methods Enzymol. **276**, 13–22.
34. Earnest, T. (1995) Conceptual Design Report for ALS Beamline 5.0, Lawrence Berkeley Laboratory PN941209-2.
35. Svensson, L.A., Ståhl, K., Cerenius, Y., Oskarsson, Å., Albertsson, J., and Liljas, A. (1997) A new beamline for crystallographic measurements at the MAX II synchrotron, Lund, Sweden, Annual Report 182.
36. Ogata, C.M. (1998) MAD phasing grows up. Nature Struct. Biol. **5**, 638–640.
37. Hendrickson, W.A. and Ogata, C.M. (1997) Phase determination from multiwavelength anomalous diffraction measurements. Methods Enzymol. **276**, 494–523.
38. Müller, A., Müller, J.J., Muller, Y.A., Uhlmann, H., Bernhardt, R., and Heinemann, U. (1998) New aspects of electron transfer revealed by the crystal structure of a truncated bovine adrenodoxin, Adx(4-108). Structure **6**, 269–280.
39. Fortier, S., Chiverton, A., Glasgow, J., and Leherte, L. (1997) Critical-point analysis in protein electron-density map interpretation. Methods Enzymol. **277**, 131–157.
40. Lamzin, V.S. and Wilson, K.S. (1997) Automated refinement for protein crystallography. Methods Enzymol. **277**, 269–305.
41. Dodson, E.J., Davies, G.J., Lamzin, V.S., Murshudov, G.N., and Wilson, K.S. (1998) Validation tools: can they indicate the information content of macromolecular crystal structures? Structure **6**, 685–690.
42. Brownstein, M.J., Trent, J.M., and Boguski, M.S. (1998) Functional genomics. Trends Biochem. Sci. **23**(Suppl.), 27–29.

16

G PROTEIN-COUPLED RECEPTORS, OR THE POWER OF DATA

Florence Horn,[1] Mustapha Mokrane,[2] Johnathon Weare,[1]
and Gerrit Vriend*,[1]

[1]BIOcomputing
European Molecular Biology Laboratory
Meyerhofstraße 1, 69117 Heidelberg, Germany
[2]Laboratoire de Génétique et de Physiologie du Développement
Université de la Méditerranée
13288 Marseille Cedex 9
France

1. WHAT TO DO WITH GENOME DATA?

Genome sequencing projects are racing at a rate of one species per month, and we can count the days till the human genome will be published in a special issue of *Nature* or *Science*. This rich source of data is spurring new research fields. The availability of all these complete genome sequences allows us to compare entire species, to look at evolution in new ways, to think about the minimal set of macromolecules needed for life, etc. From a users point of view however, data needs to be organized in databases in many different ways, in order to fully benefit from this novel data luxury. Big monolithic databases that mainly store sequences are needed for rapid sequence comparisons. One database per species is needed for species specific questions. One database per molecular class is needed to optimally harvest all information in a drug design environment. In this article we will discuss one such molecular class-specific databases: the GPCRDB. This database aims at the collection and harvesting of G protein-coupled receptor (GPCR) data. Although the project is specifically aimed at GPCRs, all principles and problems will also

* To whom correspondence should be adressed: vriend@embl-heidelberg.de

Genomics and Proteomics, edited by Sándor Suhai.
Kluwer Academic / Plenum Publishers, New York, 2000.

hold for other classes of molecules, and the software designed for this project can be used in similar projects with only minor modifications.

2. G PROTEIN-COUPLED RECEPTORS

GPCRs are a super family of integral membrane proteins. They consist of seven membrane spanning helices, three periplasmic (extra-cellular) loops, three cytosolic (intra-cellular) loops and a periplasmic N-terminal and a cytosolic C-terminal domain (Figure 1). The periplasmic loops and the N-terminal domain often are glycosylated. The

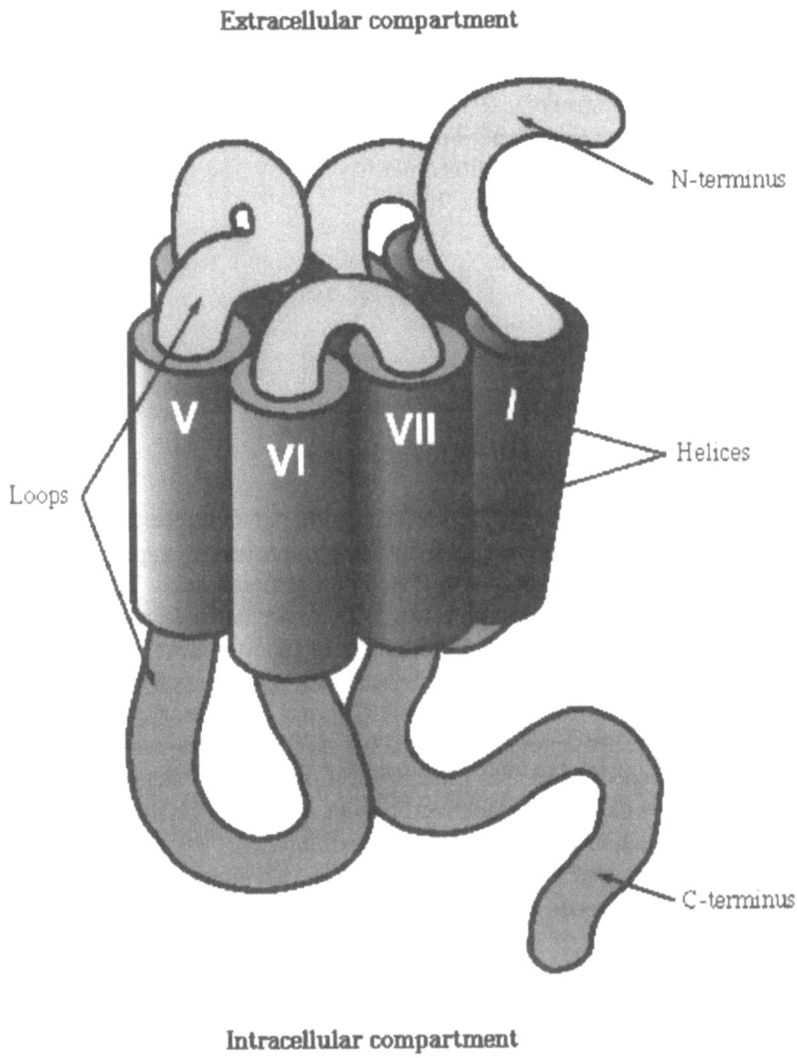

Figure 1. Schematic representation of a GPCR. (Figure kindly provided by T. Schwartz)

Table 1. Examples of GPCR-related diseases or effects

Receptors	Related diseases or effects
Bradykinin	Inflammation, asthma, pain, shock
Calcitonin	Paget's bone disease
$GABA_B$	Analgesics
CCK	Anxiolytic
FSH	Infertility
GnRH	Prostate cancer, precocious puberty
Histamine 1	Hay fever, itching, motion sickness
Serotonin	Depression migraine, post-operative vomiting
Opiod	Pain, constipation
PAF	Inflammation, asthma
Somastostatin	Tumours, glucagonoma
Vasopressin	Diabetes insipidus
Oxytocin	Induces labour and promotes lactation
Opsins	Colour blind
Metabotropic	Tinitus

C terminal domain often contains a cysteine residue that has a lipid group attached to it. This group, often a meristoyl, is embedded in the membrane, thereby making the part between the seventh helix and this cysteine quasi a fourth cytosolic loop.

GPCRs are involved in the communication between the surroundings and the cell. They detect a signal at the periplasmic side. This signal can be a protein, a peptide, a small organic molecule, an ion, or a photon that causes a structural change in a retinal group, but more exotic signaling mechanisms also exist (see Watson and Arkinstall, 1994 for review). This signal is transmitted through the transmembrane helices to the cytosolic side where a trimeric G protein becomes activated. This G protein trimer elicits the response in the cell.

Dysfunction of GPCRs results in diseases as diverse as Alzheimer's, Parkinson's, diabetes, dwarfism, colour blindness, retina pigmentosa and asthma. GPCRs are also involved in depression, schizophrenia, sleeplessness, hypertension, impotence, anxiety, stress, renal failure, several cardiovascular disorders and inflammations. Table 1 gives a short summary of some of the better known receptors and their associated disease states.

G protein-coupled receptors are of enormous importance for the pharmaceutical industry because they are the target for about 50% of all existing medicines (Gudermann et al., 1995). Many well-known medicines such as beta blockers and anti-histamines, or the drugs like opium and cannabis act on GPCRs. Every major pharmaceutical industry has a GPCR research program aimed at the design of drugs for the treatment of GPCR related diseases. And still we know very little about these receptors.

Numerous natural ligands, agonists and antagonists of GPCRs (mostly the amine receptors) are used for therapeutic purposes. For instance, muscarinic agonists are used in the treatment of glaucoma and tachycardia, while agonists and antagonists of adrenoceptors have been developed for the treatment of asthma, of high blood pressure, of other cardiovascular disorders and of anxiety. Serotonin 5HT-1D antagonists are used against migraine, and histamine H1 antagonists against allergic and anaphylactic reactions, hay fever, itching, and motion sickness. The dopamine agonist L-dopa is the agent of choice in the treatment of Parkinson's disease, and a dopamine antagonist is used to treat schizophrenia and Huntington's disease in its early stages. Other GPCR ligands have been shown to be analgesic, anti-inflammatory and anti-asthmatic agents.

Designing therapeutic drugs is the aim of all pharmaceutical industries. The design of one new drug can cost up to 500 million dollars and it is therefore not surprising that any tool that can make this process faster or cheaper is welcomed. Combinatorial chemistry, high throughput screening and computational approaches are the most important drug design developments of the last decade. Structure-based computational drug design promised spectacular results when introduced about ten years ago (Hol, 1986). Although this technique was a total failure in so far as only very few drugs actually got designed using these approaches, this technique has found its place in the drug designer's tool chest because it is rather effective in removing unlikely candidates from the high throughput screening experiments, and it is useful when designing combinatorial chemistry experiments. Lacking high resolution experimental structure data, chemists study models built *ab initio*, or based on bacteriorhodopsin as a (poor) template (Hibert et al., 1991).

Not all GPCR research is aimed directly at the design of drugs. For example, molecular biology, cell biology and related fields study the regulation, localization, membrane insertion, phosphorylation, interactions with periplasmic and cytosolic factors and a whole array of other aspects of GPCRs. There are also theoretical approaches to these aspects of GPCRs. Oliveira et al. (1994), for example, used the massive body of experimental data and combined this first into a general model for the interaction between receptors and G proteins and later into a general functional model for signal transduction (Oliveira et al., 1999). Reynolds and colleagues (Gouldson et al., 1997) used a combination of fact mining in the experimental literature and sequence analysis techniques to detect that a number of GPCRs is only active in a dimeric form. There are many more examples of increased understanding of the life of a GPCR thanks to computational work. What all these computational studies have in common is that they rely on readily available data.

About 1700 GPCR protein sequences are presently available in the publicly accessible databases, and about a dozen new sequences become available every month (e.g. the Swiss-Prot database; Bairoch and Apweiler, 2000). The continuation of this rapid increase in the number of available sequences seems guaranteed when in the next few years the major genome sequencing efforts move from small bugs to higher species, especially mammals. An uncountable number of mutations have been made and characterized, and about 5000 of these found their way into publicly accessible databases (e.g. the GRAP database; Kristiansen et al., 1996). The number of compounds tested for GPCR binding and/or efficacy in test systems must be in the millions or even billions and about 12,000 dissociation constants are now accessible via the World Wide Web (WWW) (http://www.gpcr.org/ligand/ligand.html). The small amount of structural data stands in marked contrast to this wealth of sequences, ligands and mutations. At present only low resolution electron diffraction data is available for two rather special GPCRs, the frog and bovine opsins (Schertler et al., 1993; Schertler et al., 1998). High resolution data is available for bacteriorhodopsin (e.g. Unger et al., 1995; Luecke et al., 1998; Takeda et al., 1998; Pebay-Peyroula et al., 1997), but as bacteriorhodopsin is not G protein coupled, the value of these coordinates for GPCR research, i.e. as a template for homology modelling, is doubtful at best (Schertler, 1998).

In this article, we describe the design and usage of a heterogeneous data handling system that allows for the combination of many different forms of computational and experimental data related to GPCRs. This system, available via the WWW, can be seen as a pilot study for a new generation of information systems. Classical monolithic databases contain one type of data and have a search engine attached to provide a query system. To allow for research on one class of molecules, these databases should be incorporated in an information system that permits the four major tasks: querying, browsing, retrieval and inferencing. The rapid growth of all underlying databases, partly caused by all kinds of genome projects, makes these next generation information systems an absolute necessity if we do not simply want to drown in the data streams.

3. THE GPCRDB INFORMATION SYSTEM

The GPCRDB combines three kinds of experimental data: sequences, mutation data and ligand binding data. Multiple sequence alignments, phylogenetic trees and results of correlated mutation analyses (see below) are also made available to the user. Sequence comparisons are very important because they help to extrapolate experimental knowledge from one GPCR to another.

Visualization of data is important for the user of the system. In the GPCR field two-dimensional representations by means of snake-like diagrams (see Figure 2 for an example) are used to visually combine a sequence with other types of information such as mutations. Obviously, remarks related to the snake-like plots cannot be applied to proteins that do not form helix bundles.

At present, we automatically extract all sequence data from the Swiss-Prot database. This has the advantage of providing us with non-redundant sequences which have been well annotated by expert database curators. In the January 2000 release of the GPCRDB, more than 1700 GPCR sequences were available, divided into 6 classes and 200 families and sub-families based on the pharmacological nature of their ligand and sequence similarity (see the complete list of families at http://www.gpcr.org/7tm/multali/multali.html; Oliveira et al., 1993; Kolakowski, 1994). With more than 1000 proteins, including fragments, class A (or rhodopsin-like) receptors represent about 90% of all the known GPCRs.

In the GPCRDB, each sequence is linked to the tinyGRAP mutant database via a snake-like diagram (when mutant data is available). In order to increase the accessibility of mutation data, we have implemented a data submission system, where experimental scientists in the GPCR field can enter their own mutation data.

Ligand binding data was obtained from P. Seeman who has manually collected drug dissociation constants for neuroreceptors and transporters (Seeman, 1993). In total about 12,000 constants are available for 10 GPCR families.

In an ideal world, experimentalist would always deposit experimental results in the appropriate databases, or at least make them available one way or another in a computer-readable form. Unfortunately, the scientific community has decided that only three dimensional coordinates of molecules and protein and nucleic acid sequences need to be made available before publication of the corresponding article. We once again urge the scientific community to impose a few more constraints on itself and not let so much valuable information become lost by merely publishing it in a journal that is, considering today's prices of scientific journals, accessible to just a small number of colleagues in related fields. However important, this is not the main topic of this article and will not be discussed further.

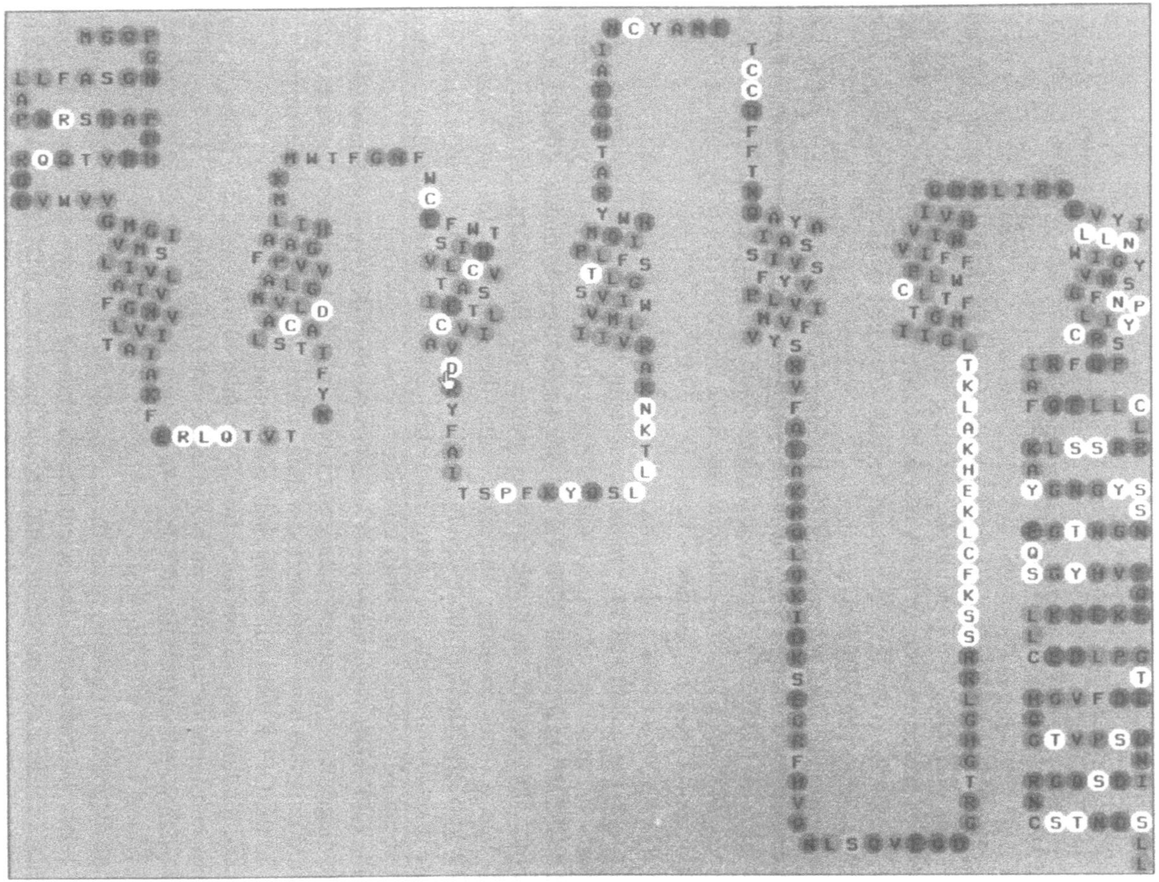

Figure 2. Snake-like diagram of the human β2 adrenergic receptor. Amino-acids are coloured based on their biochemical nature. White-coloured amino-acids have been mutated and are hyperlinked to the TinyGRAP mutation database.

3.1. The Underlying Data

The major experimental data components of the GPCRDB are sequences, mutations and ligand binding data. We will discuss some aspects of the data and how the GPCRDB is updated.

3.1.1. Sequences. All nucleic acid sequences are deposited in one of the three major sequence deposition centers, the EMBL (Bauer et al., 2000), Genbank (Benson et al., 2000) and DDBJ (Tateno et al., 2000).

An automatic procedure that runs at the EBI identifies GPCR nucleotidic sequences in the EMBL sequence databank. The coding regions are translated into a protein sequence and annotated both by automatic processes and by experts. An important delay can occur between the entry of newly cloned GPCRs in TrEMBL (translated EMBL entries) and their integration in Swiss-Prot. Here we have to make a choice. Do we want good quality data and minimal redundancy, or do we want the GPCRDB to be as complete as possible? One of the partners in the loose association of the GPCR-related databases (i.e. the GCRDb; see Table 2) is maintaining a list of all GPCR related sequences to be found in all protein and genome databases. We therefore decided to first concentrate on the optimal use of the high quality Swiss-Prot derived sequences. However, the EBI based data-mining mechanisms will soon be incorporated in the GPCRDB. At present the GPCRDB is updated in batch mode (about one update every two months). Once new GPCRs entries are added in Swiss-Prot, we produce a new release of the GPCRDB. The sequence classification and the secondary data validation still require manual intervention, and we cannot see how these steps can be automated. Nevertheless, we expect to detect GPCRs in nucleotide databanks more rapidly, using a scanning method in order to increase the availability and the completeness of GPCR proteins, and provide not only a high quality information system, but also an up-to-date overview of all available information.

3.1.2. Mutant Data. As there is a lack of experimental structure data, mutation data is the most important source of information when questions are raised that normally require three dimensional coordinates for an answer. Many mutation databases exist and are available via the WWW, for example from the EBI (http://www.ebi.ac.uk/). The GRAP database has collected GPCR mutations published from 1987 to mid-1995. At present this database contains 2000 single and multiple-point mutations that are well annotated. Annotated means that detailed information is available regarding quantitative and qualitative effects of mutations on agonist and antagonist binding. The effects on receptor-G protein coupling and signal transduction properties are also described. The GRAP database is no longer updated and the mutant information is now stored in another database, tinyGRAP, that only contains the basic data on receptor mutant and literature reference (Edvardsen and Kristiansen, 1997). The last release of tinyGRAP contains about 7600 mutants that have been published in more than 1020 papers (Beukers et al., 1999).

All mutations in the GRAP database and its derivative tinyGRAP were entered by specialists. However, we have made available an automated submission system to everyone on the WWW at http://www.gpcr.org/7tm/mutants/mutants.html (Figure 3). The data submission system for the mutant database has been designed to be a practical mechanism for the experimental scientists to submit their mutant data themselves. It is an author-in system. The current set of fields is fairly comprehensive, and miscellaneous details can also be given, which may be useful for specific experimental conditions. At

Table 2. Internet resources of GPCR-related information

Web sites	Data content	Location (http://)
GPCRDB: Information system on G Protein-Coupled Receptor (Horn et al., 1998b)	GPCR proteins, multiple sequence alignments, phylogenetic trees, drug dissociation constants, mutation data	www.gpcr.org/7tm/
GCRDb: G protein-coupled Receptor Database (Kolakowski, 1994)	GPCR sequences, multiple alignments, phylogenetic trees, GPCR-linked diseases in OMIM	www.gcrdb.uthscsa.edu/
Swiss-Prot (Bairoch and Apweiler, 2000)	List of GPCR proteins	www.expasy.ch/cgi-bin/lists?7tmrlist.txt
GRAP and **TinyGRAP** (Boukers et al., 2000)	Mutant data for class A and B receptors	www-grap.fagmed.uit.no/GRAP/homepage.html
ORDB: Olfactory Receptor Data Base (Skoufos et al., 2000)	Sequences of olfactory receptors	ycmi.med.yale.edu/senselab/ordb
CORD: Center for opioid Research and Design	Structural and schematic models, chimeric receptor studies, point mutation data, opioid ligands	www.opioid.umn.edu/
MRS: Molecular Recognition Section (van Rhee and Jacobson, 1996)	Alignments, mutation analysis	mgddk1.niddk.nih.gov/MutationAnalysis.html
Swiss-Model 7TM Interface (Guex and Peitsch, 1997)	Tool for the modelling of the helices of GPCRs	www.expasy.ch/swissmod/SWISS-MODEL.html
GPCR Pattern Recognition (Attwood et al., 2000)	GPCR fingerprints, visualisation of alignments	www.biochem.ucl.ac.uk/bsm/dbbrowser/GPCR/
Viseur (Campagne et al., 1999)	Visualisation, management and integration of GPCR-related information	www.lctn.u-nancy.fr/viseur/viseur.html

present, every entry entered via this form is checked by hand by an expert, but it is envisaged that more advanced software will be added which will do most or all such manual validation in an automatic fashion.

A new mutant is submitted using this single WWW page. This page lists all the fields that may be entered. The use of a single page instead of a step-through system has several benefits. It is transparent in that the submitter immediately knows what information must be gathered. It is more flexible as it allows editing of the entire entry at any time. It is fast as the form can be filled in off-line, with only the final submission needing a return to the server, with the added bonus that a static WWW page has caching

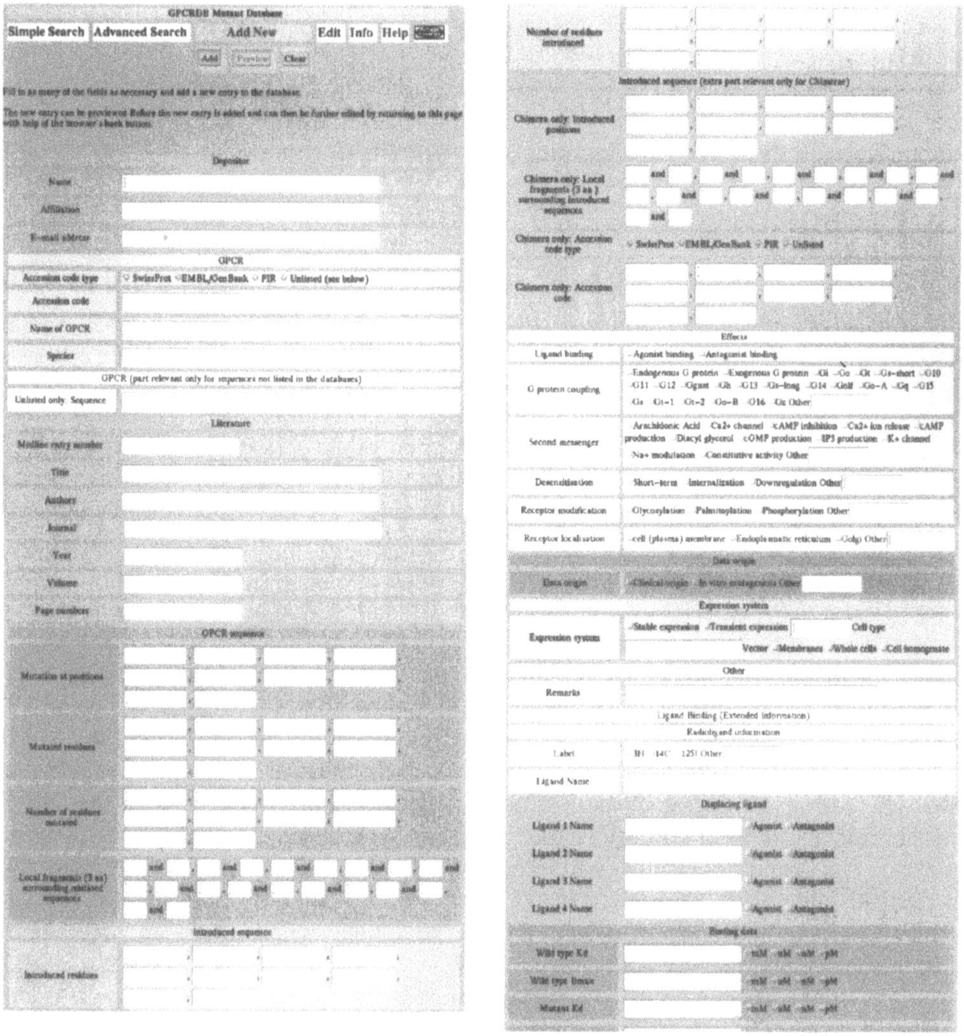

Figure 3. Mutant data submission form as part of the automated author-in system. The top left menu shows query and administration options.

benefits. The resulting page is very large, so to prevent visual fatigue, colour was used to breakup the large number of fields. To aid submission, all deposited entries can be searched and new entries can be seen in their final form before submission. Technically, this uses a non-indexed method so that searches can be done on newly deposited data. The system is simple but flexible so that it can be adapted when needed. It was designed to be generally useful and can be easily integrated with browsing systems like the Sequence Retrieving System (SRS; Etzold et al., 1996).

Maintenance of the GRAP database had to be terminated a few years ago because it was too much work for a small research group to document so many mutations so extensively. The tinyGRAP database that came in its place provides less information but as it can be maintained by about one full time person, it is nearly complete. If the submission burden was spread amongst experimentalists—each depositing mutation information via our WWW based system—it would probably be possible to maintain the GRAP database.

The mutation data can be inspected using either the query system that is a part of the tinyGRAP database, or the Viseur program (Campagne et al., 1999), or one can browse through the snake-like diagrams in the GPCRDB (Figure 2).

3.2. GPCRDB Database Design

We have now been working for more than three years on the GPCRDB. Much time was spent on so-called GPCR specific problems such as the use of experimental data or correlated mutations (see below) in order to optimize the profiles used for the sequence alignments. We also spent time on the design of inference engines. But we were unpleasantly surprised by the large amount of time that was needed for file conversions, detection and correction of errors in the data etc. A good example is the beautiful book by P. Seeman (Seeman, 1993). This book holds about 40.000 binding constants for ligand-receptor combinations. Dr Seeman has been so kind as to make this data available to the GPCRDB. However, when we received the data, it came on floppy disks written in a format that we could not read, on a computer that we did not have and by a program that we did not know. After getting the floppies converted by external specialists (thanks to Merck Darmstadt) about a hundred references were missing and had to be typed by hand. The ligand data was typed in a format mainly intended for human readability and a lot of software had to be written to make the tables computer-readable. The WWW allows for easy linking of data, and modern information systems use the HTML (HyperText Markup Language) hyperlink facilities to improve the user's browsing possibilities. The ligand data collection, though, was started in the days before the WWW existed, and the data organization was such that hyperlinking required a lot of reformatting. The ligand data is now well integrated in the GPCRDB and the fifty users that inspect this data every week seem to be happy with it, but we wonder if any of those users understand how much work went into the pages they are browsing. This was a practical lesson in open-standards for us.

The GPCRDB is a multi-country, multi-research group project, and it was decided to aim at a situation in which all data is only available where it is actually produced. Unfortunately, international networks are not always or everywhere fast enough yet to achieve this goal, and therefore we still keep most of the data on one machine. The GPCRDB has been designed, however, to be a multi-computer information system, and we expect that all data will be decentralized in less than five years.

The following sections will describe the database update procedure. It is envisaged that this description can be used as a guideline for the production of other molecular class-specific information systems.

3.2.1. Sequence Data Import. We import the GPCR protein sequences from the Swiss-Prot database. A list of all the available GPCR entries provided by Swiss-Prot makes this import easy. In the Swiss-Prot release 36.0 this list contains 1093 GPCRs, including bacteriorhodopsins. Every entry is imported. It would have been faster to only import the new GPCR entries, but a complete import permits us to detect any change in the former entries and, consequently, to optimize data quality (this extra work will no longer be needed once we can rely on a fast enough Internet network so that the sequences can stay where they are annotated and do not need transfering to the machine that hosts the GPCRDB).

Each "old" Swiss-Prot entry is compared to the current set of entries in the GPCRDB. The most important point is checking the pair identifier/accession number (ID and AC lines respectively). In the GPCRDB we name the entries by their ID as it is more explicit than the AC. However, it can be changed during the Swiss-Prot curation process (merged entries, new nomenclature, etc.). Only the AC line is definitive. Consequently, it is crucial to check whether the ID/AC couples are conserved. We also compare the date of sequence update to detect any sequence modification.

3.2.2. Sequence Cleaning. The entries with IDs that are no longer used are deleted and replaced by entries with a correct ID. Every new ID assignment is indicated in the GPCRDB.

GPCR fragments are stored in a special directory so as not to decrease the quality of the alignments. Even so, as they hold useful information, they are available in the GPCRDB. The current release includes 184 fragments.

3.2.3. Sequence Treatment. The new and the sequence-updated entries follow several steps. They are first classified using alignment against the family profiles and all the sequence-derived data (multiple sequence alignments, phylogenetic tree, models and correlated mutation analysis) of the given family and super-families is produced. The HTML pages that allow the user to access and browse the data are also created (Figure 4). We use the WHAT IF software (Vriend, 1990) for profile alignments, multiple sequence alignments, phylogenetic tree generation and for the family administration.

3.2.4. Sequence Classification. The sequence sorting is a crucial step on which the quality and the integrity of the sequence-derived data depend. A source file contains a list of profiles, each one being associated to one family or sub-family, i.e. the most specialized family level. Families are encoded in such a way that the hierarchical family classification can easily be retrieved. The same coding is used for the directory names. For example, code 001 is associated with class A, 002 with class B, etc., and code 001_001_002 represents the second sub-family of the amine receptors, which is defined as the first family of class A.

Each line of the profile list contains the name of a profile, the code of the family with which it is associated and the acceptance cut-off. WHAT IF aligns each new sequence to all profiles, retains the ten best scores, compares the convolution percentage (a measure of similarity) with the profile to the indicated cut-off and copies the sequence into the corresponding family directory and the ascendant directories. The file with profiles, cut-offs and family names still has to be maintained manually and needs to be updated every time a new family is discovered. Table 3 illustrates the sequence sorting process for the following example:

WHAT IF finds the highest convolution percentage for the sequence 'ABCD_HUMAN' with the profile of family 001_001_001_002 (muscarinic receptor vertebrate type 2). If the percentage of similarity is above the cut-off, the program copies the sequence into the following directories: 001_001_001_002/, 001_001_001/, 001_001/ and 001/. The latter three directories correspond to muscarinic, amine and class A receptor families respectively.

If this sorting method does not produce a result, the sequence is copied to a temporary directory where it stays until it gets classified manually. If adequate information is available to classify a sequence, either a new entry can be added to the profiles list to indicate this sequence's (sub-)family, or the acceptance cut-off can be decreased for one

Figure 4. Example of one of the 173 sub-families specific pages. Underlined text and icons are hyperlinked for navigation through the GPCRDB.

profile. Occasionally a new profile needs to be created. When the new entry shares no significant homology with any known GPCR, we add it to the orphan family. The analysis of phylogenetic trees and multiple sequence alignments, as well as the annotation on the function of the GPCRs (determined experimentally), help us to choose the appropriate destination of those GPCRs that cannot be classified easily.

The curator relies on literature and help from specialist users to convert (groups of) orphan receptors into a new family when experimental or computational evidence for the function of those (ex-) orphan receptors becomes available.

3.2.5. Sequence-Derived Data

3.2.5.1. Multiple Sequence Alignments. There is very little homology between the different GPCR families. The length and the sequence of the connecting loops vary a lot and aligning multiple GPCRs families is quite a delicate process. There is about 20%

Table 3. Data source for the sequence sorting process

a) Family name	Family code
Class A—Rhodopsin like	001
Amine	001_001
Acetylcholine (muscarinic)	001_001_001
Acetylcholine Vertebrate type 1	001_001_001_001
Acetylcholine Vertebrate type 2	001_001_001_002
Acetylcholine Vertebrate type 3	001_001_001_003
Acetylcholine Vertebrate type 4	001_001_001_004
Acetylcholine Vertebrate type 5	001_001_001_005
Acetylcholine Insect type 1	001_001_001_101
Acetylcholine Other	001_001_001_201
Adrenoceptors	001_001_002
Alpha Adrenoceptors	001_001_002_001
Alpha Adrenoceptors type 1	001_001_002_001_001
....	

b) Profile	Cut-off	Family code
muscar1m.prof	0.68	001_001_001_001
muscar2.prof	0.73	001_001_001_002
muscar3.prof	0.70	001_001_001_003
muscar4m.prof	0.84	001_001_001_004
muscar5.prof	0.70	001_001_001_005
muscar1d.prof	0.60	001_001_001_101
muscar4x.prof	0.79	001_001_001_201
a1aadren.prof	0.58	001_001_002_001_001
a1badren.prof	0.73	001_001_002_001_001
a1dadren.prof	0.80	001_001_002_001_001
....		

c) Profile	Family code	%similarity	%identity	Cut-off
1 muscar2.prof	001_001_001_002	74	78	73
2 gprx.prof	001_999_008	28	43	60
3 gust.prof	001_005_101	28	42	63
4 neurop1m.prof	001_002_014	40	41	70
5 unk23.prof	001_999_999	22	25	55

a) Top of the list of the GPCR families and their code. b) Top of the list of profiles used for the automatic sequence sorting. Only the most specialized family level is associated with one (or more) profile(s). c) Extract of the sorting method results for the virtual sequence "ABCD_HUMAN": it will be added to the acetylcholine muscarinic receptor type 2 family as its % of similarity is above the corresponding acceptance cut-off of 73%.

Table 4. Conserved positions for the 3 main GPCR classes[a]

Helices		I	II	III	IV	V	VI	VII
Class A	100%			R340				
(796)	≥95%	N130	L220		W420	P520	P620	P730
	≥90%		D224	C315				
Class B	100%	S120	H220	W332	P420	N510		Q718
(58)				E336				G719
				L340				C727
								N730
								V733
	≥95%	G116	R216	C315	G416	L521	L615	V722
		L124		Y339		K524	G620	
	≥90%		C214	Y324	W417		L613	Y726
			L229	N329	G418		P617	F728
							L618	E732
Class C	100%	G120	L210	C310	Q410	Y518	I611	L721
(22)		T124	Y212	G318	Q421	L522	F613	K728
			L215			C526	M615	
						A530	W622	
	≥95%	V127					I620	
		F131					P627	
	≥90%	A117					F625	I731
								I732

[a]Only the well conserved positions in the helices are indicated. The number of sequences for each family is indicated in brackets.

average pairwise similarity in an all-against-all GPCR sequence alignment. Fortunately, a few very conserved positions exist in each GPCR class, mainly located in the transmembrane domains. These residues are in a way the pillars of the multiple alignments and the knowledge of well conserved positions makes it possible to automatically validate the updated multiple sequence alignments. The newly produced multiple sequence alignments and profiles are automatically checked for the presence of these conserved positions and manually refined if needed. Table 4 summarizes these very conserved positions. Throughout the GPCRDB, we use the residue numbering scheme suggested by Oliveira et al. (1993). The residues are numbered so that the 100s digits indicate the helix number, and the most conserved residue in every helix has a round number.

The alignments are made for whole classes and families, but also for sub-families, sub-sub families, etc. The alignments are presented as HSSP files (Sander and Schneider, 1991) and as MSF files (Devereux, 1989), which allow the user to choose between two standard output formats, one of which displays the sequences horizontally (MSF), and the other vertically (HSSP). A coloured view of the multiple alignments is also available thanks to the Mview viewer (Brown et al., 1998).

3.2.5.2. Phylogenetic Trees. Phylogenetic trees are a good visualization tool for relationships between sequences in a family. This information can help to answer several different kinds of questions, e.g., questions related to ligand design. The WHAT IF program uses a neighbour-joining algorithm. Although the resultant phylogenetic trees represent, as accurately as possible, the pairwise identities between the sequences rather than their evolutionary relationships, there is a striking resemblance with phylogenetic trees based on an accepted-mutation parsimony algorithm (Kolakowski, 1994). The

Residue position	>sequence number> 1	2	3	4	5	6	7	8	9	10	11	12
7	A	A	S	T	T	S	L	L	I	V	S	P
8	L	L	L	D	D	D	L	L	L	L	D	D
119	K	R	R	R	R	R	R	K	K	K	K	R
120	E	E	E	I	I	I	E	E	E	E	I	I
121	S	S	S	T	T	*S*	S	S	S	S	T	T
122	K	E	V	T	L	L	I	V	E	E	T	T

Figure 5. Hypothetical example of correlated mutational behaviour. The residues at positions 7, 8 and 119–122 are shown for 12 (aligned) hypothetical sequences. Positions 8, 120 and 121 show correlated mutational behaviour to different extends. The correlation is strongest in the 8–120 pair where always L-E or D-I are observed. The correlations 8–121 and 120–121 are a bit weaker because of the aberrant behaviour of sequence 6 at position 121 (in italic).

phylogenetic trees presently available in the GPCRDB are made up of between 3 and 20 representative sequences per family.

3.2.5.3. Models. The GPCRDB server holds atomic coordinates of 3D models of GPCRs. Different modellers used different alignments, different modelling techniques, and different template structures to build these models. Consequently, a wide variety of models has been proposed. As it is at present not possible to decide which models are right and which are wrong, we have decided to store every suggested model in the GPCRDB. The models are grouped per depositor. For each model one can either download the coordinates or view them using a WWW helper application like Rasmol (Sayle and Milner-White, 1995).

3.2.5.4. Correlated Mutation Analysis. Correlated mutation analysis (CMA) is a computational method used to identify pairs of sequence positions which have remained conserved or mutated in tandem during evolution. Figure 5 shows a hypothetical example of a group of sequence positions which display correlated mutational behaviour. The idea behind the search for such pairs of residues is that when a mutation occurs at a functionally important site, the protein either becomes non-functional or may acquire its original or a different function due to a compensatory mutation at another position. Residues detected by the CMA method are often involved in the function of the protein, or in intermolecular interactions (Kuipers et al., 1996; Oliveira et al., 1995). The automatic detection of "important" residues is a relevant aspect of the GPCRDB effort.

3.2.5.5. Snake-Like Diagrams. Once the sequence-derived data is updated, three kinds of snake-like diagrams are generated. These plots are produced with the Viseur program and allow the user to view the GPCRs in two dimensions. The probable location of the seven transmembrane domains is extracted from the multiple alignments. The helix ends in the multiple sequence alignments were determined by Oliveira et al. (1995) using the methods described by Cronet et al. (1993).

The first plot is made for all new entries and for all those with an updated sequence (Figure 2). Amino acids are coloured based on their biochemical nature, except for the white-coloured positions, which are hyperlinked to mutant data in the tinyGRAP database. These links are based on a source file provided by the tinyGRAP people which associates each receptor accession number with one or several mutated positions.

The second snake-like diagram combines all the mutant data available for the members of one given GPCR family. The correspondence between the residue numbering of each sequence and that of the family consensus sequence is extracted from the positions of the residues in the multiple sequence alignment. These mutated positions are also hyperlinked to the tinyGRAP database. This family snake is provided for each GPCR family. This kind of plot is useful as it permits the transfer of mutant information from one given organism to another when their sequences are similar. However, we have noticed that, because chimeric constructions are often used in mutation studies, several individual and family snakes are completely white, meaning that all the sequence positions are associated with mutant information. In this case, it becomes impossible to instantly visualize point mutations. We are thinking about another schematic encoding of the mutation information in order to allow users to easily distinguish the different kinds of mutations.

The last snake-like diagram is also made for each family and sub-family, but displays the results of CMA. It is only calculated for families with more than eight members in order to give significant results. Each group of correlated positions is coded by a different color and hyperlinked to the family's multiple sequence alignment. The CMA results are also provided in a text file.

3.2.6. Database Inter-Operability. A recent addition to the GPCRDB consists of database cross-reference entries. A user-friendly view lists all the available pointers to local and remote information on each receptor present in the GPCRDB (Figure 6). Each pointer is hyperlinked to the corresponding data. This is done automatically by reading the Swiss-Prot entries and querying several remote databases as well as the GPCRDB itself. With one click the user can access all information available on the WWW for his or her favorite receptor.

A generic text file is first created in a computer-readable format (one type of data per line identified by a two-letter code). This data file allows for the generation of different views. A data representation in tabular form is then generated for use via the WWW.

This work on database cross-references pointed out several technical problems that WWW database developers should think about in order to improve the database interoperability.

WWW databases usually mention the corresponding Swiss-Prot AC or ID lines in their entries. Certain databases are sometimes one or two Swiss-Prot releases behind, and thus do not have entries that correspond to the most recently added or updated GPCR sequences. As all the cross-references are based on the Swiss-Prot identifiers, it is not rare when hyperlinked cross-references lead to nothing, i.e. dead links. Another delicate point is the change of nomenclature. This can lead to severe mistakes when one is using two different versions of the same GPCR entry. A typical example is that of the alpha-1 adrenoceptors: the alpha-1a subtypes have been renamed alpha-1c and the alpha-1d alpha-1a. Of course, the IDs of the corresponding GPCRs have also been renamed. Consequently, the content of one given adrenoceptor entry can be completely different between two databases that use a different Swiss-Prot release. For example, it could well happen that the ID "A1AA_HUMAN" (i.e. human alpha-1a adrenergic receptor) does not correspond to the same GPCR in the GPCRDB and in another database which is not regularly updated. A naming scheme that is based on accession numbers rather than file names prevents many, but not all of these problems.

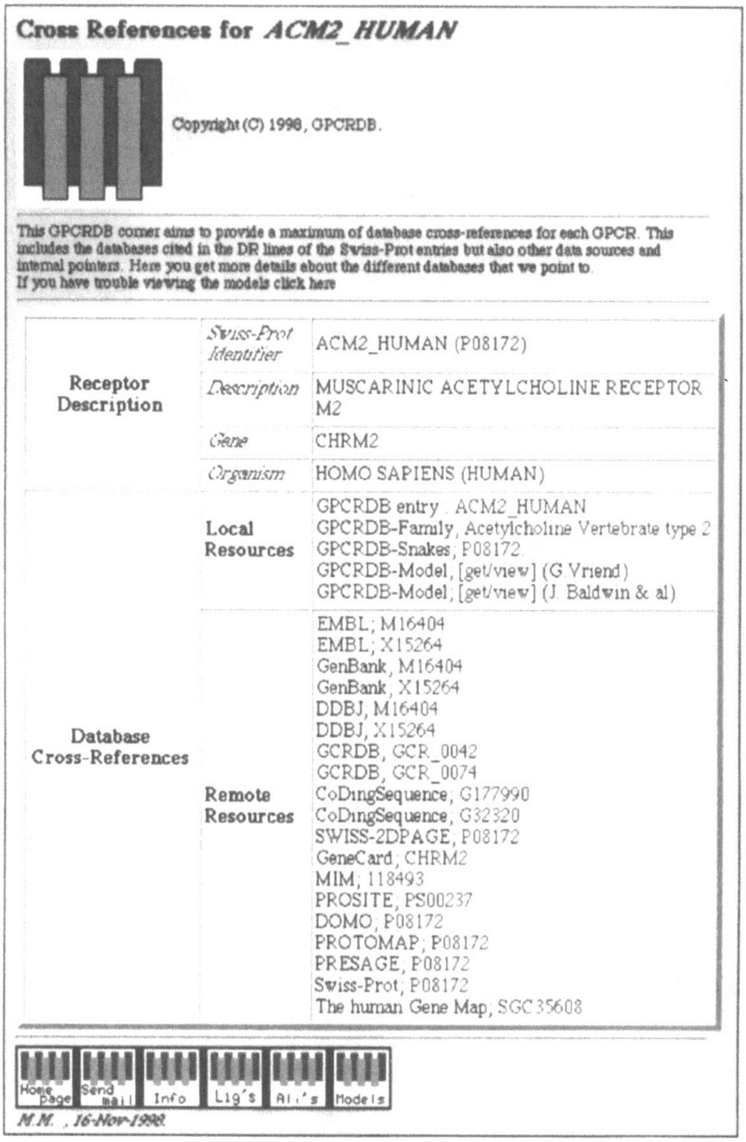

Cross References for *ACM2_HUMAN*

Copyright (C) 1998, GPCRDB.

This GPCRDB corner aims to provide a maximum of database cross-references for each GPCR. This includes the databases cited in the DR lines of the Swiss-Prot entries but also other data sources and internal pointers. Here you get more details about the different databases that we point to.
If you have trouble viewing the models click here

Receptor Description	*Swiss-Prot Identifier*	ACM2_HUMAN (P08172)
	Description	MUSCARINIC ACETYLCHOLINE RECEPTOR M2
	Gene	CHRM2
	Organism	HOMO SAPIENS (HUMAN)
Database Cross-References	**Local Resources**	GPCRDB entry . ACM2_HUMAN GPCRDB-Family, Acetylcholine Vertebrate type 2 GPCRDB-Snakes; P08172. GPCRDB-Model; [get/view] (G.Vriend) GPCRDB-Model; [get/view] (J. Baldwin & al)
	Remote Resources	EMBL; M16404 EMBL; X15264 GenBank, M16404 GenBank, X15264 DDBJ, M16404 DDBJ, X15264 GCRDB, GCR_0042 GCRDB, GCR_0074 CoDingSequence, G177990 CoDingSequence, G32320 SWISS-2DPAGE, P08172 GeneCard; CHRM2 MIM; 118493 PROSITE; PS00237 DOMO; P08172 PROTOMAP; P08172 PRESAGE, P08172 Swiss-Prot; P08172 The human Gene Map; SGC35608

Home Page | Send mail | Info | Lig's | Ali's | Models

M.M. , 16-Nov-1998

Figure 6. Example of the database cross-reference tables. For each GPCR, one can access all the available data through the World Wide Web.

Another bottleneck in the establishment of database cross-references is due to the lack of homogeneity or standardization of molecule identifiers. It sometimes requires extensive brain storming to find the correspondence between a GPCR and its corresponding entry in a remote database. The use of SRS is often useful for finding this correspondence, but mainly in the large monolithic but not in all the niche databases. This heterogeneity often implies working out elaborate computational methods in order to retrieve GPCR-related data from WWW databases.

4. DISSEMINATION FACILITIES

The GPCRDB has been conceived to provide fast and easy access to all information related to GPCRs. It should be an information tool that makes it easier for the user to think about GPCRs, and it should make suggestions for future research. For this purposes we have implemented, (and are still implementing) the four basic information system tools: browsing, retrieval, query and inferencing.

4.1. Browsing

The GPCRDB organization is based on the pharmacological classification of receptors and access to the data is obtained via a hierarchical list of known families in agreement with this classification. For one specific family, one can access the individual sequences, the multiple alignments, the profiles used to perform the latter, the snakes and a phylogenetic tree. Each type of data is displayed in a WWW page with hyperlinks to other data where appropriate. Figure 4 shows the WWW page for the muscarinic receptor family as an example. Extensive hyperlinking to other databases further increases the ease of browsing.

4.2. Retrieval

Often a user wants to work on certain data at home, independent of the GPCRDB environment. Therefore most data can be retrieved in its native form using the "save as" option of the WWW browsers or via anonymous FTP from www.gpcr.org, data being stored in the 7tm/ directory.

4.3. Query

A query system allows users to make simple queries via keywords as well as advanced queries, such as the search for a sequence pattern in a helix or a loop, by means of logical and regular expressions. The user can also refine the search by combining different queries. We are presently implementing a fault tolerant query system that will automatically adjust queries that lead to no hits. This adjustment can be linguistic (e.g. "human" corresponds to "*homo sapiens*") or relaxing (e.g. "in helix III" corresponds to "near helix III" or "PPP" to "PP").

In addition, a BLAST server at the EBI allows the user to scan one sequence pattern against all the sequences stored in the GPCRDB.

5. HARVESTING THE DATA

The main reason for setting up the GPCRDB was to be able to answer questions regarding GPCRs. In the following sections we will review 2 examples of research performed using the GPCRDB. Both studies extensively use the correlated mutation analysis (CMA) method. This method earlier has been used to determine residues that are important for ligand recognition in olfactory receptors (Singer et al., 1995), for determining which residues in the G protein are important for receptor binding (Oliveira et al., 1995), or for determining which residues in the receptor are involved in G protein selectivity (Horn et al., 2000).

5.1. The Interaction of Class B G Protein-Coupled Receptors with Their Hormones

We applied the correlated mutation analysis method for highlighting residues involved in the interaction between Class B or glucagon-like receptors and their ligands (Horn et al., 1998a).

In common with many G protein-coupled receptors, dysfunction in members of the Class B receptors can elicit a wide spectrum of disease-related activities. Consequently, they are potential targets in many different areas of pharmacological research. Unlike the class A or rhodopsin-like receptors, for which at least some structural similarity to bacteriorhodopsin has been detected, absolutely no structural information is available for the Class B GPCRs. They all have a large extra-cellular N-terminal domain (more than 120 amino acids) in which six cysteines are well conserved. These GPCRs are activated by peptides, many of which are about 30 amino acids long (Watson and Arkinstall, 1994).

We performed a computational study that exploits the experimental work performed by evolution in order to indicate those residues that are potentially involved in the binding of ligands with the Class B GPCRs. A CMA study revealed residues that show a strong correlation when analysing the receptors and the ligands at the same time. Residues that are detected by the CMA method normally have a functional role. As the only common function between the ligand and the receptor is binding to each other, we hypothesized that these residues that are detected in the CMA analysis are involved in direct contacts with each other. Only the class B ligand and receptor families which share enough sequence similarity have been analyzed in this study, i.e. the glucagon, glucagon-like peptide 1, growth hormone releasing factor, pituitary adenylate cyclase activating polypeptide, secretin and vasoactive intestinal peptide families.

Thirteen groups consisting of ligand and receptor residues, and thus potentially contacting residues, were detected by the CMA sequence analysis technique. We combined the CMA work with the results of many ligand and receptor mutation studies. Figure 7 summarizes the results. Our results are in agreement with a model for the receptor activation, in which the middle and C-terminal part of the ligands are in contact with the N-terminal part of the receptors, while the N-terminal residues of the ligands (roughly speaking residues 1–8) are located between the helices, perhaps as a short beta-hairpin loop. A detailed analysis of the available experimental data, a short summary of the literature used, and all computational details underlying this study and more detailed figures are available from the GPCRDB database (http://www.gpcr.org/7tm/articles/B_CMA/CMA.html).

5.2. Residues Important for Ligand Specificity in Amine Receptors

The correlated mutation analysis has also been used to enhance our understanding of the ligand binding on amine receptors (Kuipers et al., 1996).

Many receptors, such as serotonin (5-HT), adrenergic, muscarinic, dopamine, and histamine receptors, interact with ligands that contain a positively charged nitrogen atom. Several of the endogenous agonists (neurotransmitters) for these receptors are shown in Figure 8. The structure of acetylcholine, which activates muscarinic receptors, is rather different from other aminergic neurotransmitters. The aromatic ring system, which is present in most endogenous amine agonists, is replaced by a polar non-aromatic acetyl group in acetylcholine. Furthermore, acetylcholine contains a quaternary ammonium

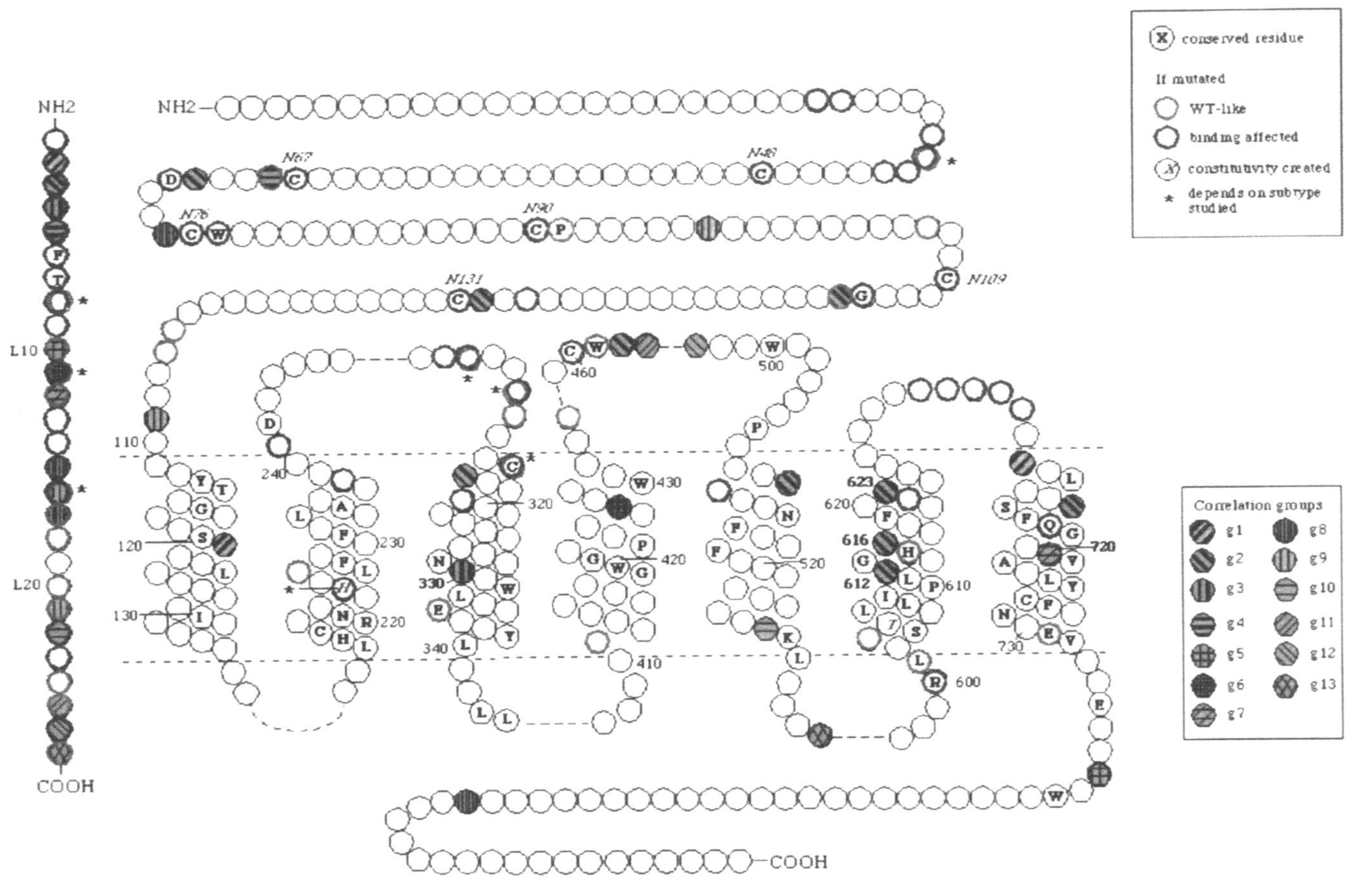

Figure 7. Schematic representation of the Class B receptors and ligands. The residue numbering is based on the consensus sequence generated from ligand and receptor multiple sequence alignments. Some key residues are numbered for easy reference. Bold numbers indicate that the residue was found in the CMA. Thin dashed lines indicate loops not included in the figure.

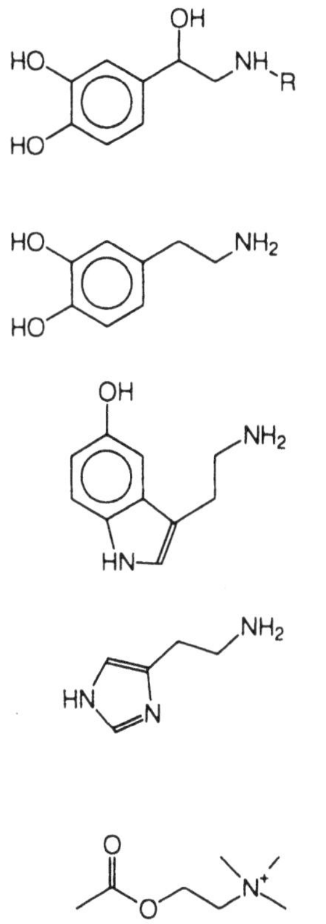

Noradrenaline R=H
Adrenaline R=CH3
Dopamine
Serotonin
Histamine
Acetylcholine

Figure 8. Natural agonists for amine receptors.

group with three ethyl substituents instead of a primary or secondary amine group as in other aminergic neurotransmitters.

This raises the question whether any residues exist that correlate with these structural differences. We compared 32 muscarinic receptor sequences with 144 other biogenic amine receptor sequences. The seven residues that discriminate best between these two classes are shown in Table 5. They are all located in or close to the putative ligand binding pocket for biogenic amine receptors. Residue positions 327 and 330 are in the middle of helix III close to the conserved aspartate residue (Asp-322) that interacts with the ligand's ammonium group. Residue position 231 is near Asp-322 in most GPCR models. The conserved mutations in positions 621 and 622, in the middle of helix VI, are of particular interest. At position 621, a highly conserved Phe in serotonin, dopamine and adrenergic

Table 5. Residues that discriminate between muscarinic and other biogenic amine receptors[a]

	231	233	327	33	621	622	722
Muscarinic	S	N	N	V	Y	N	C
Other amine	V	P	T	I	F	F	G
	(5I)	(1S)		(1H)	(1A)		

[a]The sequences used can be obtained from the GPCRDB. Here, 32 muscarinic receptors were compared with 144 other amine receptors. Four of the seven residue positions are not 100% conserved in the non-muscarinic aminergic receptors. The alter native residues and their frequency of occurence are indicated between brackets.

receptors, is replaced by an equally conserved Tyr in muscarinic and histamine receptors. The more polar character of Tyr, and its capability of forming hydrogen bonds, is in good agreement with the more polar character of their endogenous agonists. The importance of residue 621 for muscarinic agonist affinity was confirmed by the mutation of Tyr-621 Phe in M3 muscarinic receptors, which decreases agonist binding (Wess et al., 1991). The highly conserved Phe-622 in all other amine receptors is an Asn in all muscarinic receptors. This mutation agrees with the structural differences between the endogenous agonists of these receptors. Asn-622 is probably capable of forming hydrogen bonds with the polar acetylcholine, whereas Phe-622 has an aromatic-aromatic interaction with the other amine neurotransmitters. The importance of Phe-622 for agonist affinity was confirmed by mutation studies in the b-adrenoceptor and the 5-HT_{2A} receptor (Strader et al., 1989; Choudhary et al., 1993). Residue 722 is adjacent to positions claimed to bind acetylcholine in muscarinic receptors (Wess et al., 1992).

6. HINTS FOR INFORMATION SYSTEM DESIGN

Thousands of so-called "niche" databases or "boutique" databases, or other information systems can be found by browsing the WWW. Many of these systems are not updated regularly and therefore barely worth using. The normal life cycle of a niche database is as follows:

A scientist interested in X has, after years of writing grant applications, received money to create a database for X. A student is hired, and with great enthusiasm the programming is started (of course without first talking to database specialists, and without making a proper database schema). After a while (often too early) database X is made accessible via the WWW, and thanks to users who ask questions and thanks to the enormous energy of the student, system X grows and shows many colourful and ingenious features. But Doomsday is rapidly approaching. After three years, the student becomes a doctor and leaves the group. The professor writes a nice final report, but does not get the grant extended. That would be the end of the story, unless somebody else were able to take over the management of the database. This "somebody else" could be a company interested in marketing database X, or one of the large biocomputing institutes like EMBL, EBI or NIH. However, in order for the management of a database to be transfered from one group to another, several requirements must first be met:

- The raw data must be available as keyword-driven flat files.
- Every unit of information has a unique identifier associated with it.
- Standards must be adhered to whenever possible.
- Software should be written-portable, in a commonly used language.
- The WWW based view should be totally independent of the data.
- Search engines must be able to work as stand-alone programs.
- Everything should be simple, devoid of fancy features.
- Individual aspects of the system (i.e. the query engine and the browse facility) should be independent of each other.

Most databases, however, are either poorly designed or only of interest to a small audience. This makes it difficult to pass on the database to another team after funding runs out. And that is why there are so many databases on the WWW that are years behind with regard to updating. We therefore strongly suggest that niche database systems are made available via the WWW only if long-term maintenance can be guaranteed.

ACKNOWLEDGMENTS

The GPCRDB is an EU sponsored project (PL 950224). We thank K. Aberer, A. Bairoch, M. Beukers, R. Bywater, F. Campagne, G. Casari, Ø. Edvardsen, U. Göbel, L. Holm, F. Kolakowski, W. Kuipers, L. Oliveira, F. Rippmann, A. Tramontano, A. Valencia and A.P. IJzerman for stimulating discussions. We thank J. Eckert and K. Krmoian for technical assistance. We also thank Novo Nordisk Copenhagen Denmark, Solvay Weesp the Netherlands, E. Merck Darmstadt Germany and IRBM Rome Italy for testing aspects of the GPCRDB.

REFERENCES

Attwood, T.K., Croming, M.D., Elowen, D.R., Lewis, A.P., Mabey, J.E., Scordis, P., Selley, J.N., and Wright, W. (2000) PRINTS-S: the database formerly known as PRINTS. Nucleic Acids Res. **28**, 225–227.

Bairoch, A. and Apweiler, R. (2000) The SWISS-PROT protein sequence data bank and its supplement TrEMBL in 1999. Nucleic Acids Res. **28**, 45–48.

Baker, W., Vanden Broel, A., Camon, E., Hingamp, P., Sterk, P., Stoesser, G., and Tuli, M.A. (2000) The EMBL Nucleotide Sequence Database. Nucleic Acid Res. **28**, 19–23.

Baldwin, J.M. (1993) The probable arrangement of the helices in G protein-coupled receptors. EMBO J. **12**, 1693–1703.

Baldwin, J.M., Schertler, G.F.X., and Unger, V.M. (1997) An alpha-carbon template for the transmembrane helices in the rhodopsin family of G-protein-coupled receptors. J. Mol. Biol. **272**, 144–164.

Benson, D.A., Karsch-Mirrachi, I., Lipman, D.J., Ostell, J., Rapp, B.A., and Wheeler, D.L. (2000) GenBank. Nucleic Acid Res. **28**, 15–18.

Beukers, M.B., Kristiansen, K., IJzerman, A.P., and Edvardsen, O. (1999) TinyGRAP database: a bioinformatics tool to mine G protein-coupled receptor mutant data. Trends Pharmacal. Sci. **20**, 475–477.

Brown, N.P., Leroy, C., and Sander, C. (1998) MView: a web-compatible database search or multiple alignment viewer. Bioinformatics **14**, 380–381.

Campagne, F., Jestin, R., Recerasak, J.L., Bernassau, J.M., and Maigret, B. (1999) Visualisation and integration of G protein-coupled receptor related information help the modelling: description and application of the VISEUR program. J. Comput Aided Mol. Des. 13, 625–643.

Choudhary, M.S., Craigo, S., and Roth, B.L. (1993) A single point mutation (Phe340-Leu340) of a conserved phenylalanine abolishes 4-[125I]iodo-(2,5-dimethoxy)phenylisopropylamine and [3H]mesulergine but not [3H]ketanserin binding to 5-hydroxytryptamine2 receptors. Mol. Pharmacol. **43**, 755–761.

Cronet, P., Sander, C., and Vriend, G. (1993) Modelling of transmembrane seven helix bundles. Prot. Eng. **6**, 59–64.

Devereux, J. (1989) The GCG Sequence Analysis Software Package, Version 6.0. Genetics Computer Group, University of Wisconsin Biotechnology Center, 1710 University Avenue, Madison, Wisconsin, USA, 53705.

Edvardsen, ϕ. and Kristiansen, K. (1997) Computerization of mutant data: the tinyGRAP mutant database. 7TM journal **6**, 1–6.

Etzold, T., Ulyanov, A., and Argos, P. (1996) SRS: information retrieval system for molecular biology data banks. Methods Enzymol. **266**, 114–128.

Gouldson, P.R., Snell, C.R., and Reynolds, C.A. (1997) A new approach to docking in the beta 2-adrenergic receptor that exploits the domain structure of G-protein-coupled receptors. J. Med. Chem. **40**, 3871–3886.

Gudermann, T., Nurnberg, B., and Schultz, G. (1995) Receptors and G proteins as primary components of transmembrane signal transduction. Part 1. G-protein-coupled receptors: structure and function. J. Mol. Med. **73**, 51–63.

Guex, N. and Peitsch, M.C. (1997) SWISS-MODEL and the Swiss-PdbViewer: an environment for comparative protein modeling. Electrophoresis **18**, 2714–2723.

Hibert, M.F., Trumpp-Kallmeyer, S., Bruinvels, A., and Hoflack, J. (1991) Three dimensional models of neurotransmitter G-binding protein coupled receptors. Mol. Pharm. **40**, 8–15.

Hol, W.G.J. (1986) Protein crystallography and Computer graphics—toward rational drug design. Ang. Chem. **25**, 767–777.

Horn, F., Bywater, R., Krause, G., Kuipers, W., Oliveira, L., Paiva, A.C.M., Sander, C., and Vriend, G. (1998a) The interaction of class B G protein-coupled receptors with their hormones. Receptors Channels **5**, 305–314.

Horn, F., Vander Wenden, E.M., Oliveira, L., IJzerman, A.P., and Vriend, G. (2000) Receptors coupling to G proteins: Is there signal behind the sequence. Proteins, accepted.

Horn, F., Weare, J., Beukers, M.W., Hörsch, S., Bairoch, A., Chen, W., Edvardsen, Ø., Campagne, F., and Vriend, G. (1998b) GPCRDB: an information system for G protein coupled receptors. Nucleic Acids Res. **26**, 277–281.

Kolakowski, L.F. Jr (1994) GCRDb: a G-protein-coupled receptor database. Receptors Channels **2**, 1–7.

Kristiansen, K., Dahl, S.G., and Edvardsenm Ø. (1996) A database of mutants and effects of site-directed mutagenesis experiments on G-protein coupled receptors. Proteins **26**, 81–94.

Kuipers, W., Oliveira, L., Paiva, A.C.M., Rippman, F., Sander, C., and IJzerman, A.P. (1996) Analysis of G Protein-Coupled Receptor Function. In "Membrane Protein Models" (Ed. Findlay, J.), Bios Scientific Publishers Ltd: Oxford, pp 27–45.

Luecke, H., Richter, H.-T., and Lanyi, J.K. (1998) Proton transfer pathways in bacteriorhodopsin at 2.3 Angstrom resolution. Science **280**, 1934–1937.

Oliveira, L., Paiva, A.C.M., and Vriend, G. (1993) A model for G-protein coupled receptors. J. Comp. Aided Mol. Des. **7**, 649–658.

Oliveira, L., Paiva, A.C.M., and Vriend, G. (1995) Correlated mutations analysis of G protein α-chains to search for residues linked to binding. In "Peptides: Chemistry, Structure and Biology" (Eds Kazonaya, P.T.P. and Hodges, R.S.), Mayflower Scientific Ltd.

Oliveira, L., Paiva, A.C.M., Sander, C., and Vriend, G. (1994) A common step for signal transduction in G protein-coupled receptors. Trends Pharmacol. Sci. **15**, 170–172.

Oliveira, L., Paiva, A.C.M., Vriend, G. (1999) A low resolution model for the interaction of G proteins with GPCRs. Prot. Enging. **12**, 1087–1095.

Pebay-Peroula, E., Rummel, G., Rosenbusch, J.P., and Landau, E.M. (1997) X-ray structure of bacteriorhodopsin at 2.5 Angstroms from microcrystals grown in lipid cubic phases. Science **277**, 1676–1681.

Sander, C. and Schneider, R. (1991) Database of homology-derived protein structures and the structural meaning of sequence alignment. Proteins **9**, 56–68.

Sayle, R. and Milner-White, E.J. (1995) RASMOL: biomolecular graphics for all. Trends Biochem. Sci. **20**:374–376.

Schertler, G.F.X. (1998) Structure of rhodopsin. Eye **12**, 504–510.

Schertler, G.F.X., Villa, C., and Henderson, R. (1993) Projection structure of rhodopsin. Nature **362**, 770–772.

Seeman, P. (1993) Receptor Tables, vol.2: Drug dissociation constants for neuroreceptors and transporters. Toronto: SZ Research.

Singer, M.S., Oliveira, L., Vriend, G., and Shepherd, G.M. (1995) Potential ligand-binding residues in rat olfactory receptors identified by correlated mutation analysis. Receptors Channels **3**, 89–95.

Skoufes E., Marenko, L., Nadkarni, P.M., Miller, P.L., and Shepherd, G.M. (2000) Olfactory receptor database: a sensory chemoreceptor resource. Nucleic Acid Res. **28**, 341–343.

Strader, C.D., Sigal, I.S., and Dixon, R.A. (1989) Structural basis of beta-adrenergic receptor function. FASEB J. **3**, 1825–1832.

Takeda, K., Sati, H., Hino, T., Kono, M., Fukuda, K., Sakurai, I., Okada, T., and Kouyama, T. (1998) A novel three-dimensional crystal of bacteriorhodopsin obtained by successive fusion of vesicular assemblies. J. Mol. Biol. **283**, 463–474.

Tateno, Y., Miyazaki, S., Ota, M., Sugawara, H., and Gojobori, T. (2000) DNA data bank of Japan (DDBJ) in collaboration with mass sequencing teams. Nucleic Acid Res. **28**, 24–26.

Unger, V.M. and Schertler, G.F.X. (1995) Low resolution structure of bovine rhodopsin determined by electron cryo-microscopy. Biophys. J. **68**, 1776–1786.

Unger, V.M., Hargrave, P.A., Baldwin, J.M., and Schertler, G.F.X. (1997) Arrangement of rhodopsin transmembrane alpha-helices. Nature **389**, 203–206.

van Rhee, A.M. and Jacobson, K.A. (1996) Molecular Architecture of G Protein-Coupled Receptors. Drug Devel Res. **37**, 1–38.

Vriend, G. (1990) WHAT IF: A molecular modelling and drug design program. J. Mol. Graph. **8**, 52–56.

Watson, S. and Arkinstall, S. (Eds.) (1994) The G-protein linked receptor facts book. Academic Press.

Wess, J., Gdula, D. and Brann, M.R. (1991) Site-directed mutagenesis of the m3 muscarinic receptor: identification of a series of threonine and tyrosine residues involved in agonist but not antagonist binding. EMBO J. **10**, 3729–3734.

DISTRIBUTED APPLICATION MANAGEMENT IN BIOINFORMATICS

M. Senger,[1] P. Ernst,[2] and K.-H. Glatting[2]

[1]EMBL Outstation—European Bioinformatics Institute
Wellcome Trust Genome Campus
Hinxton, Cambridge, CB10 1SD, United Kingdom
[2]DKFZ—German Cancer Research Center
Department of Molecular Biophysics
Im Neuenheimer Feld 280
D-69120 Heidelberg, Germany

ABSTRACT

This paper describes approaches to achieve distributed access to analysis tools in the life sciences domain. The underlying technologies are the Web, CORBA, and the use of descriptive meta-data. The need for standarisation, extensibility and portability is underlined. The two separate applications presented here (W2H and AppLab) are both available freely.

INTRODUCTION

Bioinformaticians are dependent upon many vast databases and hundreds of applications to analyse their data. These analysis tools use sophisticated algorithms and data access methods but often suffer from a lack of factors necessary to provide a scaleable, flexible and user-friendly distributed application environment. These factors include, but are not limited to:

- Unified and intuitive user interfaces
- Platform independence and portability
- Using well-defined standards to provide extensibility
- Flexibility and customisation
- Protocols allowing co-operation between analysis components
- Secure access to sensitive data

Genomics and Proteomics, edited by Sándor Suhai.
Kluwer Academic / Plenum Publishers, New York, 2000.

Other more technical factors include:

- Load balancing for CPU consuming algorithms
- Low-cost maintainability

A well-defined distributed environment can address most of the issues above. In the next sections, we will present several approaches, based on different underlying technologies, which we have implemented and been using for the last several years.

WEB TECHNOLOGY—W2H

The issue of providing a unified user-interface to a large number of bioinformatics analysis tools arose first with the GCG sequence analysis package (Devereux et al., 1984) and other derived services, e.g. EGCG (Rice et al., 1995) and HUSAR (http://genome.dkfz-heidelberg.de).

The command-line interface to GCG has limited user-friendliness, and the graphical X-Window interface (WPI, SeqLab) supplied by GCG can often not be used on PC or Macintosh computers due to the lack of an X-server. Web technology (also adopted recently by GCG in the SeqWeb interface) as used by W2H brings immediately the advantages of:

- Platform independence on the client side
- A unified interface well known to the users using conventional web browsers
- A real client-server distribution, with optimisation of the network round-trips

W2H[1] presents more than one hundred sequence analysis tools in a unified environment (Figure 1). A typical scenario starts with a user choosing one or more sequences that are to be analysed. They continue by choosing an application and then entering analysis-controlling parameters in a dynamically generated HTML form (Figure 3). The basic cycle concludes with the running of the application and the browsing of the results (Figure 4). The W2H interface is free to both commercial and non-commercial organisations. However, the GCG programs to which the W2H package provides an interface are licensed software (Devereux et al., 1984). However, the W2H implementation is based upon the use of meta-data, which makes it possible to add non-GCG applications easily. As a result, W2H allows:

- Flexibility and extensibility
- Customisation through the specification of site-defined output data interpreters, and user-defined personal preferences

System and Methods

From its inception in 1996, W2H was designed with the following requirements:

- Platform independence: As there are an extremely large variety of computer resources in the bioinformatics domain, it is necessary to find a common denominator for a user interface. An obvious solution is to follow the trend of using the Web browsers that are available on many platforms.
- Quick access to the server: A real client/server architecture is required to save time-consuming networking. A solution that allowed the delegation of part of

[1] http://industry.ebi.ac.uk/w2h

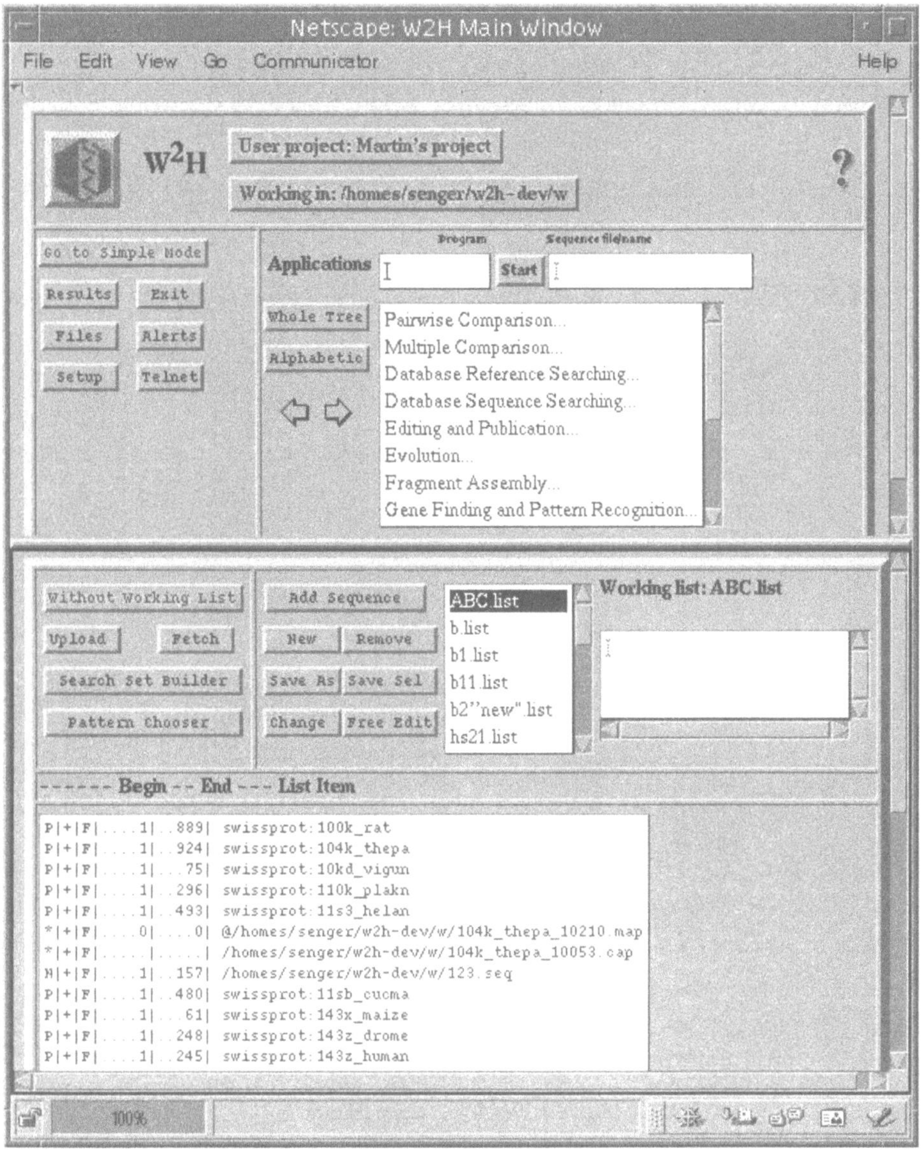

Figure 1. W2H main window—advanced mode.

the functionality to the client had to be found. The scripting language JavaScript allows the separation of verification from other features, and to have client browsers process this directly.

- Secure access to the server and data: GCG programs and other services have to be run on a remote server whilst remaining under the control of the users who invoked them. Privacy for a user's data on the remote server and during network transfer is also a consideration.
- State-holding server: The users of W2H have to be able to start GCG or other programs, interrupt their sessions and then be able to continue from where they

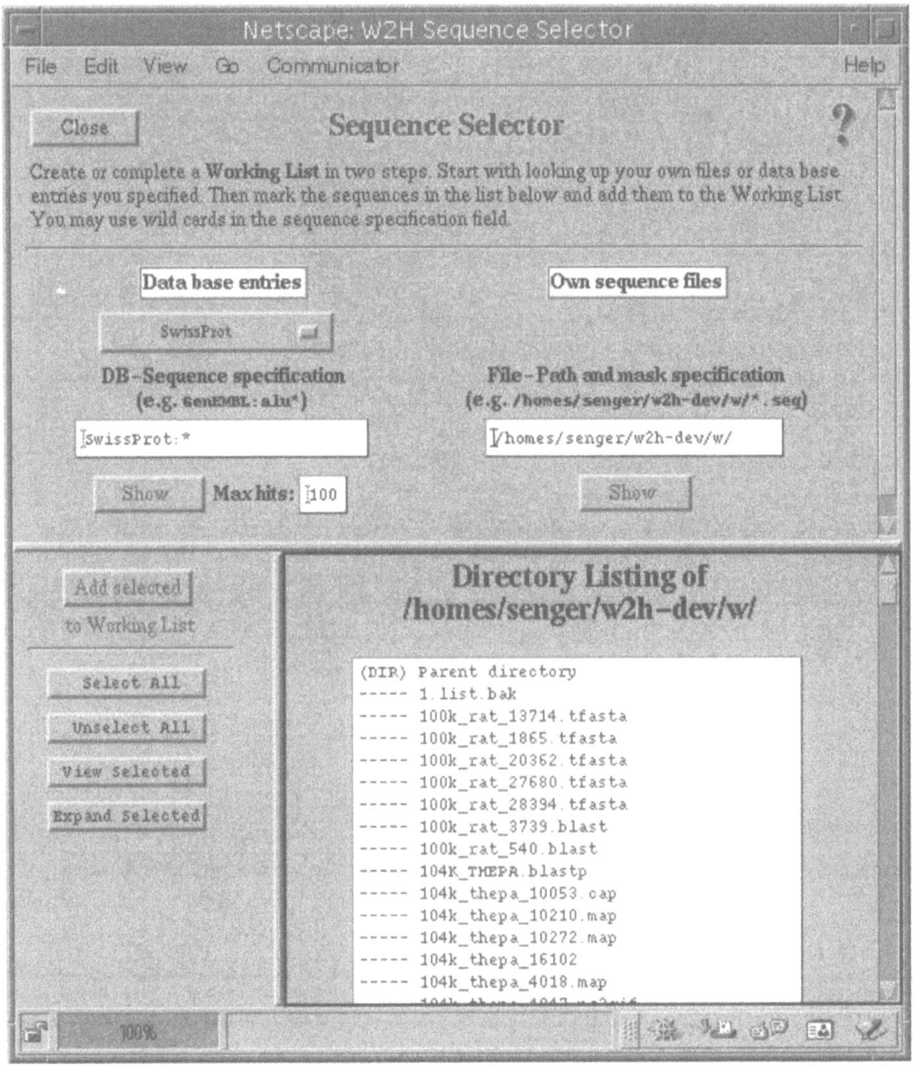

Figure 2. W2H—sequence selector.

had left off at a later time. A set of current user settings and preferences should be stored and reused later. Last but not least, the problem of re-starting the GCG name service, which can be quite slow, had to be addressed.

- Dependency on GCG installation: An underlying premise is that there should be no need to change anything in the GCG installation to be able to run the W2H interface. The approach of using a special W2H configuration and dynamically created HTML files is used.

To be able to fulfil all these requirements, the following server and client architecture was designed.

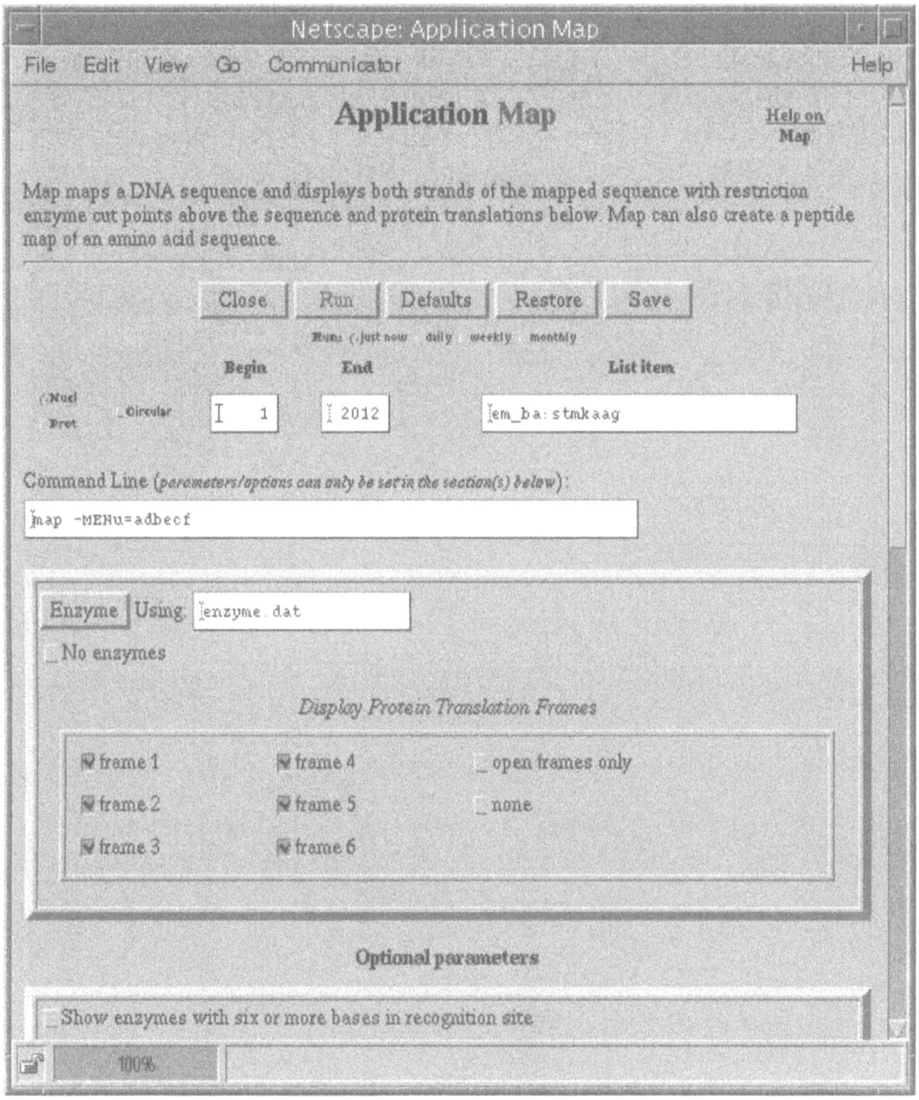

Figure 3. W2H—an application entry form.

Server Architecture Design. W2H uses the HTTP protocol to talk with the HTTP server and to start CGI scripts on the server side. However, there are at least two special issues to be considered.

Firstly, the applications started on the server side can take a long time to complete. This means that there must be a mechanism which is able to run a CGI script, and then cut the connection between the browser and HTTP server while still maintaining and monitoring the executing application.

The second issue is related to the GCG naming service. All GCG tools are required to run inside an environment built on the top of shared memory blocks. It is often time consuming (up to several tens of seconds) to establish this environment and it would be inefficient to repeat this for each client request.

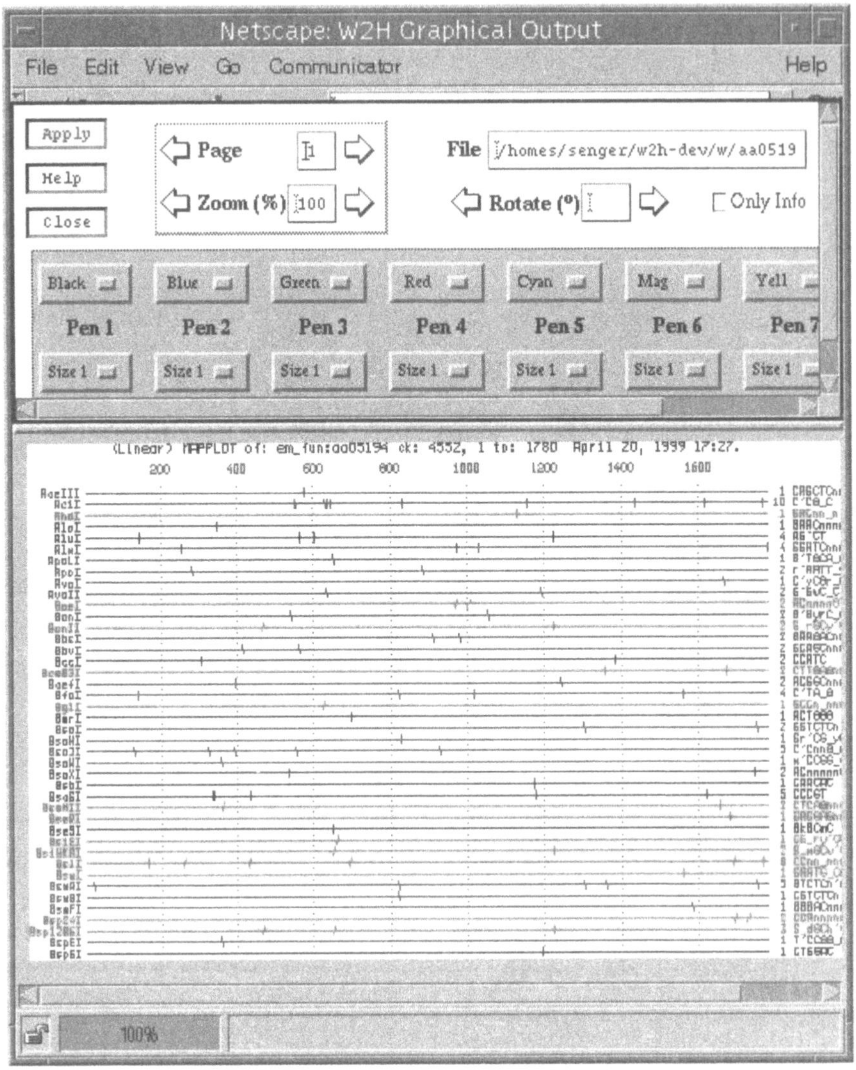

Figure 4. W2H—example of a graphical output.

Keeping state information on the server side solves both issues. W2H achieves this by keeping the following data for each user on the server side:

- GCG session environment: The GCG package allows the use of a special environment variable to keep access to an established GCG name service. Colet (Colet and Herzog, 1996) first implemented usage of this variable in the HTTP context. The result from the point of view of the user is that the first connection to W2H server can be slow, but all subsequent accesses are as fast as the network allows. Moreover, W2H provides fault-tolerance by determining if the access to the GCG name service is still valid; if not then it restarts the name service in a transparent manner.
- A table of processes: W2H creates and maintains a record about each application invoked. The table contains a list of data sources, process identification and

log information. This table ensures that if the connection is cut, the next request can reconnect, look into the table, check if the processes are still running and report it in a dynamically created HTML document.

- A table for tracking all output files. W2H can constantly report the current status of processes and created data by investigating the table of processes and using client-pull technology.
- User preferences: Both application-dependent and general preferences are kept for each user. These include the last used working directory, a current working list of sequences to be analysed, a set of GCG specific global options (such as the length of sequence documentation used), a preferred graphical device, etc. When a W2H session starts, these preferences are mirrored on the client side using Netscape cookie technology.

Besides from keeping the state information, each started application also has its own W2H script waiting and controlling the application. In this way a comprehensive log file can be maintained, and the user can always get fresh information about what is happening. A similar approach can be used to group the application into batches.

Client Architecture Design. The client must be as independent from the server as possible. This saves network round-trips and gives a faster response for the end user. W2H uses JavaScript to achieve this. JavaScript is very close to HTML forms but with extensions that listen for user events and requires less code to be transferred across the network compared to Java applets. The main domains for JavaScript are:

- Parameter checking: The proper formats, allowed range, and sequence type dependencies are checked. It would also be useful to check dependency rules between parameters but this would require much more dynamic behaviour of the HTML layout than is provided by current browsers.
- Creation of the command line: This is not a crucial requirement because the command line is created and used on the server side. However, this provides an immediate image of the parameters that will be used without having retrieve this from the server.
- Confirmation messages: W2H creates small HTML windows to confirm requested actions before sending the request to the server (for example, before deleting output files).
- Help cards: The W2H comprehensive help system uses JavaScript to organise the help pages into smaller pieces (help cards) which are all transferred within one HTML document, saving network load and making the help facility much more interactive.

Security Issues. Since the operation of a web server raises several security issues, the W2H design has to consider carefully how to protect the server machine from unauthorised use, and how to prevent unauthorised access to user data by other users. The authentication mechanism of the HTTP protocol ensures that only authorised users get access to the server machine. Data protection is accomplished using a special server script which changes the HTTP server process started by an authenticated user, to a user-identified process so that all the processes can be related to individual users and the resulting data can be stored in these users' own disk space on the server side. Using the UNIX access rights mechanism, this user's disk space and data can then be protected from access by unauthorised users.

An additional security aspect that needs consideration is secure data transfer over the network. Fortunately, current web technologies allow encrypted datatransfer (completely transparent to web applications) between the user's web browser and the remote server. It is therefore sufficient for W2H to use the Common Gateway Interface (CGI) for network communication. An optional data encryption can thus be provided by using standardised mechanisms in browser and server, known as HTTPS. The transferred data are encrypted and unencrypted "on the fly" by introducing an encrypting SSL protocol layer (Secure Socket Layer[2]) between the HTTP and the TCP protocol layers. This technology is supported by most popular web browsers and HTTP/HTTPS servers.

For special environments, like workshops, conferences and company intranets, there is a special W2H mode (Intranet mode) with less security constraints that allows access to W2H without having separate UNIX accounts for all users on the server side. In these situations, security is considered either not too important (e.g. for conferences), or is provided by an independent mechanism outside of W2H (e.g. a site firewall).

Meta-Data Files. The GCG package comes with a pre-defined set of configuration files for all included applications. These can be considered as meta-data files describing usage and properties of any non-interactive, command-line driven application. W2H uses these files to generate on-the-fly HTML documents with forms for entering values for command-line parameters to the applications. Every application has its own HTML document but all of them follow the same style. The GCG configuration files contribute to the W2H design.

The meta-data files are human-readable text files containing the description of the entire application (name, type of sequences it deals with, etc.), the description of the output files (types, how to name files, etc.), the detailed description of all command-line parameters, including:

- A parameter description (prompt)
- A method describing how to place the parameter on the command-line
- A default value for the parameter
- Additional data for parameter validation
- Layout hints on how to display the parameter on the screen
- A definition of any dependency rules between parameters

The syntactical rules (i.e. grammar) of these files had to be determined, and a parser was designed to process them. It is easy to extend these meta-data files as there is only one layer to the W2H parser that processes them.

Open Architecture. W2H was developed primarily as an interface to GCG tools but its architecture allows the addition of new, non-GCG applications. A new application must be described by a meta-data file, and generally be wrapped in a shell script so it behaves similarly to other GCG programs. Both requirements can be achieved easily and templates for the shell wrapping are provided.

For the display of output data, W2H can be extended to use a set of result viewers. Depending on the application that produced the results, and on the output specification (as defined in a meta-data file), W2H decides which viewer or output interpreter to use.

---

[2] http://home.netscape.com/eng/security/

Implementation

The W2H interface is quite comprehensive, providing advanced features like sequence selector (Figure 2), search set builder, enzyme chooser, and access to sequence databases in addition to the GCG analysis programs. It is available in two modes: the fully featured "advanced mode" and the more problem-oriented "simple mode" (Figure 5). Both modes are interchangeable. An important feature is the possibility to upload client sequences both as files and as "copy and paste" areas to the server where they are to be processed.

However, There are Still Non-Addressed Issues . . .

W2H is being used by many users now and has achieved quite a success. However, from the point of view of a well-designed distributed architecture as described at the beginning of this article, there are still several open issues:

Server-Side Platform Independence. This was not considered as an important issue because of the underlying GCG package running on one server only. But with possibility of adding non-GCG applications, the topic becomes important. There is a need to be able to re-distribute computational efforts between more application servers and ideally, to be able to control the load balancing between them too.

Poor Standardisation. W2H uses CGI interfaces without any application-specific standards. For better and deeper integration, we need a well-defined interface to analysis tools, which should be independent of the implementation.

Furthermore, the meta-data approach, which proved to be very successful, could be based on a more publicly aware format. The XML[3] format seems to be predestined for this.

Little Support for Analysis Co-Operation. W2H supports only two basic forms of component collaboration:

- Simple piping of results from one tool to another. W2H attempts to mimic the current behaviour of GCG SeqLab interface. However, this piping is defined in meta-data and is static.
- Hooking together user-defined data post-processors is powerful, flexible and quite dynamic, but it was meant primarily for displaying output data, not for chaining several applications together.

CORBA TECHNOLOGY—APPLAB

The life science domain became very active recently in attempts to define specifications and standards to achieve better compatibility and integration between domain components. This approach reflects the richness and quantity of biological data, as well as the need for better data exchange between analysis tools.

In 1997, a Life Science Research (LSR) Domain Task Force was established and started to work in the frame of the Object Management Group (OMG). The OMG is a

---

[3] http://www.w3c.org/XML

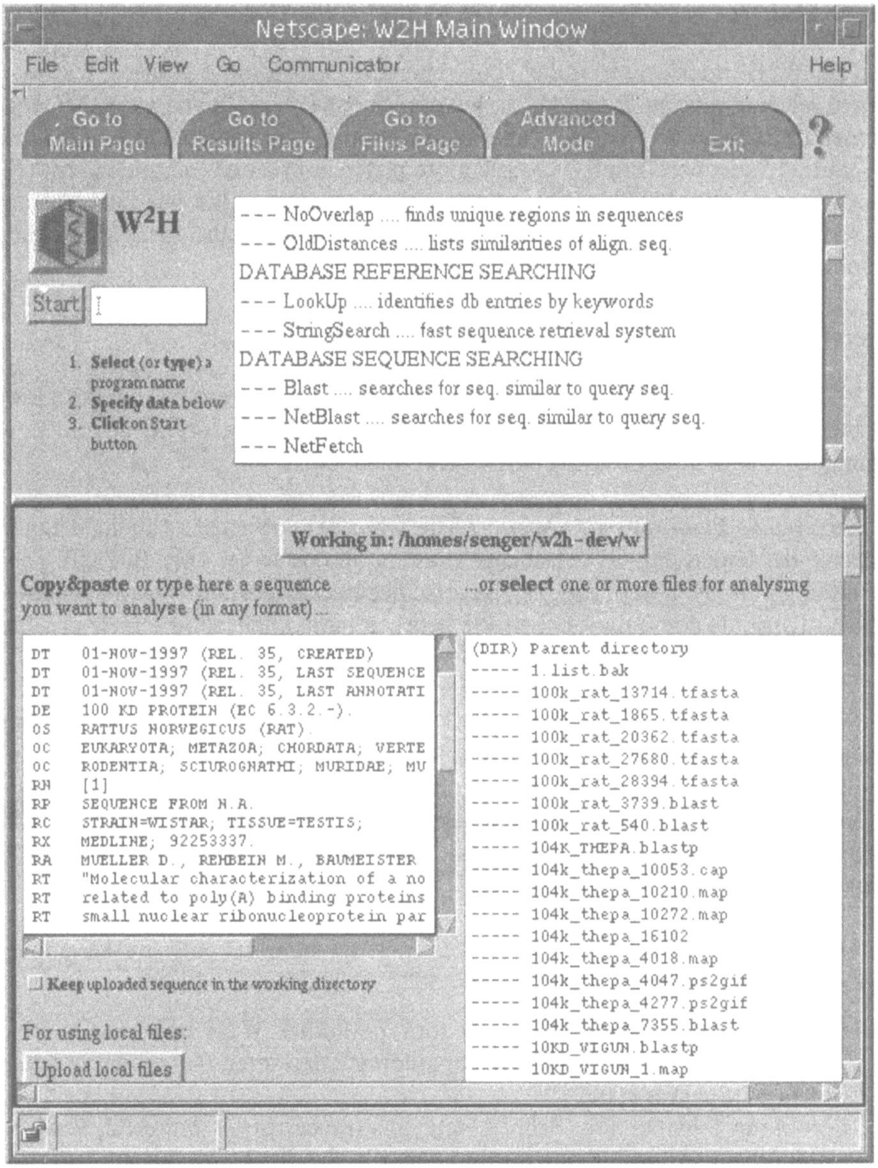

Figure 5. W2H main window—simple mode.

consortium of more then 800 industrial, governmental and academic institutions and is committed "... to develop technically excellent, commercially viable and vendor independent specifications for the software industry ...".[4] It defines and uses CORBA[5] and the Object Management Architecture (OMA) to achieve this. The LSR defines domain standards in many areas,[6] including, but not limited to:

---

[4] http://www.omg.org/omg/background.html

[5] Common Object Request Broker Architecture (http://www.omg.org/corba/beginners.html)

[6] http://www.omg.org/homepages/lsr

- Bibliographic services
- Cheminformatics
- Clinical trials
- Gene expression
- Genomic maps
- Sequence analysis
- Workflow and frameworks

To address missing features in W2H and focus on standardisation issues, we started the AppLab project[7] as a parallel development to W2H of a CORBA-based interface to GCG and other similar applications. From the beginning (1998) it was meant as a broader project to solve:

- Standardised access to analysis tools
- Server-side platform independence
- Component collaboration

CORBA provides a language independent communication protocol (IIOP) and interface definition language (IDL). CORBA can provide a technology to design and develop interfaces that can be used generally to access most (if not all) analysis tools.

System and Methods

There are both similarities and differences between W2H and AppLab. The concepts that proved useful in W2H were re-used in AppLab. For example, extensibility, use of meta-data and automatic code generation. The extensibility was again one of the most important issues. Development of security protocols was of less concern with AppLab because of the existing CORBA Security service from OMG that standardised this part of the project.

Java™[8] Code Generator. AppLab uses Java as its primary language. The code is generated from the meta-data descriptions and hides completely all the CORBA calls. It means that no CORBA knowledge is required from the users to use the system, including those users who are adding new applications into the system.

The code generators require the use of meta-data only during the initial preparation stage when the code is generated to produce the CORBA interface to an application. There is no need to use meta-data during run-time. However, the meta-data are available throught the whole lifecycle, but their format is no longer critical because they have been converted into Java code.

Standard Access to the Analysis Tools. The IDL definitions represent the only connection between a server (where the analyses are executed) and a client (where the GUI resides). The IDL is general and can be used for defining all analyses. Alternatively, an IDL interface could inherit from a very general IDL analysis interface and so create analysis-specific definitions.

The AppLab IDL was used as a prototype for submission to the OMG's "Request for Proposals" (RFP) on "Biomolecular Sequence Analysis".[9] In the second half of 1999,

---

[7] http://industry.ebi.ac.uk/applab
[8] Java is a trade-mark of Sun Microsystems Inc.
[9] http://www.omg.org/homepages/lsr/rfs.html

the revised submission to this RFP should be accepted by LSR and OMG groups. The sequence analysis implementations based upon the standard should be available within 12 months of acceptance of the RFP. AppLab is a good candidate to be one of those implementations.

The latest version of AppLab made a further step towards standardisation by adopting XML as the format for describing application meta-data. This approach allows programmers to use a number of developing XML tools for creating, verifying and maintaining meta-data files.

Server-Side Platform Independence. The main benefit of using CORBA is that more servers can perform the same task because all of them use the same interface. The servers can be implemented in different programming languages, using different algorithms or using specialised hardware, but they all communicate with clients in an identical way.

Component Collaboration. Using JavaBeans™[10] technology allows collaboration between AppLab components and smooth integration with other Java applications. Software developers who need to call external applications from their Java programs can use AppLab to generate fully featured software components with a simple and unified interface. The exchange of data between AppLab components takes place using Java events. This data exchange takes place on the client side using only references to these data objects, while the real (and possibly) large data flowing between components can travel from application to application only on the server side.

The importance of these component collaborations grows even more important with the development and production of more Java-based visualisation widgets and techniques.

Implementation

AppLab code consists of a set of Java libraries, and several Perl scripts used only on the server side. The additional Java code is generated from meta-data files for each application individually (Figure 6). The AppLab system includes a notion of project management (Figure 7), and provides a simple browser of available applications. Both can be replaced by more sophisticated techniques (e.g. the CORBA trader or naming service could substitute for the application browser, and framework environments could replace the project management).

AppLab can be used by applications to call other programs at runtime for data analysis. An example is the Genome Builder application developed at the EBI (Muilu, 1999) that uses AppLab to call the CAP3 program and build assemblies of EST sequences.

However, There are Still Not Fully Addressed Issues . . .

AppLab represents a fully distributed computing system based on open standards. However, because of current incompatibilities between CORBA implementations from different vendors and incomplete support for Java in many Web browsers, there are

[10] Java Bean is a trade-mark of Sun Microsystems Inc.

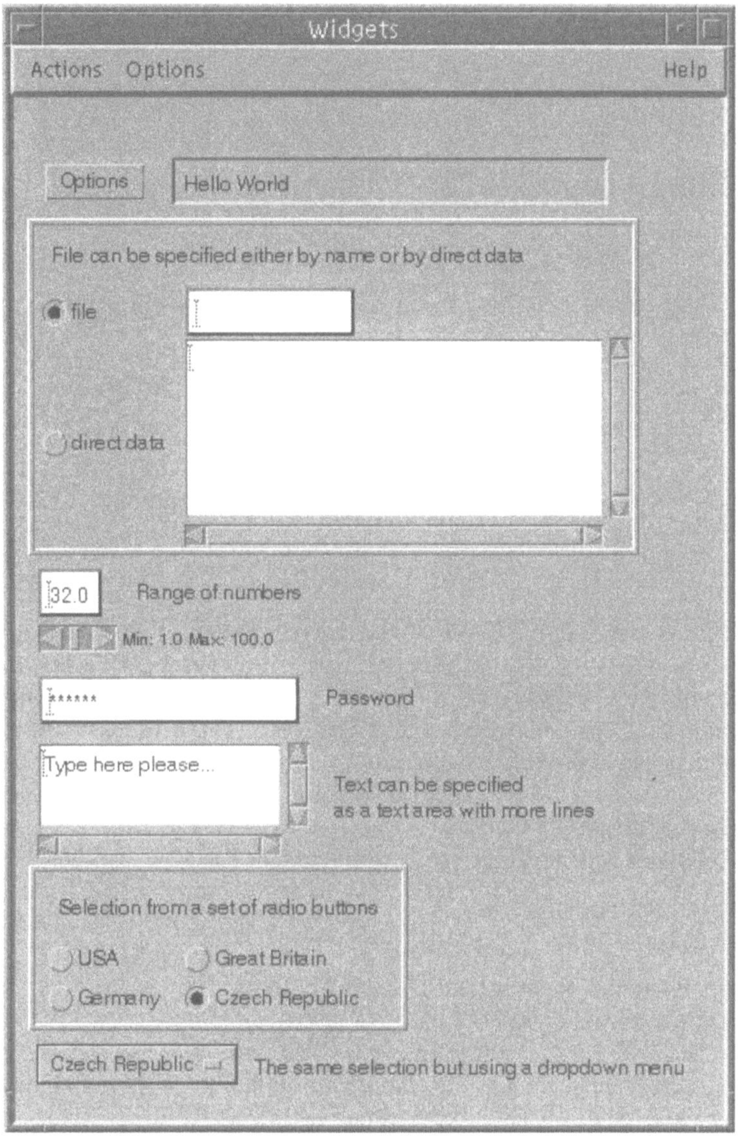

Figure 6. AppLab—an example of a generated hypothetical application.

technical limitations in using AppLab components on top of Web technology. We believe that this will be solved in the near future.

Currently, AppLab provides a component-based environment using JavaBeans, which is obviously limited to components written in Java. In the future, we will address component collaboration based on CORBA technology, which will provide interoperability between components regardless of their language of implementation. Fortunately, the OMG is now in the process of adopting specifications for "CORBA Components"[11]

[11] URL is not yet fixed—try document search at http://www.omg.org

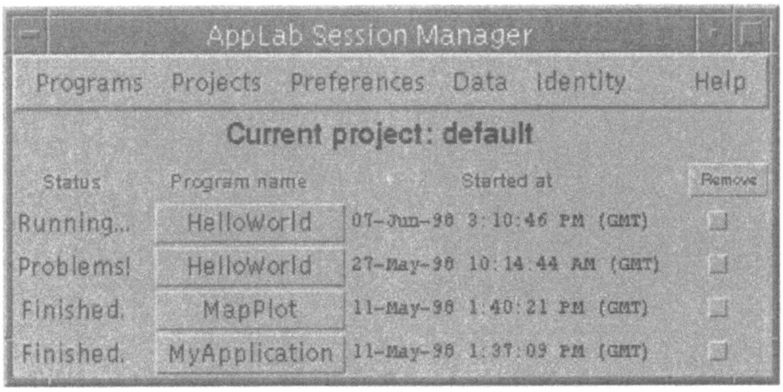

Figure 7. AppLab—project manager window.

that address exactly this issue. We intend to use this specification once it has been issued and accepted.

THE FRAMEWORK ISSUE—PROBLEM-ORIENTED TASKS

The collaboration between analysis tools is needed in both W2H and AppLab. While AppLab has a better starting position because of its JavaBeans nature, W2H has a richer user-interface and seamless Internet access. AppLab developers must work on re-using or creating the existing JavaBeans environments (JavaStudio, Visual Café, Sun's BeanBox, etc.) to develop the GUI.

The component collaboration can be seen from at least two aspects:

- Connectivity aspect
- Task aspect

Connectivity is the more technical issue, and deals with event handling and the interfaces between components. The areas to be explored include CORBA Components, Enterprise JavaBeans™[12] and the InfoBus.[13]

The task aspects describe how the components talk to each other behind the scenes. A problem-oriented task is usually represented as a network of applications. A task is not fulfilled by simply running a single application, but by running several applications in parallel and/or sequentially. Goal is to have a system in which it is possible to define a task in a standard fashion, i.e. choosing appropriate component applications and assembling them into a task description. To make stand-alone applications to collaborate, the following issues have to be addressed:

- Prepare input in different formats
- Analyse/comprehend the results of the individual applications
- Manage control flow and data flow through the network of applications
- Merge the individual information sources (application outputs) into a conclusive task result

---

[12] Enterprise JavaBeans is a trade-mark of Sun Microsystems Inc., http://www.javasoft.com/products/ejb
[13] http://java.sun.com/beans/infobus

A task description can be created by a task administrator or by an advanced user. An interactive task manager to create such descriptions will also be employed.

SUMMARY

It was shown that distributed application management gains from using general and domain-specific specifications and standards. Well-defined standards together with the meta-data approach allow extensibility, flexibility, portability, and finally help developers to build user-friendly environments. Available visualisation widgets can be used as additional components within the whole architecture providing a strongly customised look and feel.

Both Web and CORBA technology with their related protocols (HTTP, IIOP) are valuable and complementary. They can even be used together and support each other, e.g. a CORBA server receiving client requests through a Web interface. The slightly weaker standards used for the Web are compensated by better support in current Web browsers.

A framework of collaborating components is a critical requisite for achieving more complex tasks in the future.

ACKNOWLEDGMENTS

The W2H interface is being developed as a collaborative project between the European Bioinformatics Institute (EMBL-EBI), Hinxton, UK (within the BioStandards project in the Industry Support Programme) and the German Cancer Research Centre (DKFZ), Heidelberg, Germany. The developers from DKFZ are very active with the task-oriented approach. The AppLab project and its framework are developed within the BioStandards project at EBI. [The authors would like to thank Alan Robinson of the EBI for his reading of the manuscript during its preparation.]

REFERENCES

Devereux, J., Haeberli, P., and Smithies, O. (1984) A comprehensive set of sequence analysis programs for the VAX. *Nucleic Acids Res.* **12**, 387–395.

Rice, P., Lopez, R., Doelz, R., and Leunissen, J. (1995) EGCG 8.0. *Embnet. news* **2**, 5–7.

Colet, M. and Herzog, A. (1996) A Web interface to the GCG biological sequences analysis software. *Comput. & Graphics* **20**, 445–450.

Senger, M., Flores, T., Glatting, K.H., Ernst, P., Hotz-Wagenblatt, A., and Suhai, S. (1998) W2H: WWW interface to GCG sequence analysis package. Bioinformatics **14**, 452–457.

Life Science Research Domain Task Force (1998), Pocket Guide, http://www.omg.org/homepages/lsr/FAQ.html.

Muilu, J. (1999) Genome Builder—integration of databases and legacy applications using CORBA. http://industry.ebi.ac.uk/~muilu/Gbuilder.

18

IS HUMAN GENETICS BECOMING DANGEROUS TO SOCIETY?

Charles J. Epstein

Department of Pediatrics
University of California
San Francisco
California 94143–0748

1. THE UNABOMBER MANIFESTO

Several years ago, a man who was called the Unabomber sent a bomb to an American geneticist—who was probably the only person in America to be targeted for death just for being a geneticist. This was neither the first nor last bomb that he sent, but to explain his actions, the Unabomber issued a manifesto. This is what he had to say about genetics in the section of the manifesto entitled "The 'bad' parts of technology cannot be separated from the 'good' parts" (FC, 1995):

> Suppose . . . that a cure for diabetes is discovered. People with a genetic tendency to diabetes will then be able to survive and reproduce as well as anyone else. Natural selection against genes for diabetes will cease and such genes will spread throughout the population . . . The same thing will happen with many other diseases susceptibility to which is affected by genetic factors (for example, childhood cancer), resulting in massive genetic degradation of the population.
>
> The only solution will be some sort of eugenics program or extensive genetic engineering of human beings, so that man in the future will no longer be a creation of nature, or of chance, or of God (depending on your religious or philosophical opinions) but a manufactured product. If you think that big government interferes in your life too much NOW, just wait till the government starts regulating the genetic constitution of your children. Such regulation will inevitably follow the introduction of genetic engineering of human beings, because the consequences of unregulated genetic engineering would be disastrous.

He then goes on in the following vein:

> The only code of ethics that would truly protect freedom would be one that prohibited ANY genetic engineering of human beings, and you can be sure that no such code will ever be applied in a technological society

Genomics and Proteomics, edited by Sándor Suhai.
Kluwer Academic / Plenum Publishers, New York, 2000.

... Inevitably, genetic engineering will be used extensively, but only in ways consistent with the needs of the industrial-technological system ... And, as nuclear proliferation has shown, new technology cannot be kept out of the hands of dictators and irresponsible third-world nations. Would you like to speculate about what Iraq or North Korea will do with genetic engineering.

A potent image, indeed!—Iraq or North Korea and genetic engineering. I would suspect that the leaders of both countries have other things on their minds just now.

There has been considerable speculation about the state of Unabomber's mind when he was engaged in his campaign of bombings which, fortunately, ended with his capture and conviction, but this is not something I want to get into at this time. However, I think that it is important to realize that the sentiments expressed in his manifesto are *not* out of line with much that has been said or written by minds less demonic than his. The Unabomber—or Theodore Kaczynski as we now know him to be—was playing to fears and concerns that already existed. The only thing that really distinguishes him from so many other critics of technology is that he killed people in cold blood, and attempted to kill others, to draw attention to his message. However, it really did not really require the Unabomber to alert us to the fact that there is a problem. The daily press and weekly and monthly publications are constantly telling us that there are matters of concern. Consider, for example, two periodical covers from a few years back. On one, from *Technology Review* (Technology Review, 1996), a DNA helix replaces the serpent in the Garden of Eden, and on the other, from *The Economist* (Economist, The, 1996), a man is shown ensnared in fetters composed of DNA helices. If we take the DNA helix as representing an understanding of our genes or knowledge of our genetic constitution, it doesn't take too much imagination to see the equation that is being suggested: there is something malevolent about DNA—genetic knowledge is dangerous; it will enslave us. Kaczynski couldn't have portrayed it better himself.

These images represent somewhat extreme views of what genetic knowledge might portend. For the most part, criticisms of human genetics have been particularly leveled at certain clinical applications, particularly prenatal diagnosis and carrier screening, which have been referred to variously as racist, sexist, insensitive, and/or just plain misguided. Gene therapy has certainly not been immune. Quite the opposite! It has conjured up a whole host of concerns and fears of its own. I once witnessed people in wheel chairs coming to a meeting of the NIH Recombinant DNA Advisory Committee—the RAC—to assert that gene therapy constitutes a form of discrimination against the disabled. Why? Because by wishing to treat or prevent a genetic disease we are somehow sending the message that having a genetic condition, whatever it is, is bad—that the person with the condition should not exist.

2. CONCERNS ABOUT HUMAN GENETICS

In the latter part of 1993, some months after the Unabomber sent his bomb to the geneticist, I wrote these words in my final editorial, which was entitled "Seven Momentous Years" for the *American Journal of Human Genetics*:

[Of greater] concern ... is the public's very real fear of what progress in genetics might bring. It cannot be ignored. The scientific hubris and resulting chaos portrayed in *Jurassic Park*, the history of the eugenics movement in America, the Nazi racial purification schemes which culminated in the Holocaust, and the anti-gene therapy stance of Jeremy Rifkin have all had a negative influence on public thinking about genetic research and what it might lead to (Epstein, 1993, p. 1164).

If anything, these sentiments are even more true today than they were then. Let me give you a very recent example. An article with the headline, "Concern among Jews is heightened as scientists deepen gene studies," appeared in the April 22 (1998) issue of the *New York Times* (Stolberg, 1998). The article described the increasing apprehension being expressed by leaders of the Ashkenazi Jewish community about genetic investigations of Ashkenazi Jews. As the article explains, this community, like many other ethnic groups, is being extensively studied because it is still relatively homogeneous genetically. This genetic homogeneity greatly facilitates studies aimed at the mapping, cloning, and identifying genes which cause or predispose to a variety of genetic conditions. The Ashkenazi population does not have more genetic diseases or mutant genes than other groups, but, as with other groups, it does have some that are present in higher than average frequencies—for example, Tay-Sachs disease, Gaucher disease, torsion dystonia, and the so-called Ashkenazi *BRCA1* and *BRCA2* breast cancer gene mutations.

Why are the leaders of the Jewish community so sensitive to genetic research? Here are some quotes:

- No other group carries the psychological scars of the Holocaust, a calculated extermination attempt rooted in the notion that Jews were genetically inferior.
- There is a historical context to this that I don't think that you can ignore . . . People are anxious.
- We are getting a bad reputation. All of the bad genes you talk about are Jewish genes.
- Just raising the topic of genetics and Jews will "fan the flames of fear."

Although the article frames the issue in Jewish terms, and there is certainly a historical reason for doing so, the problem being raised is not unique to the Jewish community and affects all of us, Jews and non-Jews:

- Who, after all, wants to hear themselves described as carrying genetic defects or mutant genes?
- The use of the word mutation gets to our very soul. It's the whole question of stigma and our own view of ourselves.

However, the issue of stigma goes far beyond how we think of ourselves or how one potential marriage partner thinks about another who carries a deleterious gene detected in one of many screening programs—a gene for sickle cell disease, or beta thalassemia, or cystic fibrosis, or, yes, Tay Sachs disease in Ashkenazi Jews. It gets to the concerns raised by the disabled at the RAC meeting mentioned earlier, and ultimately it gets to concerns, which are particularly acute in the United States, about discrimination by insurance companies and employers. This is a problem with which we have already had extensive experience in presymptomatic detection programs for Huntington disease, an incurable dominantly inherited degenerative condition of the nervous system which has its onset usually in middle adult years and ultimately leads to death. Using molecular diagnostic tests, it is possible to determine who does and who does not have the mutant gene with a very high degree of accuracy. For a variety of reasons of their own, persons at risk choose to have themselves tested, and it is at this point that the specter of stigma and discrimination raises its ugly head. What if their insurance company finds out? Would they be denied medical or life insurance? Or, what if their employer finds out—would they lose their jobs? The result of all of this, of course, is that the persons coming for presymptomatic testing in the United States generally pay for the counseling, the medical and psychological evaluations, and the molecular studies themselves, and the evaluations and results are held in confidence and excluded from the general medical records.

Now, you might think, Huntington disease is fairly rare, so why should we be concerned? Well, the issue really isn't very different for some much more common situations—in particular, the presymptomatic identification of mutant or variant genes predisposing to breast and ovarian cancer. Ultimately, we will have to be concerned about a whole host of conditions, both common and rare, which will be shown to be genetically caused or influenced, and *all* of us will be affected.

But, let me return to "the flames of fear" that were mentioned in the newspaper report about the Ashkenazi Jews. An instructional videotape and an accompanying teacher's guide aimed at high school and college students were produced a few years ago by a group working at my own university. What I want to consider is the title: "Winding Your Way Through DNA: Promises & Perils of Biotechnology: Genetic Testing" (University of California, San Francisco, 1996). The thing that troubles me is the word "perils."

The videotape presents two genetic stories. One is concerned with the presymptomatic testing of a young woman for Huntington disease. She turns out in the end to carry the gene, and, as the narrator puts it, ". . . taking a test has removed the uncertainty of her situation with devastating clarity." The other story is about the treatment of familial hypercholersterolemia in a mother and her daughter, and for them the major issue appears to be insurability in the face of an identified predisposition to genetic disease.

So, what are the "perils of biotechnology" and of genetic testing spoken of in the title? A promotional description of the tape suggests an answer:

> Advances in biotechnology are allowing doctors to use genetic testing to identify more and more genetic conditions. The information provided by the genetic tests not only helps expectant couples learn about the health of their developing fetus but also confirm the presence of genetic conditions for children and adults. These findings pose ethical, legal and social dilemmas about how this information should be used.

Are these, then, the "perils"—the ethical, legal and social dilemmas or issues raised by genetic testing—what those of us in the American genetics community have come to know by the acronym of ELSI? As I have already said, there is certainly no question that presymptomatic testing for Huntington disease, or breast cancer, or the short QT syndrome, or Alzheimer disease raises a large number of important and difficult issues—privacy and confidentiality, insurability, employability, stigmatization, responsibility to other family members, and more. The problems are real, they are here now, and they require serious attention! But, are these problems "perils," a term, which to my ear, at least, has a very ominous sound? Perhaps that was, in fact, the intention—to raise public consciousness about the ELSI issues by equating biotechnology or genetic testing with danger (one of the definitions of the term "peril"). Or was the purpose of the title merely to attract people to use the videotape. "Promises and Perils" does have a nice alliterative ring about it, and I would suspect that whoever coined it wasn't really trying to make it sound ominous. Fair enough!

3. REGULATORY AND LEGISLATIVE APPROACHES TO THE CONTROL OF HUMAN GENETICS

But let's pursue the issue a bit further with another example in which I feel that the semantics and what lay behind them are of real importance. In the United States, all proposed federal regulations are published in the daily *Federal Register*. Therefore, it was interesting to read in the *Federal Register* of March 14, 1996, that

The FDA [the US Food and Drug Administration] could regulate as Class III devices only those ASR's [analyte specific reagents] used in tests intended for use in overtly healthy people to identify a genetic predisposition to a dementing disease, or to fatal or potentially fatal medical disorders (for example, cancer or Alzheimer's disease) in situations where penetrance is poorly defined or variable and latency is long (5 years or longer) (US Food and Drug Administration, Department of Health and Human Services, 1996, p. 10486).

Now, what, you may ask, are Class III reagents? Looking through the FDA document, we find them variously referred to as

- ASR's presenting a high risk to public health.
- [Reagents whose] use presents particularly high risks.
- Serious health risks [are] associated with their use or in the class of test in which the ASR is being used. These include active ingredients used in tests intended to diagnose potentially fatal contagious infections (for example, HIV or tuberculosis) or intended to safeguard the blood supply.

So, why did the Immunological Devices Panel make its recommendation to the FDA concerning human DNA reagents? In its report, the Panel cited two types of risks. First, there are the:

General risks: variable quality, inappropriate labeling, use by persons without adequate qualifications; clinicians ordering test may be unaware that the clinical performance characteristics of the tests have not been independently reviewed by FDA.

True enough, but then there are what are supposedly considered unique risks:

The panel also identified a subset of ASR's whose use posed unique risks to public health because of the substantial clinical impact of the information generated using these devices.

So now we have it—the high risk to the public health, in the category of risks posed by reagents designed to safeguard us against infectious diseases and to protect the blood supply, derives from "the substantial clinical impact of the information." INFORMATION!!! Information is risky!!! Frankly, the notion boggles the mind!

Fortunately, the proposed regulation was not adopted, but, once again, there is a kernel of truth behind the terminology. As more and more disease genes and predisposition-to-disease genes are being cloned, new DNA diagnostic reagents are being introduced daily into research and clinical practice. Many of these are indeed what the FDA calls "home brews", tests developed in research laboratories rather than in the traditional pharmaceutical manner, and there are, of course, issues with regard to quality of the reagents, accuracy of tests, and qualifications of the clinicians using them. This is true for all reagents used in medical testing. It is also true that that the results from many of the genetic tests which are performed do have a substantial clinical impact. But, I do not believe that defining genetic information as a unique risk to the public health is the way to approach the problem. Doing so serves only to increase public apprehension about genetics and geneticists and will, I believe, inhibit both research and practice. That the public needs protection is without question, but implementing an unduly stringent FDA regulation is not the way to provide it.

Before I leave the issue of regulation, let me know discuss another attempt at governmental regulation of genetics. A few years ago, the so-called Domenici bill, S. 1898,

"To protect the genetic privacy of individuals, and for other purposes," was introduced into the United States Senate (US Senate, 1996). The main thrust of this bill was the regulation of genetic research, in particular research on stored samples capable of providing DNA for analysis. This has been an extremely contentious issue in the United States, which I do not want to discuss here. However, what I would like to point out are some of the premises on which the legislation was based—the so-called findings of Congress. There were ten in all, but I want to cite just three to give you the flavor:

- Genetic information has been misused resulting in harm to individuals.
- The improper use and disclosure of genetic information can lead to significant harm to the individual, including stigmatization and discrimination.
- The potential for misuse with respect to genetics is tremendous since genetics transcends medicine. It has the potential to penetrate many aspects of life including employment, insurance, forensics, finance, education, and even one's self-perception.

As you can imagine, the Domenici bill raised quite a furor in the human genetics community when it was introduced. Had it been enacted, it would have spelled serious trouble for virtually all types of human genetic research. Fortunately, it was not.

4. CLONING

Thus far I have focused primarily on problems associated in some way with genetic testing and related research. However, the concerns of the public go far beyond this application of genetics. Think for a moment about what was the biggest scientific news—more particularly biological news—during the past two years. Why, it was Dolly, of course, Dolly, the sheep supposedly cloned in Scotland by Dr. Ian Wilmut (Wilmut et al., 1997). *Science* referred to the cloning of Dolly as the "breakthrough of the year" in 1997 (Science, 1997). Although there was initially some concern about whether Dolly was an anomalous result and not really a clone derived from an adult cell, this issue has been largely laid to rest with the birth of Cumulina, the mouse cloned from cumulus cells (Wakayama et al., 1998).

However, my purpose here is not to discuss the science of cloning but, rather, the public response (in the United States) to the cloning. At first came the light hearted jibes by political cartoonists These were rapidly followed by expressions of fear and concern coupled with pronouncements from the President ("let's not 'play God'") (Marin Independent Journal, 1997), a ban on funding for research on human cloning, and the deliberations of the National Bioethics Advisory Committee (NBAC) which resulted in the recommendation that a moratorium be place on human cloning.

Following all of the initial responses came the announcement of additional cloned animals, some of which were transgenic, and then, a year after it all began, two things occurred almost simultaneously. First, there was evidence for the beginning of a reversal in public opinion—at least in some quarters. Research on human cloning was held out as having significant potential benefits in the future. And, then, somewhat ominously, there came an attempt to make all research on cloning, whether for the purposes of reproduction or not, a crime. Another bill was introduced into the United States Senate: S.1611—a bill to prohibit all human cloning (US Senate, 1998). This bill, which its authors tried to sneak through with parliamentary maneuvers aimed at blocking debate, was soundly defeated, and a more moderate bill is now somewhere in committee.

However, the NIH has placed a moratorium on research on human cloning, and the future of this endeavor is still undecided.

Looking back on it, the cloning of Dolly and the public response to it is quite instructive. The immediate response was one of awe and wonder, followed rapidly by fear and concern—fear based on projecting the worst possible outcomes and concern over whether ethical and moral boundaries would be transgressed. Of course, it wasn't sheep that anyone was worried about, it was humans—even though only one viable sheep had been produced after hundreds of attempts and there was at the time no evidence that the same could be done in humans. The whole public debate, the political, ethical, and religious pronouncements, the attempted restrictive legislation, and even the political cartoons were all based on just one, and at the time, still unconfirmed cloning of a sheep from an adult cell. That this one animal could excite such an outpouring of talk and action clearly attests to the sensitivity, indeed hypersensitivity, of both the public and the scientific community to the possibility of a new and potentially powerful way to manipulate the human genome.

Although Dolly and her kind were farm animals, their situation could be directly extrapolated to humans. However, the fears and concerns about genetic research evoked by Dolly are not limited to human genetics. Consider an article from *Nature,* the heading of which tells us that "the Swiss have embarked on a national debate about the use of transgenic animals, threatening devastation of biological science and industry in their country. Are they a barometer of wider public antipathy?" (Referendum's challenge to transgenic research, 1997). In this instance the discussion is about all types of transgenic animals, right down to the lowly mouse. Again, the message is clear enough: any type of genetic manipulation is taboo. Fortunately, the Swiss referendum was defeated. But, even plants are not immune, as headlines from two recent reports in *Nature* and *Science* point out. The first article ["Transgenic corn ban sparks a furor"] speaks about a French government decision to ban the growing of transgenic corn—although, curiously enough, not its importation and consumption—because the growing of transgenic plants was regarded as an environmental hazard (Balter, 1997). The second ["Agricultural biotech faces backlash in Europe"] deals with similar matters (Williams, 1998).

5. THE HUMAN GENOME PROJECT

The human genome project was recently very much in the news these days in the United States, especially with Craig Venter's challenge to the genome project establishment. The many potential benefits of this research have been widely discussed. However, it is also important to consider the concerns that genomic research has engendered. You can get a sense of what they are from some recent newspaper article headlines:

- DNA mapping could revolutionize medicine. Genetic data may bring cure, discrimination (Goetinck, 1998a).
- Humanity confronts heredity. As genetic knowledge explodes, society is being compelled to deal with a host of issues barely imagined a decade ago (Goetinck, 1998b).

What are these issues?

First, there are the short term concerns, all of which have already been mentioned, that presymptomatic identification by genetic testing of individuals who will or are predisposed to develop disease may result in

- Testing in advance of the ability to use the information appropriately
- Problems with insurance and insurability
- Problems in employment with discrimination because of potential risk of developing a genetic disease
- Ethnic discrimination because of high frequencies of particular genes

And then there are the more long-term concerns, that genes governing behavior, intelligence, and physical attributes will be discovered and that this knowledge will be used to

- Identify persons with traits thought to be undesirable
- Select for children with traits thought to be desirable
- Manipulate the genome of unborn children
- Manipulate the germ line
- Clone genetically modified/enhanced people

These are the things that I believe society is really concerned about.

6. THE PREVALENT DISTRUST OF SCIENCE

A few years ago, John Maddox, the former editor of *Nature*, published a commentary on "The prevalent distrust of science" (Maddox, 1995). I would like to share with you some of what he had to say.

> Distrust of science is still alarmingly prevalent, which conflicts with reasonable expectation. Is not the century now drawing to a close most of all remarkable for the technology that now fills our world and for the understanding of that world that has been won since, say, the discovery of the electron in 1897 [or, I would add, since the rediscovery of Mendelian genetics in 1900]?
>
> During that long period, the improvement in the human condition has been immense . . . In general, science and technology have helped to make us healthy, wealthy and wise in a manner and to a degree not foreseen except by a few visionaries such as H.G. Wells . . . What is now being learned of human genetics . . . will be dramatically reflected in the health of our populations in the decades ahead.

I think that most, if not all of us would agree with this assessment of the contributions of science and technology and, in our own little world, of genetics. So, why, Maddox goes on—

> So why, given all of these benefits of health, wealth, and wisdom, to which science has made such important contributions, does there persist the deep distrust of science we see around us? . . . The standard answer is that science and scientists have in the past made exaggerated claims of what innovation will do for the world at large, so that scientists are no longer trusted. . . . The nuclear power saga of the 1950s may be one illustration, molecular genetics is at risk of becoming another.

It is interesting that Maddox couches what he calls the "standard answer"—exaggerated claims—in terms of the past, but, when it comes to genetics, he predicts trouble in the future. Well, it is obvious from all that I have said that the future has already arrived. A few years ago, a Panel to Assess the NIH Investment in Research on Gene Therapy, the so-called Orkin/Motulsky Committee, presented its report to the NIH Director's Meeting (Orkin & Motulsky, 1995). Among the Committee's findings were the following:

> Expectations of current gene therapy protocols have been oversold. Over-zealous representation of clinical gene therapy has obscured the exploratory nature of the initial studies, colored the manner in which findings are

portrayed to the scientific press and public, and led to the widely held, but mistaken, perception that clinical gene therapy is already highly successful. Such misrepresentation threatens confidence in the field and will inevitably lead to disappointment in both medical and lay communities. Of even greater concern is the possibility that patients, their families, and health providers may make unwise decisions regarding treatment alternatives, holding out for cures that they mistakenly believe are "just around the corner."

These are strong words, but to the extent that scientists themselves bear some of the responsibility for the mistrust of the science of genetics because of our own hyperbole and inflated claims, the prescription is relatively straight forward. As stated by the Orkin/Motulsky Committee:

The panel urges gene therapy investigators and their sponsors—be they academic, governmental, private, or industrial—to be more circumspect regarding the aims and accomplishments of clinical protocols when discussing their work with the scientific community, the public, and the media.

However, Maddox, in the article referred to earlier (Maddox, 1995) points out that there is more to the problem of the public perception of science than just hype.

The general distrust of science has other and more primitive roots. To the extent that science and its applications bring improvements in our lot, they also imply change, and change is never welcome for its own sake. Then this knowledge that I've been extolling is often unwelcome.

Maddox then tells the following story:

. . . a panel of parliamentarians gathered to discuss the legislative position on genetics and genetic manipulation in their countries. A woman member of the German Bundestag, and a representative of the Green Party, spoke clearly and intelligently and said this: "You must understand that we Greens believe that to represent the nature of human beings by a description of their genes undermines their dignity as human beings. We shall oppose in the Bundestag any legislation that condones research in human genetics.

Maddox springs to the defense of the geneticists in a way that I think we would all applaud.

This implacable position is arresting. It also succeeds in misrepresenting the position of the research community. Broadly speaking, geneticists themselves are deeply suspicious of genetic determinism—the assertion that a person is determined almost exclusively by the genes there happen to be in his or her genome. To their credit, geneticists have also been among the first to draw attention to the respects in which the rapid development of their field is likely to create social problems, chiefly by the use of genetic diagnosis as a basis for discrimination between individuals, mainly in employment and insurance. But evidently the geneticists will win no credit from the German Greens for their perceptiveness.

Why do I tell you this? Well, it wasn't because the Greens might be part of a coalition government here in Germany, because I just learned that last week. That will certainly give us something to think about. Rather, I tell you this because I believe that situations are arising in which opposition to genetic research and genetic testing may be based on similar types of premises.

An editorial entitled "Crimes against genetics" appeared in *Nature Genetics* just about the same time as the Maddox article (Crimes against genetics, 1995). This title is, of course, a clever play on words, since the article dealt with the controversy surrounding a meeting held about a month earlier on the subject of "The Meaning and Significance of Research on Genetics and Criminal Behavior". The editorial broadly summarized the two sides of the dispute that surrounded the conference in the following way:

- On the side of the organizers, the goal was "to explore the implications of current genetic research of violent, antisocial and criminal behavior . . . to help to identify and aid those most likely to fall victim to sociological circumstances."
- For the opponents, there was the fear "that these studies will lead only to the enslavement of the underclasses as social changes are abandoned in favor of easy-answer drug treatments or harsh restrictions on those deemed genetically irredeemable."

I am perhaps not the right person to analyze these two positions and to decide which, if either, is right and which is wrong. As a matter of fact, I think that there is considerable merit to some of the arguments about behavioral research on both sides of the issue. However, what troubles me is that there is or is starting to be a breakdown in our ability to engage in rational discourse about what genetic research is all about. For reasons that are certainly grounded in the history of the applications and misapplications of genetics, there is a movement to proscribe, to prohibit certain areas of genetic research—because the findings or, perhaps more accurately, the potential applications of the findings, are believed to be so frightening because of the possibilities for abuse. The editorial to which I referred, in paraphrasing the remarks of one of the speakers concerning the behavioral research controversy, puts it this way:

> . . . the public also sees scientific information, regardless of the soundness of the methods, as powerfully legitimizing, and, furthermore, the public's perception of genetic findings is that they are immutable. Thus the mere perception of reality (rather than the realities themselves) can provide impetus for the enactment of inequitable laws . . .

There is a bit of a paradox here—although the public fears what genetics can do, it may uncritically accept what they think the geneticists are saying. And, it is not even the reality of what has been found. It's just the mere perception of reality.

Before concluding, there is one last point that I would like to bring up. I have been discussing what the public thinks and what is fears, but the fact of the matter is that the public doesn't know very much all. Consider the results of a recent poll in which people were asked what they knew about tests for genetic diseases (Goetinck, 1998c). 40% said they knew "not much" or "nothing at all"; 42% said they knew "some." These are not very encouraging findings.

7. CONCLUSION

As the present century is drawing to a close and a new one is about to begin, human genetics and its applied clinical science, medical genetics, are more powerful, rewarding, and exciting than ever. Progress *has* been enormous, and I believe that geneticists have every reason to be proud of what they have been able to accomplish in a remarkably short time.

But, having said this, let me return to the question I posed in the title of this talk. Is genetics becoming dangerous to society?

My own answer is no—genetics is not becoming dangerous to society. However, I have presented several examples of situations that suggest that there is a public belief or fear that it is, and when I say "public", I include some scientists and geneticists as well. These include:

- the Ashkenazi Jewish community's response to genetic research
- the videotape on the "promises and perils" of genetic testing
- the proposed classification of DNA reagents as being risky to the public health
- the finding in the Domenici report that genetic information has been misused
- the furor over the cloning of Dolly
- the banning of transgenic animals and plants
- the uproar over studying whether there is a genetic basis to criminal behavior.

And this list does not include all of the science fiction extrapolations to the future which conjure up the fearful prospects of being able not only to predict personality traits, physical characteristics, and intelligence, but actually to determine them.

Although it is my belief that genetics is not dangerous in the sense implied in many of these examples, there is, nevertheless, a reality about people's concerns about stigmatization and discrimination, about problems with insurance, about being able to learn things about themselves that might be frightening to contemplate, and about having to make difficult decisions because of this knowledge. And, even though I have referred to it as science fiction, there *is* a sense, if not of actual fear, at least of unease about just how much manipulation of the human genome will ultimately become possible and who will be able to manipulate it.

How, then, should we deal with all of this? Two thousand years ago, the famous sage Rabbi Hillel said:

If I am not for myself, who is for me? If I am only for myself, what am I? If not now—when? What does this mean in the context of human genetics at the dawn of the twenty-first century?

If I am not for myself, who is for me?

Speaking first as a human geneticist, I believe that the human genetics community needs to inform the public about all of the positive aspects of what it has done, of what it is now doing, and of its future goals. Geneticists need to be strong advocates for their profession, but must avoid claiming or promising too much themselves or allowing others to make such claims in their names or on their behalf. They need to avoid conjuring up unfounded public fears and apprehensions by what they say. They need to work for regulations and legislation that, while preserving personal rights, enhance rather than unnecessarily restrict their ability to carry out research and to treat patients.

I do not say all of this just to be self serving and to preserve interests rooted in ego or money. Rather, it is said out of the conviction that genetics has and will continue to make enormously valuable contributions to society's ability to prevent and treat human disease and suffering and the belief that this ability should be defended and preserved.

If I am only for myself, what am I?

However, as Rabbit Hillel recognized, it is not sufficient for geneticists and other scientists to make the case only for what they themselves are about, as valuable to society their activities may be. The genetics community must be ever mindful of the facts that it does not function in isolation and that it has responsibilities that transcend the purely professional. Geneticists need to educate the public about what genetics is all about and about what geneticists can and cannot do. They need to listen very carefully to the fears and concerns that have been expressed and need to respond in a positive fashion. They must continue to be and, if anything, become more involved in the social and ethical debate that increasingly surrounds everything they do. Geneticists need to be cognizant of the fact that they constitute just one element in the societal debate—which, hopefully, will be a rationale one—about the human applications of genetic knowledge. Important decisions about these applications certainly *will not* and *should not* be theirs alone to make.

If not now—when?

The tension between scientific advance and societal concerns is not new, and it is certainly not unique to genetics. But, the rapidity with which genetic information is being accumulated and new applications are being put forward makes the situation particularly acute for human genetics. The challenge facing both human geneticists and the rest of society is to find the proper balance between the hopes and fears of society and the goals and interests of science—the discovery of new knowledge and the improvement of health and curing of disease. This challenge goes well beyond the weighing of issues at a conceptual level and extends to quite practical and important matters of control and regulation. It is a challenge that all of us must face, and it is one that must be faced now!

ACKNOWLEDGMENT

An earlier version of this article appeared in the *American Journal of Human Genetics* (Epstein, 1997).

REFERENCES

Balter, M. (1997) Transgenic Corn Ban Sparks a Furor. Science 275, 1063
Crimes against Genetics (1995). Nature Genet. 11, 223–224
Economist, The (1996) 340 (7983): cover.
Epstein, C.J. (1993) Seven Momentous Years. Am. J. Hum. Genet. 53, 1163–1166
Epstein, C.J. (1997) Toward the 21st Century. Am. J. Hum. Genet. 60, 1–9
F.C. (1995) Industrial Society and its Future (June 1995). New York Times, September 19
Goetinck, S. (1998a) DNA Mapping Could Revolutionize Medicine. The Dallas Morning News, July 19
Goetinck, S. (1998b) Humanity Confronts Heredity. The Dallas Morning News, July 20
Goetinck, S. (1998c) Few Understand Potential of Genetic Revolution. The Dallas Morning News, July 19
Maddox, J. (1995) The Prevalent Distrust of Science. Nature 378, 435–437
Marin Independent Journal (1997), March 4
Orkin, S.H. and Motulsky, A.G. (1995) Report and Recommendations of the Panel to Assess the NIH Investment in Research on Gene Therapy. Presented at the NIH Director's Advisory Committee Meeting of December 7
Referendum's Challenge to Transgenic Research (1997) Nature 389, 103
Science (1997) 278, December 19: cover
Stolberg, S.G. (1998) Concern among Jews is Heightened as Scientists Deepen Gene Studies. New York Times, April 22
Technology Review (1996) 99(6): cover
University of California, San Francisco (1996) Promise & Perils of Biotechnology: Genetic Testing. Teacher's Guide and Videotape. Pyramid Media, Santa Monica
US Food and Drug Administration, Department of Health and Human Services (1996) Proposed rules. Docket No. 96N-0082. Fed Reg 61, 10484–10488
US Senate (1996) A Bill to Protect the Genetic Privacy of Individuals, and for other Purposes. 104th Cong., 2d sess., SR 1898
US Senate (1998) Prohibition on Cloning of Human Beings Act of 1998. 105th Cong., 2d sess., SR 1611
Wakayama, T., Perry, A.C.F., Succotti, M., Johnson, K.R., and Yanagimachi, T. (1998) Full-term Development of Mice from Enucleated Oocytes Injected with Cumulus Cell Nuclei. Nature 394, 369–374
Williams, N. (1998) Agricultural Biotech Faces Backlash in Europe. Science 281, 768–771
Wilmut, I., Schnieke, A.E., McWhir, J., Kind, A.J., and Campbell, K.H. (1997) Viable Offspring from Fetal and Adult Mammalian Cells. Nature 385, 810–813

CONTRIBUTORS

Josep F. Abril
Departament d'Informàtica Mèdica
Institut Municipal d'Investigaó Mèdica (IMIM)
c/ Doctor Aiguader 80
08003 Barcelona
Spain

Pankaj Agarwal
Dept. of Bioinformatics
SmithKline Beecham Pharmaceuticals
UW2230, 709 Swedeland Road
PO Box 1539
King of Prussia
PA 19406
USA

Sally A. Amundson
National Cancer Institute
National Institutes of Health
9000 Rockville Pike
Bethesda, MD 20892
USA

Jerome Baudry
SBPM/DBCM
CEA-Saclay
F-91191 Gif-sur-Yvette Cedex
France

Michael Bittner
Cancer Genetics Branch
National Genome Research Institute
National Institutes of Health
Bldg 49, Room 4A52
9000 Rockville Pike
Bethesda, MD 20892
USA

H. Bornfleth
Institute for Anthropology and Human Genetics
Ludwig Maximilians University
Richard-Wagner-Str. 10/I
D-80333 Munich
Germany

E. Fidelma Boyd
Dept. Geographic Medicine and Infectious Diseases
Tupper Institute
Tufts University Medical School
Boston, MA 02111
USA

Nigel P. Brown
National Institute for Medical Research
Division of Mathematical Biology
The Ridgeway, Mill Hill
London NW7 1AA
UK

Moisés Burset
Departament d'Informàtica Mèdica
Institut Municipal d'Investigaó Mèdica (IMIM)
c/ Doctor Aiguader 80
08003 Barcelona
Spain

Carlos D. Bustamante
Dept. Organismic and Evolutionary Biology
Harvard University
16 Divinity Avenue
Cambridge, MA 02138
USA

Yidong Chen
Cancer Genetics Branch
National Genome Research Institute
National Institutes of Health
Bldg 49, Room 4A52
9000 Rockville Pike
Bethesda, MD 20892
USA

Bastien Chevreux
DKFZ-German Cancer Research Center
Dept. of Molecular Biophysics (H0200)
Im Neuenheimer Feld 280
D-69120 Heidelberg
Germany

C. Cremer
Institut für Angewandte Physik
Universität Heidelberg
Heidelberg
Germany

Thomas Cremer
Institute for Anthropology and Human Genetics
Ludwig Maximilians University
Richard-Wagner-Str. 10/I
D-80333 Munich
Germany

Serge Crouzy
DBMS
CEA-Grenoble
Avenue des Martyrs
Grenoble
France

Hajo Delius
DKFZ-German Cancer Research Center
Division of Applied Tumor Virology
Im Neuenheimer Feld 280
D-69120 Heidelberg
Germany

S. Dietzel
Institute for Human Genetics
University of Heidelberg
Heidelberg
Germany

Edward R. Dougherty
Computer Assisted Medical Diagnostic Imaging Laboratory
Dept. Electrical Engineering
Texas A&M University
Texas
USA

P. Edelmann
Institut für Angewandte Physik
Universität Heidelberg
Heidelberg
Germany

R. Eils
Interdisziplinäres Zentrum für
Wissenschaftliches Rechnen (IWR)
Universität Heidelberg
Heidelberg
Germany

Charles J. Epstein
Dept. of Pediatrics
University of California
San Francisco, CA 94143-0748
USA

P. Ernst
DKFZ-German Cancer Research Center
Dept. of Molecular Biophysics
Im Neuenheimer Feld 280
D-69120 Heidelberg
Germany

James W. Fickett
Dept. of Bioinformatics
SmithKline Beecham Pharmaceuticals
UW2230, 709 Swedeland Road
PO Box 1539
King of Prussia, PA 19406
USA

Albert J. Fornace
National Cancer Institute
National Institutes of Health
9000 Rockville Pike
Bethesda, MD 20892
USA

Juergen Frevert
Biote Con
Ggesellschaft für Biotechnologie und Consulting mbH
Potsdam
Germany

Volker Gawantka
DKFZ-German Cancer Research Center
Division of Molecular Embryology
Im Neuenheimer Feld 280
D-69120 Heidelberg
Germany

K.-H. Glatting
DKFZ-German Cancer Research Center
Dept. of Molecular Biophysics
Im Neuenheimer Feld 280
D-69120 Heidelberg
Germany

Roderic Guigó i Serra
Departament d'Informàtica Mèdica
Institut Municipal d'Investigaó Mèdica (IMIM)
c/ Doctor Aiguader 80
08003 Barcelona
Spain

Markus Hammermann
DKFZ-German Cancer Research Center
Dept. of Biophysics of Macromolecules (H0500)
Im Neuenheimer Feld 280
D-69120 Heidelberg
Germany

Jens Hanke
Max-Delbrück-Centrum für Molekulare Medizin
Robert-Rössle-Str. 10
D-13125 Berlin
Germany

Daniel L. Hartl
Dept. Organismic and Evolutionary Biology
Harvard University
16 Divinity Avenue
Cambridge, MA 02138
USA

Udo Heinemann
Forschungsgruppe Kristallographie
Max-Delbrück-Centrum für Molekulare Medizin
Robert-Rössle-Str. 10
13122 Berlin
Germany

Klaus-Peter Hofmann
Institut für Medizinische Physik und Biophysik
Klinikum Charité der Humboldt-Universität zu Berlin
Berlin
Germany

Jörg D. Hoheisel
DKFZ-German Cancer Research Center
Functional Genome Analysis
Im Neuenheimer Feld 506
D-69120 Heidelberg
Germany

Florence Horn
EMBL
Meyerhofstrasse 1
D-69117 Heidelberg
Germany

Gerd Illing
Interdisziplinärer Forschungsverbund Strukturbiologie
Berlin
Germany

W. Jäger
Interdisziplinäres Zentrum für Wissenschaftliches Rechnen (IWR)
Universität Heidelberg
Heidelberg
Germany

Javed Khan
Cancer Genetics Branch
National Genome Research Institute
National Institutes of Health
Bldg 49, Room 4A52
9000 Rockville Pike
Bethesda, MD 20892
USA

D. Kienle
Institute for Human Genetics
University of Heidelberg
Heidelberg
Germany

Konstantin Klenin
DKFZ-German Cancer Research Center
Dept. of Biophysics of Macromolecules (H0500)
Im Neuenheimer Feld 280
D-69120 Heidelberg
Germany

G. Kreth
Institut für Angewandte Physik
Universität Heidelberg
Heidelberg
Germany

Joachim Klose
Institut für Humangenetik
Virchow-Klinikum
Humboldt-Universität
Augustenburger Platz 1
D-13353 Berlin
Germany

Jörg Langowski
DKFZ-German Cancer Research Center
Dept. of Biophysics of Macromolecules (H0500)
Im Neuenheimer Feld 280
D-69120 Heidelberg
Germany

P. Lichter
DKFZ-German Cancer Research Center
Dept. of Biophysics of Macromolecules (H0500)
Im Neuenheimer Feld 280
D-69120 Heidelberg
Germany

G. Little
Interdisziplinäres Zentrum für
Wissenschaftliches Rechnen (IWR)
Universität Heidelberg
Heidelberg
Germany

Carsten Mehring
DKFZ-German Cancer Research Center
Dept. of Biophysics of Macromolecules (H0500)
Im Neuenheimer Feld 280
D-69120 Heidelberg
Germany

Paul S. Meltzer
Cancer Genetics Branch
National Genome Research Institute
National Institutes of Health
Bldg 49, Room 4A52
9000 Rockville Pike
Bethesda, MD 20892
USA

Boris Mirkin
DKFZ-German Cancer Research Center
Dept. of Molecular Biophysics (H0200)
Im Neuenheimer Feld 280
D-69120 Heidelberg
Germany

Mustapha Mokrane
Laboratoire de Génétique et de Physiologie du Développement
Université de la Méditerranée
13288 Marseille Cedex 9
France

Christian Münkel
DKFZ-German Cancer Research Center
Dept. of Biophysics of Macromolecules (H0500)
Im Neuenheimer Feld 280
D-69120 Heidelberg
Germany

Christof Niehrs
DKFZ-German Cancer Research Center
Division of Molecular Embryology
Im Neuenheimer Feld 280
D-69120 Heidelberg
Germany

Harmut Oschkinat
Forschungsinstitut für Molekulare Pharmakologie
Berlin
Germany

T. Pfisterer
DKFZ-German Cancer Research Center
Dept. of Molecular Biophysics (H0200)
Im Neuenheimer Feld 280
D-69120 Heidelberg
Germany

Nicolas Pollet
DKFZ-German Cancer Research Center
Division of Molecular Embryology
Im Neuenheimer Feld 280
D-69120 Heidelberg
Germany

Ercole Rao
Institute of Human Genetics
University of Heidelberg
Im Neuenheimer Feld 328
D-69120 Heidelberg
Germany

Gudrun A. Rappold
Institute of Human Genetics
University of Heidelberg
Im Neuenheimer Feld 328
D-69120 Heidelberg
Germany

Jens Reich
Max-Delbrück-Centrum für Molekulare Medizin
Robert-Rössle-Str. 10
D-13125 Berlin
Germany

Otto Ritter
DKFZ-German Cancer Research Center
Dept. of Molecular Biophysics (H0200)
Im Neuenheimer Feld 280
D-69120 Heidelberg
Germany

Benoit Roux
Département de Chimie
Université de Montreal
Succursale centre ville
Montreal
Canada

Stanley A. Sawyer
Dept. Mathematics
Washington University
St. Louis, MO 63130
USA

Wolfram Saenger
Institut für Kristallographie
Freie Universität Berlin
Berlin
Germany

Martin Senger
EMBL Outstation—European Bioinformatics Institute
Wellcome Trust Genome Campus
Hinxton, Cambridge CB 10 1SD
UK

Jeremy C. Smith
Lehrstuhl für Biocomputing
IWR der Universität Heidelberg
Im Neuenheimer Feld 368
D-69120 Heidelberg
Germany

Randall F. Smith
Dept. of Bioinformatics
SmithKline Beecham Pharmaceuticals
UW2230, 709 Swedeland Road
PO Box 1539
King of Prussia, PA 19406
USA

I. Solovei
Institute for Anthropology and Human Genetics
Ludwig Maximilians University
Richard-Wagner-Str. 10/1
D-80333 Munich
Germany

E. H. K. Stelzer
EMBL
Meyerhofstrasse 1
D-69117 Heidelberg
Germany

Sándor Suhai
DKFZ-German Cancer Research Center
Dept. of Molecular Biophysics (H0200)
Im Neuenheimer Feld 280
D-69120 Heidelberg
Germany

William R. Taylor
National Institute for Medical Research
Division of Mathematical Biology
The Ridgeway, Mill Hill
London NW7 1AA
UK

Katalin Tóth
DKFZ-German Cancer Research Center
Dept. of Biophysics of Macromolecules (H0500)
Im Neuenheimer Feld 280
D-69120 Heidelberg
Germany

Jeffrey M. Trent
Cancer Genetics Branch
National Genome Research Institute
National Institutes of Health
Bldg 49, Room 4A52
9000 Rockville Pike
Bethesda, MD 20892
USA

Gert Vriend
EMBL
Meyerhofstrasse 1
D-69117 Heidelberg
Germany

Johnathon Weare
EMBL
Meyerhofstrasse 1
D-69117 Heidelberg
Germany

Gero Wedemann
DKFZ-German Cancer Research Center
Dept. of Biophysics of Macromolecules (H0500)
Im Neuenheimer Feld 280
D-69120 Heidelberg
Germany

Rolf Zettl
Max-Planck-Gesellschaft zur Föderung der Wissenschaften
Ressourcenzentrum im DHGP am MPI für Molekulare Genetik
Berlin
Germany

D. Zink
Institute for Anthropology and Human Genetics
Ludwig Maximilians University
Richard-Wagner-Str. 10/I
D-80333 Munich
Germany

INDEX

MIX
Papier aus verantwortungsvollen Quellen
Paper from responsible sources
FSC® C105338

If you have any concerns about our products,
you can contact us on
ProductSafety@springernature.com

In case Publisher is established outside the EU,
the EU authorized representative is:
Springer Nature Customer Service Center GmbH
Europaplatz 3, 69115 Heidelberg, Germany

Printed by Libri Plureos GmbH
in Hamburg, Germany